Praise for *Tableau Prep: Up and Running*

It has never been more important for analysts to take control of their data preparation, and—thanks to Allchin—it has also never been easier.

—*Ryan Sleeper, Founder of Playfair Data and Author of* Practical Tableau *and* Innovative Tableau

Years of experience in data preparation packed into an accessible how-to guide for users of all-levels. Essential reading for Tableau users!

—*Kimberly Bolch, University of Oxford Tableau Student Ambassador*

Carl has used his years of consulting and training experience to produce a compelling and thorough overview of Tableau Prep Builder's functionality and the thought processes required to successfully clean, manipulate and enhance data. This is an invaluable guide for budding data preppers!

—*Luke Stoughton, The Information Lab*

Tableau Prep: Up & Running

Self-Service Data Preparation for Better Analysis

Carl Allchin

Beijing • Boston • Farnham • Sebastopol • Tokyo

Tableau Prep: Up & Running
by Carl Allchin

Copyright © 2020 Carl Allchin. All rights reserved.

Printed in the United States of America.

Published by O'Reilly Media, Inc., 1005 Gravenstein Highway North, Sebastopol, CA 95472.

O'Reilly books may be purchased for educational, business, or sales promotional use. Online editions are also available for most titles (*http://oreilly.com*). For more information, contact our corporate/institutional sales department: 800-998-9938 or *corporate@oreilly.com*.

Acquisitions Editor: Michelle Smith
Development Editor: Angela Rufino
Production Editor: Daniel Elfanbaum
Copyeditor: Rachel Monaghan
Proofreader: Piper Editorial, LLC
Indexer: Judith McConville
Interior Designer: David Futato
Cover Designer: Karen Montgomery
Illustrator: O'Reilly Media, Inc.

September 2020: First Edition

Revision History for the First Edition
2020-08-03: First Release

See *http://oreilly.com/catalog/errata.csp?isbn=9781492079620* for release details.

The O'Reilly logo is a registered trademark of O'Reilly Media, Inc. *Tableau Prep: Up & Running*, the cover image, and related trade dress are trademarks of O'Reilly Media, Inc.

The views expressed in this work are those of the author, and do not represent the publisher's views. While the publisher and the author have used good faith efforts to ensure that the information and instructions contained in this work are accurate, the publisher and the author disclaim all responsibility for errors or omissions, including without limitation responsibility for damages resulting from the use of or reliance on this work. Use of the information and instructions contained in this work is at your own risk. If any code samples or other technology this work contains or describes is subject to open source licenses or the intellectual property rights of others, it is your responsibility to ensure that your use thereof complies with such licenses and/or rights.

978-1-492-07962-0

[LSCH]

Table of Contents

Part IV. Output

Part V. Cleaning Data

Part VI. Beyond the Basics

Part VII. Managing Your Data

Preface

Data is everywhere, but for most people it's largely unusable for these key reasons:

- Some data is stored in databases hidden behind a coding language that the majority of the workforce have never been taught.
- Some data lurks on individuals' computers stored away from those who might find it useful.
- Some data is held in formats that only the developer of the system that created it could read.

So why should you care about this? Frankly, that data holds the answers to questions you have and questions you have yet to even ask. Self-service data preparation is a skill that will change what data analyses you can undertake, reduce the time you need to complete data projects, and fundamentally improve the quality of your analyses.

This book aims to teach you how to access this data and turn it into information to answer those questions using one of the most user-intuitive tools on the market—Tableau Prep Builder. Released in April 2018 to support Tableau Desktop, Server, and Online users, Tableau Prep Builder enables you to turn messy data into a format where it can be analyzed in Tableau's software. Tableau Desktop, Server, and Online are software platforms that make data easy to explore and analyze visually.

Previously, the largest gap in Tableau's flow from source system to delivering insightful analysis was manipulating the data into a format that's easy to use. Tableau, like the majority of business intelligence tools, requires the data to be "clean" and structured into rows and columns. Many analysts used to take on this manual work themselves, hence the need to automate this task and spend that valuable time on the actual analysis instead.

Tableau Prep Builder allows users to easily clean, manipulate, and output data sets that are ready for analysis. Not only that, but Tableau has also embedded a number of its visual analytical approaches within the software so that users can often find the answers to their questions in Prep Builder without having to export the data at all.

Why I Wrote This Book

If everything in Tableau Prep is so intuitive, why do you need to learn how to use it from this book? Simply put, using the tool is only one part of the challenge of preparing data. The other parts consist of:

- Understanding why to prepare data at all
- Connecting to all the data you require
- Understanding how different data types affect the cleaning operations you will need to perform
- Breaking down the challenge of preparing data to plan your approach
- Ensuring the proper changes are being made during the cleaning and manipulation operations
- Combining multiple data sets
- Determining how and where to output your resulting data

As with all software, it takes a bit of time to learn how to use each function, so this book is filled with screenshots and walkthroughs of the more complicated techniques. A lot of the knowledge shared here will help you undertake your own data preparation projects in *any* data preparation tool. These techniques will empower you to tackle data sets that previously wouldn't have been accessible to you. That is why I wrote this book: to empower you to use data, or more data, to improve your decision-making.

In my career, I have been on both sides of the data preparation cycle: as the receiver and as the provider of the outputs. As the receiver, I was often frustrated with the time it took to get my hands on the information I required. The information I did receive was often not in the form I required, or it was missing key pieces of data that became required after I originally put in the data request. As the provider, I always took care to understand the problem, the underlying reason why someone wanted the data, so I could deliver the best solution rather than just what they asked for. I was also conscious that the longer I spent on each request, the longer the queue of others waiting to get their own views of different data sets. This is why I began to teach users how to get to the data themselves. Obviously, it isn't possible for everyone to spend the time to develop SQL querying skills (don't worry if you don't know what this means) in order to access tables they didn't even understand why they needed yet. Tableau Prep Builder allows you to complete your own preparation with only a few hours of training rather than the days or weeks it would take to get going with SQL.

By reading this book and taking the time to practice the skills it covers, you should feel empowered and equipped to complete your own data preparation and deliver better analytical answers faster than ever before.

Who This Book Is For

This book is for people from different parts of the spectrum that covers working with data, such as those who are:

- New to data and the workplace. Data is a major part of most jobs now, so if you're fresh out of school or university, learning the skills this book will cover should prepare you for the future.
- New to data but an experienced professional. Supplementing your experience with the knowledge you'll gain from this book can create some amazing results. Without that experience, the data can be meaningless and lack context for you. This book will give you the skills to add data to back up your professional experience.
- Experienced in visual analytics but not data preparation. Tableau Desktop has empowered many people to conduct their own visual analytics rather than waiting for IT and reporting teams to build reports for them. Tableau Prep Builder is doing exactly the same for data preparation. This book will boost your visual analysis skills and allow you to access data sets that seemed previously impossible.
- Experienced data prepper. OK, that's not your official job title, but it's what you are in my eyes. You might be using Excel, SQL, or another scripting language. Thanks to automation and simplification, Tableau Prep Builder will enable you to work faster than you can with your current methods and tools.
- Colleagues of the experienced data prepper. Being familiar with Tableau Prep Builder will allow you to take on the simpler, more repeatable tasks of experienced data preppers so they can concentrate on the harder challenges. They will be your oracle on how to develop, so you can help them and yourself at the same time.

How This Book Is Organized

There are seven parts within the book. They have been organized to progressively build the skills and knowledge you require, while also acting as easy reference points when you need to jog your memory. After Chapter 1 looks more deeply at why self-service data preparation is important, chapters are arranged as follows:

Part I (Chapters 2–6)
: After introducing Prep Builder, this part explores how to plan your data preparation and goals for the resulting data set. The final two chapters look at connecting to both data files and databases.

Part II (Chapters 7–10)
: Understanding what data you are using and preparing is key. These chapters help you know what to look for when preparing data and introduce some of the functions you can use to work with data fields.

Part III (Chapters 11–18)
: Once you have an understanding of your data fields, this part helps you analyze the shape and profile of your data set. You'll also learn about the transformational steps within Prep Builder.

Part IV (Chapters 19–21)
: After all your hard work, it's time to output the data for analysis. This part covers how to output your data from your preparation flow to either a file or a database. This part also covers the other Tableau Prep product, Prep Conductor, which allows you to automate your workflow as well as share your flows with others.

Part V (Chapters 22–34)
: Getting to this point means that you know the basics of how to produce a simple flow. Data preparation often contains other challenges, however. To help you tackle those, this part introduces you to many more of the cleaning functions built into Prep Builder.

Part VI (Chapters 35–41)
: Knowing all the relevant techniques is one thing, but knowing when to use them is quite another. Therefore, this part describes how to use the techniques you've learned in real-world scenarios and considerations for when you're faced with more difficult scenarios.

Part VII (Chapters 42–49)
: This part centers on making your data and flow available to others by managing and documenting the output as well as focusing on the result.

These chapters will give you the knowledge and foundation to prepare your own data for analysis. But like anything in life, practice will hone your skills. To that end, some of the chapters feature data sets, examples, and challenges from Preppin' Data (*http://preppindata.blogspot.com*) to allow you to practice the techniques the chapter has covered. Jonathan Allenby and I designed Preppin' Data as a weekly challenge to allow people with a range of experience to practice their data preparation skills. These exercises are purely optional, but by practicing the technique, it's much more likely that you'll understand how to apply it when you next need to. Each exercise explains its intention and requirements, just as if it was a request from someone you know. The Input and Output data sets allow you to try to meet the challenge set in the exercise. Solutions are available on the blog, but there is no right or wrong solution if you have delivered the output requested. Finally, Preppin' Data often references a company called Chin & Beard Suds Co., which is a mock soap retailer that Jonathan and I use as an example in our exercises. This allows us to use terrible soap-based puns, for which we are unapologetic. The Preppin' Data site has had 80,000+ hits, 260+ participants, and 2,000+ challenge solutions submitted. We'd love for you to join this community of data preppers.

Acknowledgments

This book would not have been possible without a number of phenomenal people that I get to call peers, colleagues, and friends. First, the one Excel user in my life that I can't bring into the modern data age, my partner of 15 years, Toni Feather. A lot of her pragmatism exists in these pages, and by writing this, I might finally get her to use different data preparation tools.

A huge thank-you goes to my friends and colleagues at The Information Lab and The Data School in London. Without these brilliant minds and passionate people, this book would never have happened. Four years of consulting experience with the team produced a lot of the use cases you will read about in the pages to come. Tom Brown, Craig Bloodworth, and Robin Kennedy—thank you for making a truly amazing environment to learn and develop in. The Data School consultants deserve special praise too; having the luxury of teaching them over the years has allowed me to refine the "messages" that need to be conveyed, so they have massively shaped the content through the questions they ask every single day. The book began with an idea shared with Dan Farmer (one of the fantastic content editors) who helped form the early skeleton that I then fleshed out. Thank you for helping me shape this thing, Dan.

I became more focused on data preparation when one of the trainee consultants at The Data School, Jonathan Allenby, asked whether there was any way to put into practice the teachings I had just given on Tableau Prep. That prompted the creation of Preppin' Data, and the success of our blog and the level of demand for instruction on data preparation led to this book.

Those who have shaped the actual content deserve massive praise as they have helped me turn my normal teaching content into this printed form. Angela Rufino at O'Reilly has been a fantastic content editor and made sure everything made sense even to new data preppers. The technical content editors—Jonathan Drummey, Ryan Sleeper, Kimberly Bolch, and Luke Stoughton—have all added a lot to this book. Their feedback has gone beyond just editing and ensured this book will deliver value to everyone reading it.

Finally, thanks to you for reading this book. By adding more data-driven decisions to your personal and work life, you will be improving the world for yourself and those around you. I have the luxury of working with lots of sectors, and the work of those around me inspires me every day. We can make this world a much better place with better use of information and insight—you are now part of those efforts to help others.

Conventions Used in This Book

The following typographical conventions are used in this book:

Italic
: Indicates new terms, URLs, email addresses, filenames, and file extensions.

`Constant width`
: Used for program listings, as well as within paragraphs to refer to program elements such as variable or function names, databases, data types, environment variables, statements, and keywords.

`Constant width bold`
: Shows commands or other text that should be typed literally by the user.

`Constant width italic`
: Shows text that should be replaced with user-supplied values or by values determined by context.

This element signifies a tip or suggestion.

This element signifies a general note.

This element indicates a warning or caution.

Using Code Examples

Supplemental material (code examples, exercises, etc.) is available for download at *https://oreil.ly/5k_uH*.

If you have a technical question or a problem using the code examples, please send email to *bookquestions@oreilly.com*.

This book is here to help you get your job done. In general, if example code is offered with this book, you may use it in your programs and documentation. You do not need to contact us for permission unless you're reproducing a significant portion of the code. For example, writing a program that uses several chunks of code from this book does not require permission. Selling or distributing examples from O'Reilly books does require permission. Answering a question by citing this book and quoting example code does not require permission. Incorporating a significant amount of example code from this book into your product's documentation does require permission.

We appreciate, but generally do not require, attribution. An attribution usually includes the title, author, publisher, and ISBN. For example: "*Tableau Prep: Up & Running* by Carl Allchin (O'Reilly). Copyright 2020 Carl Allchin, 978-1-492-07962-0."

If you feel your use of code examples falls outside fair use or the permission given above, feel free to contact us at *permissions@oreilly.com*.

O'Reilly Online Learning

For more than 40 years, *O'Reilly Media* has provided technology and business training, knowledge, and insight to help companies succeed.

Our unique network of experts and innovators share their knowledge and expertise through books, articles, and our online learning platform. O'Reilly's online learning platform gives you on-demand access to live training courses, in-depth learning paths, interactive coding environments, and a vast collection of text and video from O'Reilly and 200+ other publishers. For more information, visit *http://oreilly.com*.

How to Contact Us

Please address comments and questions concerning this book to the publisher:

O'Reilly Media, Inc.
1005 Gravenstein Highway North
Sebastopol, CA 95472
800-998-9938 (in the United States or Canada)
707-829-0515 (international or local)
707-829-0104 (fax)

We have a web page for this book, where we list errata, examples, and any additional information. You can access this page at http://oreilly.com/catalog/9781492079613.

Email *bookquestions@oreilly.com* to comment or ask technical questions about this book.

For news and information about our books and courses, visit *http://oreilly.com*.

Find us on Facebook: *http://facebook.com/oreilly*

Follow us on Twitter: *http://twitter.com/oreillymedia*

Watch us on YouTube: *http://www.youtube.com/oreillymedia*

CHAPTER 1

Why Self-Service Data Prep?

With every organization swimming in data lakes, repositories, and warehouses, never before have their employees had such an enormous opportunity to answer their questions with information rather than just their experience and gut instinct.

This isn't that different from where organizations stood a decade ago, or even longer. What is different is who wants access to that data to answer their questions. No longer is the expectation that a separate function of the business will be responsible for getting that data; now, everyone feels they should have access to it. So what has changed? Self-service data visualization. What is about to change to take this to the next level? Self-service data preparation.

A Short History of Self-Service Data Visualization

More than a decade ago, all things data related were the domain of specialist teams. Data projects went to either Business Intelligence (BI) teams or Information Technology (IT) teams, who set up data infrastructure projects to produce reports. This was an expensive and time-consuming process that often resulted in products that were less than ideal for all concerned.

The reason this methodology doesn't work is the iterative nature of BI. Humans are fundamentally intelligent creatures who like to explore a question, learn, and then ask more questions as they become more intrigued. Humans are also great at spotting visual patterns and charting data sets to help find those patterns. With the traditional IT or BI projects, once the first piece of analysis was delivered, the project was over. The initial question was answered, but because the people doing the analysis were different from those asking the questions, any follow-up questions simply went unanswered or were cobbled together from disparate reports or different levels of aggregation.

This all changed with the rise of self-service data visualization tools like Tableau Desktop. With these tools' focus on the user, suddenly individuals were able to drag and drop data fields around the screen to form their own analysis, answer their own questions, and immediately ask their next questions in a visual way that allowed them to share their findings with others.

The previous decade has seen data visualization and analysis become increasingly important throughout the organization, and a significant part of many roles that are no longer considered solely the domain of IT or data teams. Analytical capacity has come to the business, rather than the business having to go and ask specialists to access and analyze the data. This represents a big transformation in how we work and requires us to rethink what skills people now need.

Accessing the "Right Data"

The rise, and entrenchment, of self-service data visualization into individuals' roles has surfaced needs and tensions in the analytical cycle. The analytical cycle involves:

1. Having a question posed from someone
2. Sourcing data that may help answer the question
3. Preparing the data for analysis
4. Analyzing the data
5. Forming new or additional questions (returning to step 1)

Enabling self-service requires opening access to data sources, which has traditionally been a pain point in this cycle. With the right data, optimized for use in visual analysis tools, we can now find answers as soon as the business expert can form the questions. But accessing the "right data" is not that easy. The data assets owned by organizations are optimized for storage, optimized for tools that now seem to work against users rather than with them, and regulated by strict security layers requiring coding to access the data.

Many data projects are now focused on extracting data from their storage locations. The specialists are focused on using data skills to:

- Find data in existing repositories, including Excel workbooks
- Find data in public or third-party repositories
- Create feeds of data from previously inaccessible sources and systems

The gap in the analytical cycle now sits between taking these sources and preparing them for visual analytics. There are different levels of complexity to this work (as with any skill in life), from opening an Excel spreadsheet to running application programming interface (API) queries. The challenges of data preparation with Excel or coding are different, but they highlight why another solution is necessary. Excel is very flexible and a tool many people are familiar with, but it is difficult to automate. Therefore, it often requires a lot of manual rework when data is updated. Excel's non-visual nature also makes it easier to introduce errors without noticing. With coding, having to make manual updates isn't as much of an issue because the coding is often built to rerun at the press of a button. However, in most workplaces, fewer people understand coding than Excel, making it difficult to hand over this work to colleagues. And, as in Excel, it's easy to make mistakes and not notice them because you don't see the data until the end of the transformation. So, although there is a significant need for individuals to be able to complete data preparation tasks, for a long time there has been a gap between that need and the ability to access tools that help fulfill it.

The Self-Service Data Preparation Opportunity

This gap is being addressed by new tools that enable the business experts to access data and answer their questions using self-service visual analytics. Tableau Prep Builder makes the process of data preparation easier than other tools by bringing the same logic that enabled visual analytics to the data preparation process. By using a user interface (Figure 1-1) similar to the one that data visualizers are already accustomed to, Prep Builder makes the transition to self-service data preparation a simple one, even for those trying to complete these tasks for the first time.

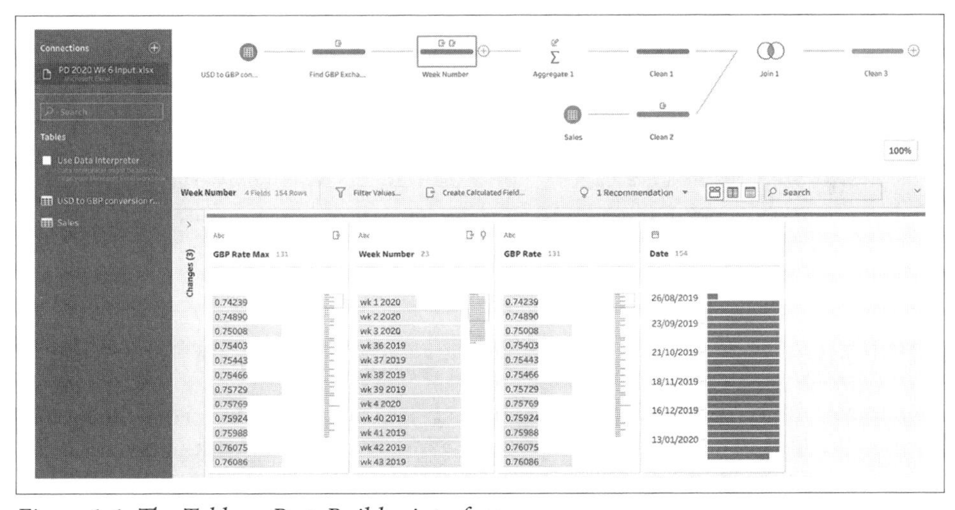

Figure 1-1. The Tableau Prep Builder interface

Data preparation isn't just the process of preparing a data set to make some charts as a one-off exercise. It encompasses many tasks, such as confirming accuracy of the data points, removing extraneous data, reshaping the data set to optimize it for easy analysis, and doing all of this efficiently so people aren't left waiting for the data to be ready so they can answer their questions. Good data preparation enables use of the data set in a timely manner, avoids wasting time on manual manipulation (thanks to tools like Tableau Prep Builder), and creates a repeatable process that anyone can use. The aim of this book is that the time you spend learning these data preparation skills will be repaid many times over as you begin to deploy them in the real world many times over. After all, the more time you spend manually preparing data means the less time you have to analyze it—if you decide to use such a messy data set at all.

To date, my career has taken me through all of the roles in the traditional analytics cycle—from a user suffering through the pain of waiting for reports to be built for me, to learning how to build the analysis for myself, to wrangling the data sets in the database, and finally, to training others in data visualization and data preparation skills.

There is still a significant gap between potential data preparers ("data preppers") and skilled ones. Learning what to do with the self-service data preparation tools, and why they are needed, is a significant undertaking but one that is worthwhile. Prep Builder will make the knowledge I have gained over many years a lot easier for you to learn and deploy straightaway.

Tableau Prep Up and Running

This book is designed for those who would benefit from being able to prepare their own data but lack the skills to do so. Over the next chapters you'll learn how to utilize the tools to tackle the tasks that are currently acting as roadblocks in your organization. The exercises will introduce commonly used techniques to solve these problems away from the pressure of the workplace. Over time, you will develop your own strategies to prepare the data exactly how you want it, whether it comes from files, databases, surveys, pivot tables, messy data, or tangled text fields. There isn't a straightforward recipe to follow, but by practicing, you'll soon be able to handle these challenges.

Summary

Hopefully, you now have a view of why self-service data preparation is vital and why learning the techniques this book covers will assist you in your work with data. How you should approach those challenges and where you should start are the subjects of the next chapter.

PART I

Getting Started

CHAPTER 2

Getting Started with Tableau Prep Builder

So you have a data set that you would like to prepare for analysis, and you want to learn Tableau Prep Builder. This chapter will take you through the basics of Tableau Prep Builder, including downloading the software, familiarizing yourself with Prep terminology, and using the software to produce a data set for the first time.

Where to Get Tableau Prep Builder

Tableau's products are available for download through the company's website (*http://tableau.com/esdalt*). Figure 2-1 shows Tableau's download page for Prep Builder, where you can select the version of the program you want to use.

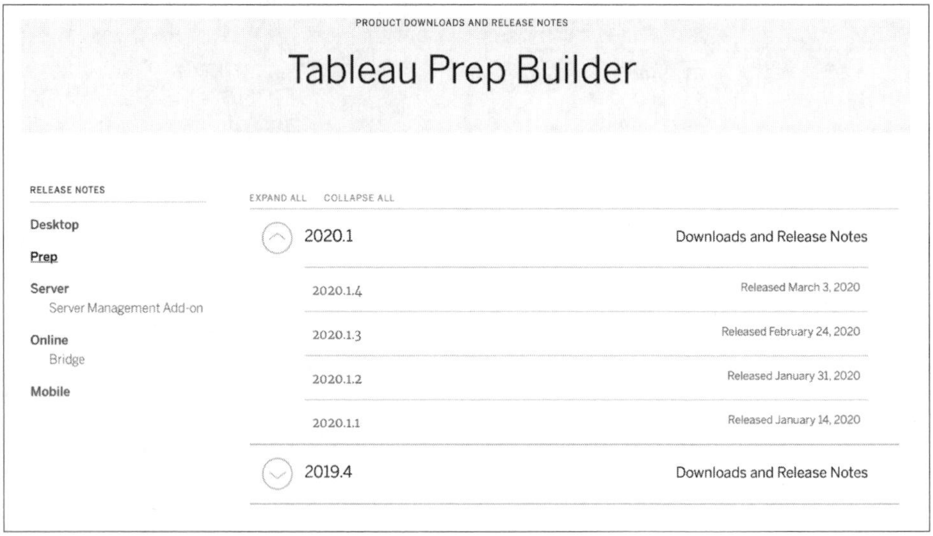

Figure 2-1. Tableau's download page

Click the version number you want to use, select Download Tableau Prep *<version number>* at the top of the screen, and choose whether you want to download the Windows or Mac version of the software. After you make your selection, the file will download to your machine. Once the download is complete, you can run the file and follow the instructions given.

After downloading Prep Builder, you'll see a new folder in your *Documents* folder, called *My Tableau Prep Repository*, which contains a lot of useful files and subfolders for storing the flows and data sets you'll create within Prep. Figure 2-2 shows the view of my Documents in Finder on Mac.

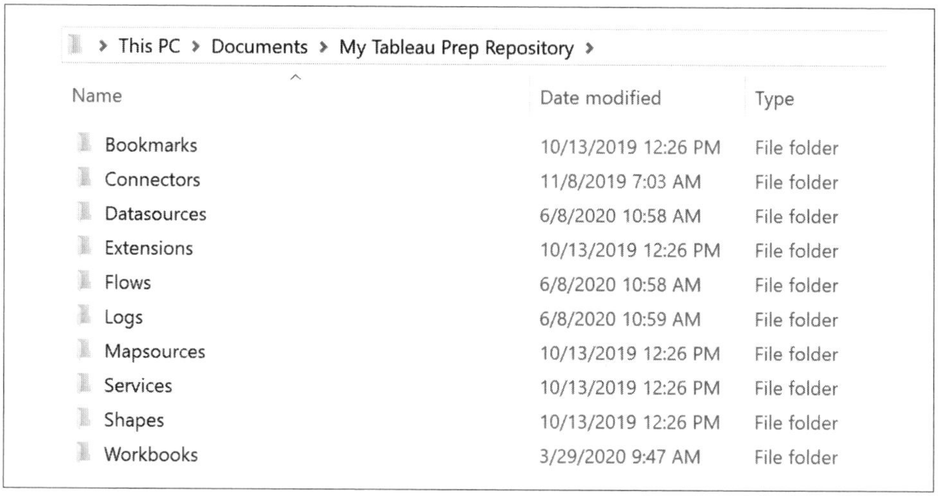

> This PC > Documents > My Tableau Prep Repository >

Name	Date modified	Type
Bookmarks	10/13/2019 12:26 PM	File folder
Connectors	11/8/2019 7:03 AM	File folder
Datasources	6/8/2020 10:58 AM	File folder
Extensions	10/13/2019 12:26 PM	File folder
Flows	6/8/2020 10:58 AM	File folder
Logs	6/8/2020 10:59 AM	File folder
Mapsources	10/13/2019 12:26 PM	File folder
Services	10/13/2019 12:26 PM	File folder
Shapes	10/13/2019 12:26 PM	File folder
Workbooks	3/29/2020 9:47 AM	File folder

Figure 2-2. My Tableau Prep Repository in Windows File Explorer

How to Get a License for Prep Builder

For most users, Tableau Prep Builder is not free. The main way to get Prep is to purchase a Tableau Creator license, which is a monthly paid subscription that packages together Tableau Prep Builder, Tableau Desktop, and a single access to Tableau Server or Tableau Online (where Tableau hosts the Server instance). Tableau Prep Conductor is part of the Data Management add-on for Tableau Server and Tableau Online that can be purchased separately. All of these licensing options apply on a per-user basis.

There are 14-day trials of all of the full Tableau tools available from the main Tableau website. Educators and currently enrolled students can also get Tableau for free after a simple verification process.

After downloading and installing the application, you will be prompted to enter your licensing information or sign up for a trial (Figure 2-3).

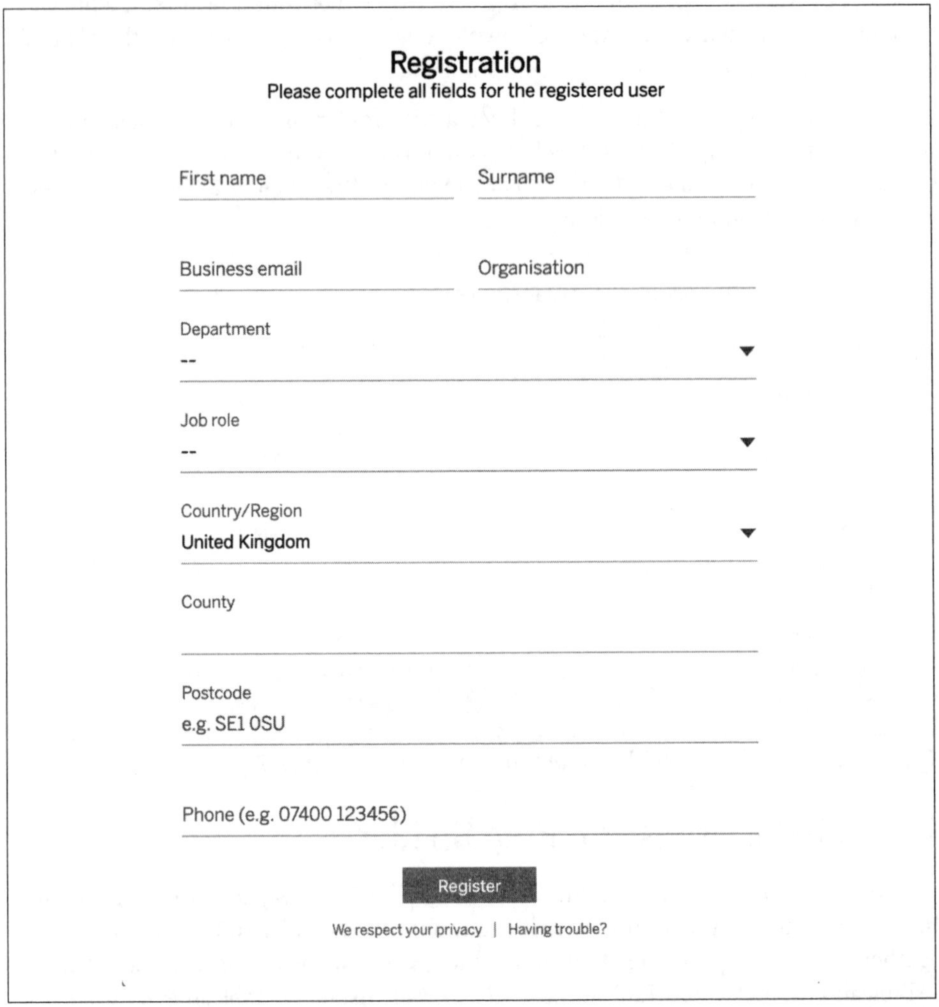

Registration

Please complete all fields for the registered user

First name

Surname

Business email

Organisation

Department

--

Job role

--

Country/Region

United Kingdom

County

Postcode

e.g. SE1 0SU

Phone (e.g. 07400 123456)

Register

We respect your privacy | Having trouble?

Figure 2-3. Prep Builder registration screen

The Tableau Prep Builder Screen

When you load Prep Builder for the first time, you will be presented with the screen shown in Figure 2-4.

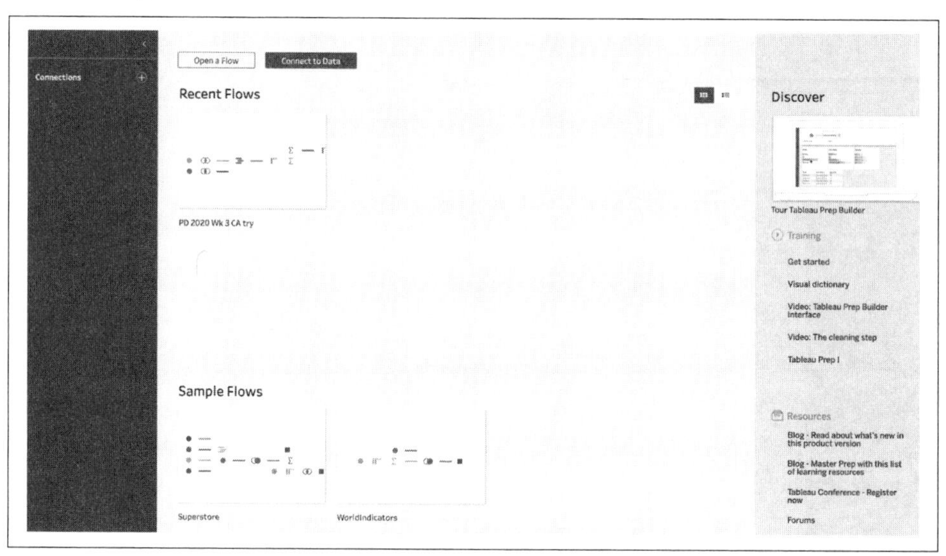

Figure 2-4. The Prep Builder initial screen

Let's walk through this screen:

Connections pane (left-hand blue pane)
: To see the list of available connections, click the plus sign (+) at the top right of this pane. The list will expand, showing the File, Server, and ODBC/JDBC connections available. ODBC (Open Database Connectivity) and JDBC (Java Database Connectivity) connections are useful where Tableau doesn't have a bespoke connector for the data source.

Recent Flows (top center)
: This area contains the latest flows that you have been working on. If this is the first time you have used Prep, this space will be blank.

Sample Flows (bottom center)
: This is where you'll find a couple of sample flows from the Tableau team to let you explore and experiment with an established, complete flow.

Discover pane (right-hand gray pane)
: This pane has links to videos, blog posts, and articles to help you learn the basics of using Prep.

Once you are connected to your data set, you will be taken to the screen shown in Figure 2-5.

Connecting to data files and databases is covered in Chapters 5 and 6 respectively.

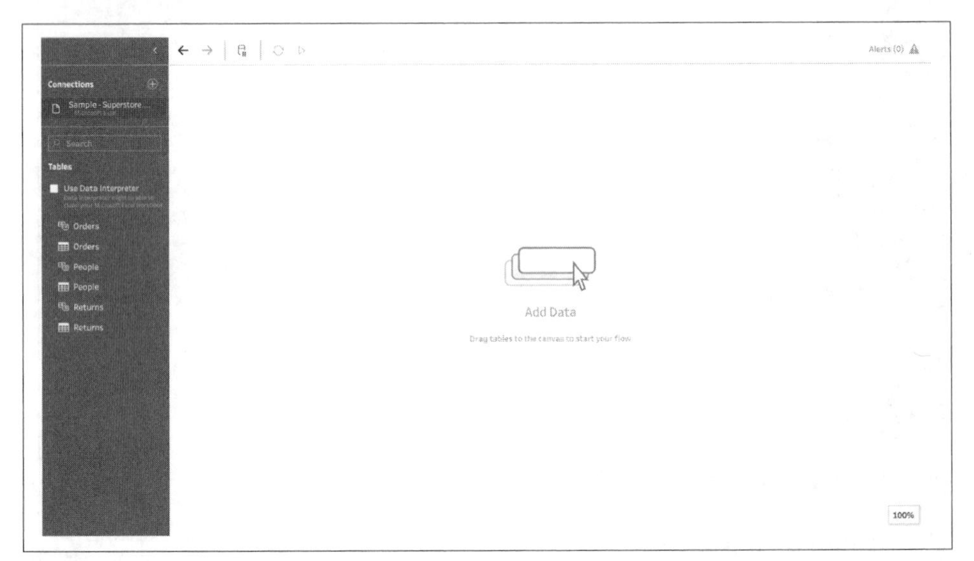

Figure 2-5. The Prep Builder main screen

This screen is split into two parts:

Connections pane (left-hand blue pane)
: As in the first screen, this is where you'll be able to edit the data connection, add another connection, or select tables to use as the input into your Prep flow. In Figure 2-6, I have chosen to load an Excel connection, so I also have the option to use the Data Interpreter to find the table(s) of data within a formatted Excel worksheet.

Canvas (center)
: This area will change considerably depending on the steps you take in your data preparation process.

Basic Steps of Data Preparation

In this section, we'll dig into a few key steps of data preparation.

Input Step

After connecting to a data source, select the input by dragging the data source from the Connections pane onto the canvas. By default, Prep Builder samples the input data set to speed up the process of building your workflow. When you run the workflow when an output is set up, all of the data will automatically be processed. The Input step is shown in Figure 2-6.

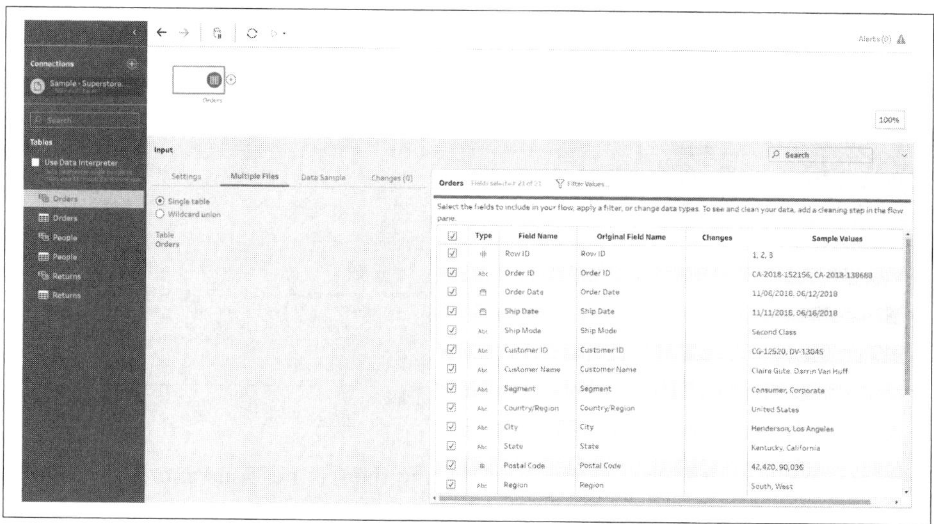

Figure 2-6. Prep Builder's Input step screen

The three key parts of this screen (excluding the Connections pane) are:

Flow pane (top)
: This area reflects the current step of the data preparation process.

Input pane (gray area)
: This is where you set up the input. You can use multiple files and choose how the data is sampled as you are building the flow. Prep keeps a log of changes you make during this step.

Data fields (white table)
: This is where you select the specific data categories to bring into the data preparation flow. You can change the names of the data fields (columns) or alter the

data types. Prep displays a small subset of sample values to help you understand what each column contains.

Clean Step

The Clean step is where the majority of the work takes place, so getting familiar with it is key to your overall understanding of Prep Builder. To begin the Clean step, click the plus sign to the right of the selected input data, as shown in Figure 2-7.

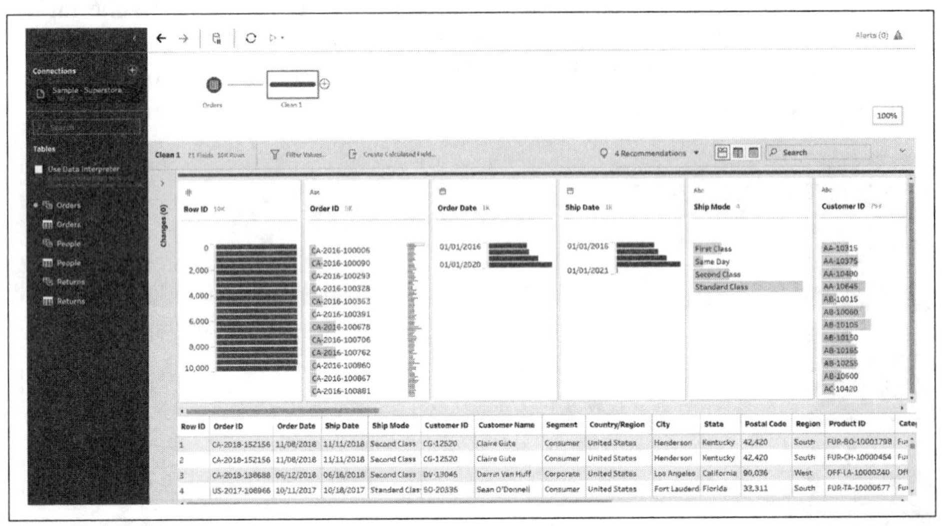

Figure 2-7. Prep Builder's Clean step screen

There are four key parts to this screen, including the Connections and Flow panes. The other two are:

Profile pane (center)
: Prep takes the sample of the data you set up during the Input step and distributes it in the appropriate data field of your data set. You can complete a number of preparation tasks within the Profile pane by selecting certain values. For example, to examine the relationships between values, select one value in a data field and see what values also appear in that row.

Data grid (bottom)
: Here you can see the records (rows) of your data set. There are three icons at the top right of the Profile pane where you can change how your data is displayed in the Data pane (Figure 2-8): the Profile pane and Data grid; just the Data grid; or the List view, which shows the metadata of the data set.

Figure 2-8. The data view options (left to right: the Profile pane and Data grid, Data grid only, and List view)

Output Step

Once you have cleaned the data, you can output the data set to make it available for analysis in other tools (Figure 2-9).

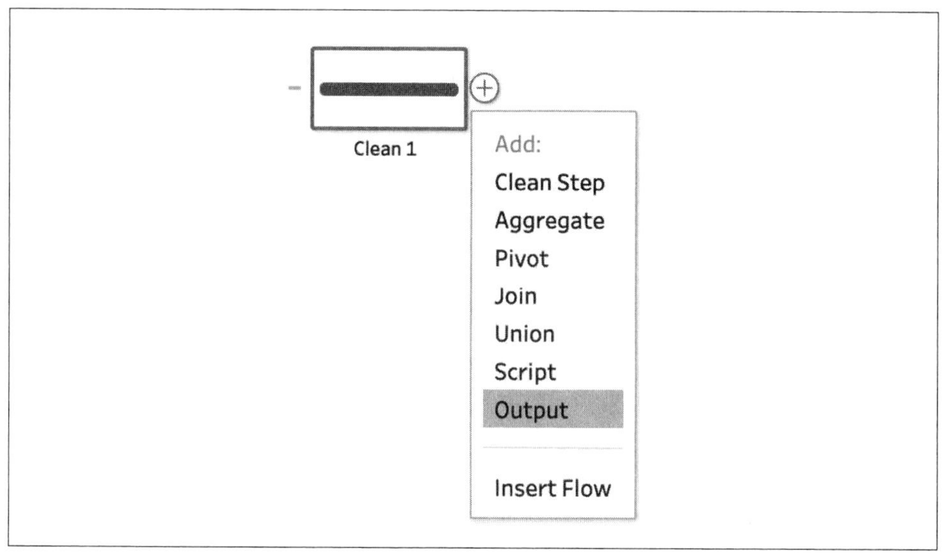

Figure 2-9. Adding an Output step

The Output step gives you a number of choices about how to output your data. Part IV of this book is dedicated to outputs, with a focus on outputting to files in Chapter 19 and to databases in Chapter 20. The default option for the output is to save to a file in your *My Tableau Prep Repository* folder.

Saving a Flow

Whether you have completed your flow or just need to pause your work to continue later, saving the flow is key. To do so, select Save or Save As from the menu at the top of the screen (Figure 2-10).

The Save option is the same as Save As if the file hasn't been saved before. If the file has been saved previously, use Save As to change the filename.

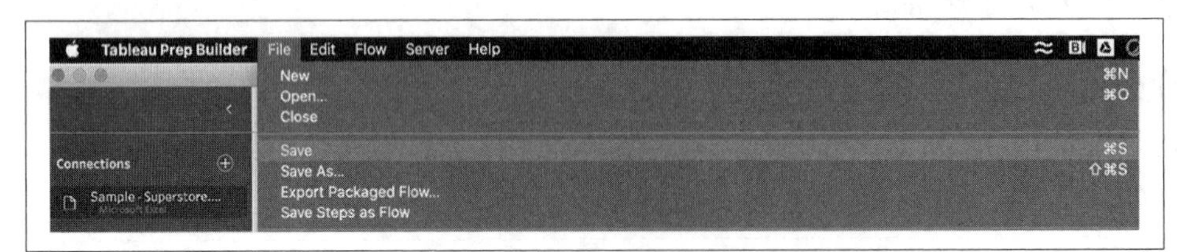

Figure 2-10. Saving a flow file

When saving the file, you'll see a screen where you have two choices for the file type (Figure 2-11):

- Tableau flow file (*.tfl*) saves the logic of the flow as well as the input and output file locations. Therefore, you'll need access to the input and output locations to make use of this file format.
- Packaged Tableau flow file (*.tflx*) saves not just the logic but also the input and output files.

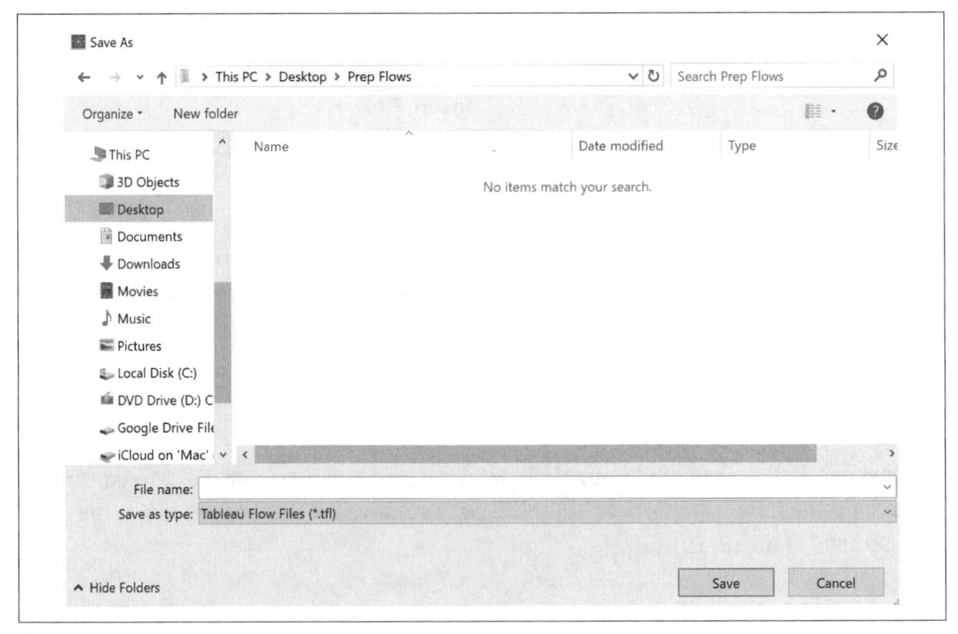

Figure 2-11. Saving to a file in Prep

Once you've selected a save option, you can process the flow by clicking the run icon (▷) either at the top of the Flow pane (to run every output) or on the output step icon (to run a single output), as shown in Figure 2-12.

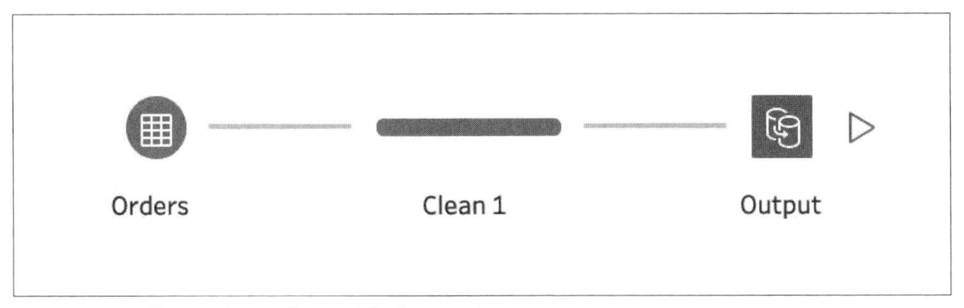

Figure 2-12. Processing a flow with the run icon

Summary

This chapter has given you the basics to start working with a simple flow in Tableau Prep Builder. Subsequent chapters will build upon this knowledge. The steps and other screen layouts you'll need to be familiar with are covered in the chapters on specific techniques:

- Pivot: Chapters 13 and 14
- Aggregate: Chapter 15
- Join: Chapter 16
- Union: Chapter 17

Now that you've gained some familiarity with the tool, you're better equipped to navigate the terminology in the other chapters and throughout your use of Prep Builder.

CHAPTER 3

Planning Your Prep

So you know your data isn't suited to the purpose for which you need it. This is probably because you have tried to analyze it but hit a roadblock early on. Maybe there are multiple data fields containing the information where you would expect just one. Maybe the data you are analyzing has gaps. Or, maybe the data you want to analyze comes from multiple sources.

What do you do after reaching this realization? How do you develop a solution when all you see are the issues in front of you?

This chapter recommends a staged approach to help you plan your data preparation, define the outcome, and build a workflow to solve your challenges. The four stages in the proposed framework are as follows:

1. Know your data (KYD).
2. Identify the desired state.
3. Determine the required transitions from KYD to the desired state.
4. Build the workflow.

To illustrate this process, we'll walk through a simple example data set, some sales data from Chin & Beard Suds Co. (Figure 3-1).

Branch	Product	01 Jan 2019	01 Feb 2019	01 Mar 2019	01 Apr 2019
Wimbledon	Liquid-Soap	173	344	427	470
Wimbledon	Soap-Bar	708	652	377	305
Lewisham_1	Liquid-Soap	276.31	804.22	655.94	789.63
Lewisham_1	Soap-Bar	359.12	647.53	401.26	291.99
Lewisham_2	Liquid-Soap	643	555	686	661
Lewisham_2	Soap-Bar	323	727	810	504

Figure 3-1. Sample data from Chin & Beard Suds Co.

Stage 1: Know Your Data

Without understanding your data set as it currently stands, you will not be able to deliver the results you need. For small data sets, sometimes it's very easy to develop this understanding and a corresponding plan. With larger data sets, the planning process can take longer, but it is arguably more important since you can hold only so much information in your own memory. Table 3-1 shows what to look for in data sets (using the sample data from Figure 3-1).

Table 3-1. Considerations for your data set

What to look for	Our example
Columns, rows, and crosstabs: how is the data structured?	There are two columns of categorical data, with each column representing a month of values.
Headers and columns: are they all there as expected?	Month headers should ideally be one column, with values listed in a separate column.
Data types: what type of data exists in each column? There should be just one data type in each column. Data types include strings (alphanumeric characters), numbers, dates, and Boolean (true/false) values.	There are two text values, but each subsequent column is a numeric value.
Granularity of rows: what does each row represent?	Each row is a different product sold in a store for a specific date.
Data points: are there any missing?	There are no data points present, so we can disregard this factor for this example.
Number of records: does the data set have the number of observations you expect?	There are three branches and two products, totaling six rows of data. With four months of data, there are 24 values.
Business/organization rules: are there any rules that the data set should adhere to?	In this instance, Lewisham should be recorded as a single branch and not two.

A quick sketch of the data set can often help you think through and answer these questions. Figure 3-2 shows how we'd sketch the sample data.

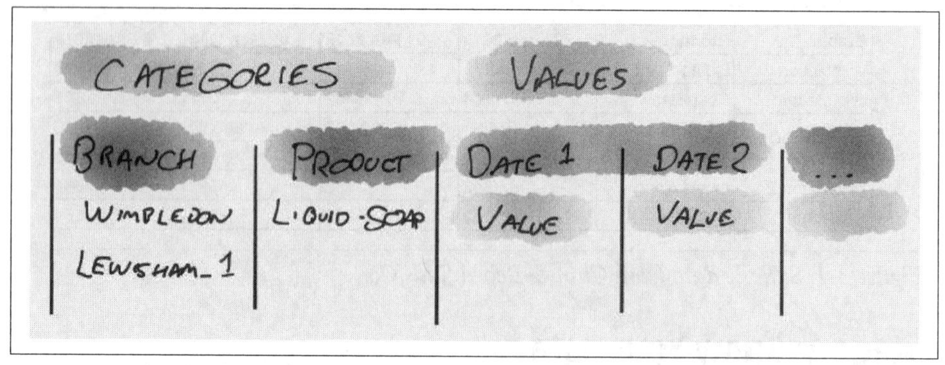

Figure 3-2. Sketch of data set highlighting categories and values

By identifying categorical data and the fields that contain the values you need for your analysis, you can see how complete the data set is, why it isn't ready for analysis yet, and what it might take to prepare the data. Notice that it's not vital to capture everything. In fact, simplifying the sketch can help you focus on the core issues rather than drowning in the details.

Stage 2: Identify the Desired State

So what is the desired state of your data?

For most modern data analysis tools or visualization software, you will need to structure your data into columns. The first row contains the column heading—the label for the data—and each subsequent row should be an individual record or instance. For example, each row could represent a product purchased from your store or the number of new employees each month (Figure 3-3).

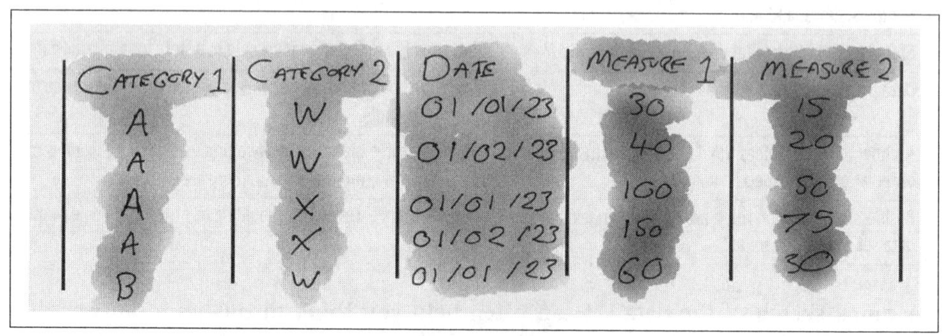

Figure 3-3. Sample desired state of a data set for analysis

Understanding what each row of your data represents is key to preparing and analyzing your data correctly. Most tools require a single input table, so all of the data fields you need for your analysis should go in one table.

Sketching Your Data Set

Simple data sets can be very easy to just doodle out. Sketching wide data sets with lots of columns, though, can be a bigger task. Start by listing out your categorical data—that is, subdivide your data into groups. Often there is a limited number of categories compared to the infinite range of numerical values, so categorizing data adds meaning by describing the numerical values in the data set. These categories are parameters that you will analyze the numbers by—for example, region, customer, product, or course. Each unique combination of these categorical data fields will establish the granularity of your desired state data set. In the example data set, the combination of branch, product, and date will set the value being recorded.

Next, add a column for each measure (measurement) you want to analyze. At this point, you should also consider values that might help the analytical tool's performance. Adding a ranking, market share, or subtotal for each category at this stage could make your data set more accessible to novice users and also more performant, allowing you to complete your analysis more quickly.

Returning to our example, Figure 3-4 shows what we need to output to complete our analysis.

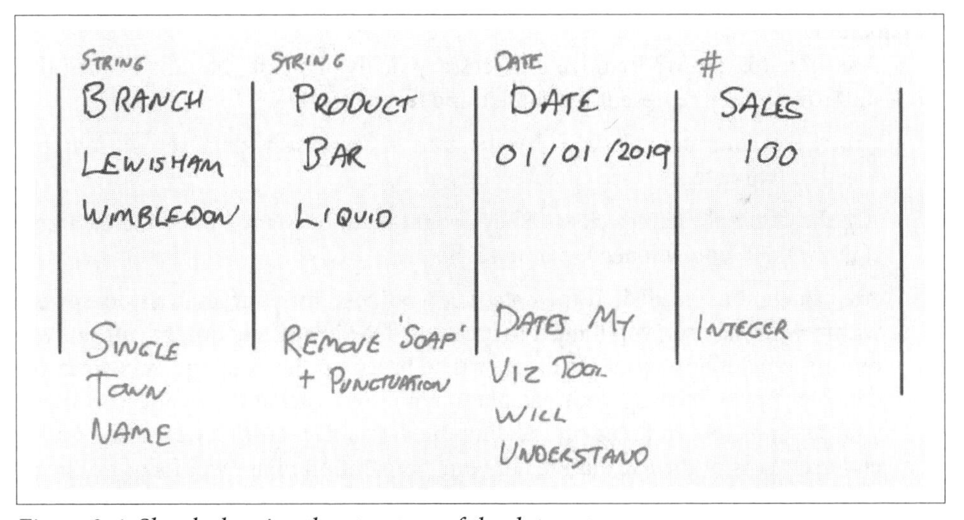

Figure 3-4. Sketch showing the structure of the data set

The categorical data fields are Branch, Product, and Date. Note how capturing the data type here is useful too (see the labels above the column headers). The output will have only one field for analysis: a simple sales value.

As you become more experienced in producing your own data sets, this stage becomes a lot easier and almost second nature. You will be able to look at a data set and know what the desired outcome should be. In the meantime, identifying the desired state can seem daunting, so sketch it!

Stage 3: Determine the Required Transitions from KYD to the Desired State

Take your hands away from the keyboard and mouse; for this stage, you won't need a computer, just your brain. By looking at the original data and identifying the desired end state, you will start to understand some of the transitions you will need—cleaning, pivoting, joining, and aggregating the data—between the two stages. (Don't worry if these types of transitions within data are completely new concepts for you; upcoming chapters will help you understand and master them to solve your own data challenges.)

Start by making a list of the transitions you think you will need to make. Doing this outside of Prep Builder will help you think through the necessary steps, rather than worrying about how you'll implement them in the software. If you don't know how to complete a step, you can deal with that at a later point. You may not end up doing them all anyway, since you're not yet building the workflow. Here are some of the questions to ask yourself at this stage:

Columns

- Are there too many? Remove unnecessary fields or pivot columns to turn them into rows if they represent the same thing (e.g., dates).
- Are there columns missing? Maybe join a secondary data set. A *join* pulls together two data sets (see Chapters 32 and 33).
- Are the data field names clear? They should be understandable to the data users. If they're not, you will need to amend them.
- Are calculations needed? If the data field you need for your analysis doesn't exist within your data set, you'll need to create it. This is possible only if you can work out the logic of how to create the new field based on the existing data within your data set. By completing your calculations in your preparation tool, you free up your analysis tool to focus on forming the data and require fewer skills from the end user. Lots of the calculations in your preparation phase will focus on the categorical data fields, as the actual measures will be aggregated during the analysis phase.

Rows

- Are there more rows than expected? Filter out unnecessary rows. Aggregate the data to be less granular.
- Are there fewer than expected? Pivot columns to create more. Add an additional data set through unioning or join a data scaffold.

- Are the values clean? For example, is there punctuation (or other extraneous characters) where there shouldn't be any? Take note of individual changes, as they will likely be separate data prep steps.
- Are there any blanks? Should they be there? If not, you'll need to filter them or find a way to fill them.

Multiple data sources

- Do you need to join multiple sources together to add all the data required for the analysis?
- Do you need to union together multiple sources to add more rows to complete a data set?

Let's return to our example to see how this works (Figure 3-5).

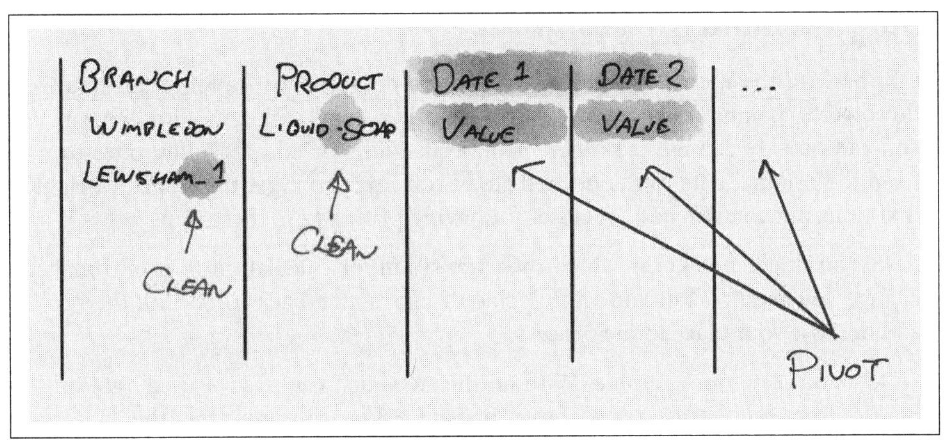

Figure 3-5. Sketch planning the transitions required in the sample data set

Very quickly we can spot the challenges in this sketch of the shape of the data. For example, we can see:

- We have two categorical data fields (Branch and Product).
- The rest of the columns are all headed by dates and contain the sales values.
  - The dates need to be pivoted to form a third categorical field. This will add another row for each month/branch/product combination. The pivoting process will create a column of values.
- Both the Branch and Product fields contain punctuation (underscores and hyphens, respectively) that will need to be removed.

For the rows of data, the analysis requires a row for each unique combination of branch (town), product, and date. This means we will need to:

- Aggregate the Lewisham_1 and Lewisham_2 sales together to output the data at the correct granularity. This will also change the number of rows in the desired state data set compared to the original data set.

We also might need to take these steps:

- Rename fields. Is each column name a clear reference to the data it contains?
- Change the data types of fields. Changing a string to a number allows for aggregation. (This will be covered in more detail in Part II of the book, Data Types.)

Stage 4: Build the Workflow

OK, back to the mouse and keyboard. Where do you start building out this workflow? Well, making a basic step-by-step plan is a good strategy. With Tableau Prep, you can quickly change the order of the transitions or add forgotten ones to move from the original data to the desired state. You might not get the workflow right the first time, but you will be a lot closer for having planned out these steps.

Also, you might not know which tool, transition, or calculation to use to make the change you require, but you will be able to take a step back to rethink the problem and not lose your place in the process.

So let's complete our example. With all the steps just captured, I used Tableau Prep Builder to create the workflow shown in Figure 3-6, from inputting the data to outputting it as a CSV (comma-separated value) file. The first step (icon) in Figure 3-6 is the Input step, where the data is imported into Prep Builder for processing. The second icon shows the step of pivoting multiple columns of data to rows of data. The pivoting process will convert all of the different dates into one column, with another column holding the respective value for that date, branch, and product combination.

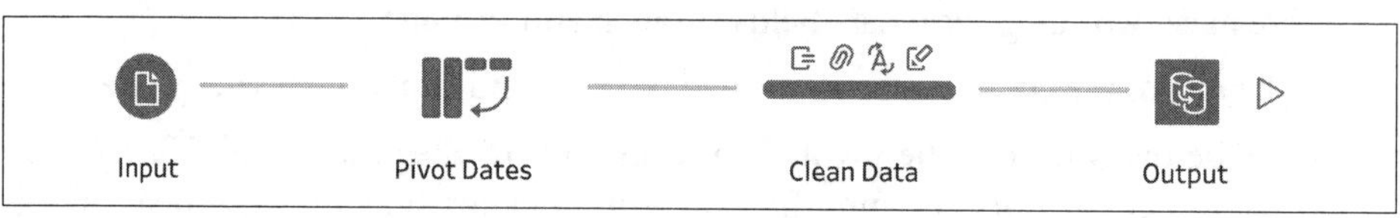

Figure 3-6. The Tableau Prep flow from input to output

The Clean step (the third icon) contains a lot of detail that is captured in the tools Changes pane (Figure 3-7), namely that we:

- Created a calculation to change the product from "Liquid-Soap" to "Liquid" and "Soap-Bar" to "Bar."
- Grouped two Lewisham stores together.
- Changed the Pivot1 Names field name to Date.
- Changed date type for the newly named Date column to a date.
- Renamed the Pivot1 Values column to Sales.

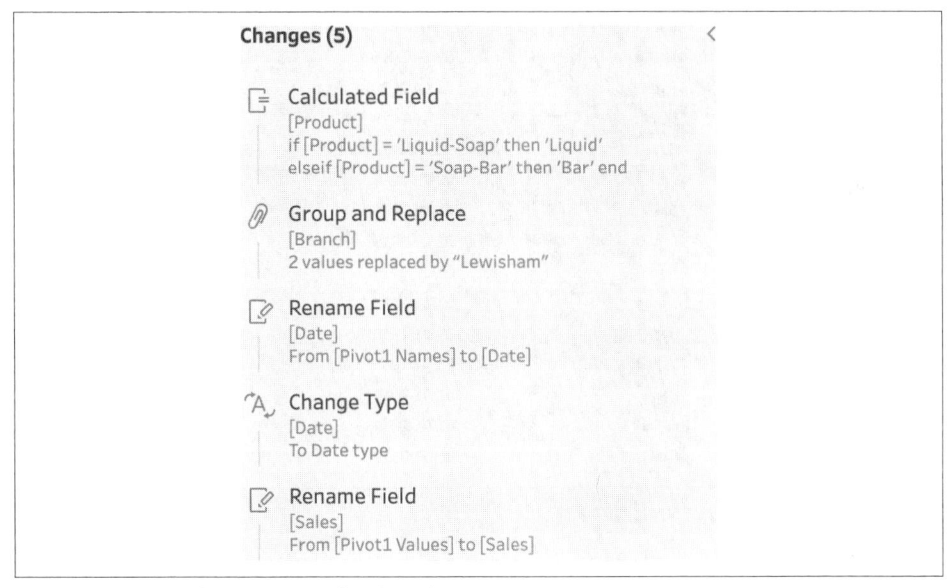

Figure 3-7. The Changes pane for the Clean step in Figure 3-5

This leaves us with a nice clean data set, ready for analysis when we output it in the fourth and final step (Figure 3-8).

We can output the data set in a few ways, but if it will be analyzed in Tableau Desktop, then a Hyper file is a good choice. Output options will be covered in more detail in Part IV.

See the book's website (*https://oreil.ly/5k_uH*) to access all the files used if you want to have a go at the exercise yourself.

Branch	Product	Date	Sales
Wimbledon	Liquid	01/03/2019	427
Wimbledon	Bar	01/03/2019	377
Lewisham	Liquid	01/03/2019	655.94
Lewisham	Bar	01/03/2019	401.26
Lewisham	Liquid	01/03/2019	686
Lewisham	Bar	01/03/2019	810
Wimbledon	Liquid	01/04/2019	470
Wimbledon	Bar	01/04/2019	305
Lewisham	Liquid	01/04/2019	789.63
Lewisham	Bar	01/04/2019	291.99
Lewisham	Liquid	01/04/2019	661
Lewisham	Bar	01/04/2019	504
Wimbledon	Liquid	01/02/2019	344
Wimbledon	Bar	01/02/2019	652
Lewisham	Liquid	01/02/2019	804.22
Lewisham	Bar	01/02/2019	647.53
Lewisham	Liquid	01/02/2019	555
Lewisham	Bar	01/02/2019	727
Wimbledon	Liquid	01/01/2019	173
Wimbledon	Bar	01/01/2019	708
Lewisham	Liquid	01/01/2019	276.31
Lewisham	Bar	01/01/2019	359.12
Lewisham	Liquid	01/01/2019	643
Lewisham	Bar	01/01/2019	323

Figure 3-8. The restructured data

Summary

By planning your data preparation, not only will you be more focused on the task of preparing data, but you'll also have a solid basis to work from when you're ready to analyze that data. Some of these techniques are challenging, especially when you apply them for the first time, but understanding and practicing them will improve your odds of success. Input and output data sets can be large and complex, so the planning process might require a significant investment of time before you can start making progress on manipulating the data. It's only normal to want to dive right in, but this planning effort will repay you in the long term by reducing your risk of hitting roadblocks or otherwise getting offtrack in your data analysis.

CHAPTER 4
Shaping Data

As discussed in the previous chapter, the first step of data preparation is understanding how the original data set is structured, followed by determining what structure the data set needs to be in—that is, the desired state—for analysis. This chapter looks at these factors and Prep Builder's functionality for reshaping data to help you quickly identify how to shape future data sets for analysis.

What to Look for in Incoming Data Sets

Let's look at a typical input data set from Excel by building a pivot table (Figure 4-1). This example uses sales of ice cream-scented soap.

Category	Measure	Jan-2019	Feb-2019	Mar-2019	Apr-2019	May-2019	Jun-2019
Mint Choc Chip	Sales	6933	9895	1871	4649	6492	8956
Mint Choc Chip	Profit	965	353	357	469	525	996
Strawberry	Sales	3832	2512	4738	8254	3816	4109
Strawberry	Profit	574	775	715	523	119	949
Vanilla	Sales	5206	1440	6397	2299	1178	7046
Vanilla	Profit	174	921	346	675	968	968

Figure 4-1. Ice Cream Scents sales and profit data

When assessing an incoming data set, it's important to identify both the dimensions (or categorical values) of the data and the measures. *Dimensions* is the term Tableau Desktop uses to refer to the columns of data that describe the records example, the regions a product is sold, in or the category that product belongs to). *Measures* refers to the numeric values of the data set that are being analyzed (such as the number of students in a college class or the tuition they are paying). If the dimensions are all in individual columns, you can move on to measures without having to think about any structural changes. Likewise, you need to assess the measures to ensure each has a separate column in the data set.

The data set from Figure 4-1 has been colored in Figure 4-2 to highlight the structure:

- First two headers: Header for dimension column
- First column values and monthly data headers: Categorical values
- Second column values: Headers for the measure column
- Everything else: Values for the measures columns

Category	Measure	Jan-2019	Feb-2019	Mar-2019	Apr-2019	May-2019	Jun-2019
Mint Choc Chip	Sales	6933	9895	1871	4649	6492	8956
Mint Choc Chip	Profit	965	353	357	469	525	996
Strawberry	Sales	3832	2512	4738	8254	3816	4109
Strawberry	Profit	574	775	715	523	119	949
Vanilla	Sales	5206	1440	6397	2299	1178	7046
Vanilla	Profit	174	921	346	675	968	968

Figure 4-2. Data set broken down into dimensions and measures

By highlighting the structure of the input data set, you can more clearly see what changes you need to make. Once you have this understanding, the process of shaping the data for analysis becomes quicker.

What Shape Is Best for Analysis in Tableau?

When you load data into Tableau Desktop, the software sets the first row of data as the headers for the columns and all subsequent rows as the data points for those headers.[1] Here are the key aspects to consider when structuring data for Desktop:

- Is there a single column for each data field? These columns will form the data fields that are then dragged and dropped in Desktop.
- Is the data field a dimension or measure? Desktop will divide all data fields into dimensions (category) and measures (the numerical values to analyze).
- Is there a single data type for each data field? A data field in Tableau (and most other tools) requires a single data type. Measures must be either an integer or float data type for analysis.

There are a few aspects that do not matter:

1 This is not the case for Data Interpreter with Excel data sources. The Data Interpreter analyzes the worksheet to return the table(s) of data the sheet contains.

Order of columns

Tableau Desktop, Server, and Online will import a data set and order the fields shown in the Data grid in alphabetical order. Therefore, there is no need to order the columns.

Order of the rows

You will analyze the rows through the visualizations you build. The charts and graphs can be sorted, but the rows of data won't be shown in the order they are imported into the tool, so you don't need to worry about their order in the data source you are building.[2]

Geographic roles

These are string fields you assign to location-based fields. Tableau Desktop commonly associates the role you apply (e.g., City or Country) to a longitude and latitude value. In Prep, however, spatial objects (aka shapes) can't be prescribed for use in Desktop, as the output doesn't maintain this metadata. The spatial role is often a string data type so that the data field can then be assigned a geographic role in Desktop. The data fields that you wish to assign geographic roles should be cleaned, but no further actions need to be taken. If your data set already contains longitude and latitude values, I recommend you use these, as they are likely to be more detailed than the longitude and latitude values Desktop associates with a geographical role.

Let's apply these factors to our ice cream scents example in Figure 3-2. We can see the Category column is in the correct state. The Category header is at the top of the column containing all the relevant dimension values.

The Measure header is in the correct location, but is it necessary? The measures are listed under each individual month. One column containing all of the different dates would be more preferable and easier to use to analyze data over time in Tableau. Therefore, in this case, the dates currently listed as headers will need to be pivoted.

The measures that are named in the Measure column would be much easier to analyze if they were individual columns. That would allow us to more simply create totals or averages within either Prep Builder or Tableau Desktop. Having one column labeled Sales and one labeled Profit would enable us to use these two columns as measures when analyzing the data in Desktop.

Now that we've identified how we need to shape the data, let's look at the steps for actually making these changes.

2 Prep Builder does not maintain the data source order when loading data. If row order matters, create a row identifier within the data source before loading your data in Prep Builder.

Changing Data Set Structures in Prep Builder

There are four key steps to changing the structure of the input data sets.

Pivot

The Pivot step is the most important one for changing the data structure. There are two types of pivot, as shown in Figure 4-3.

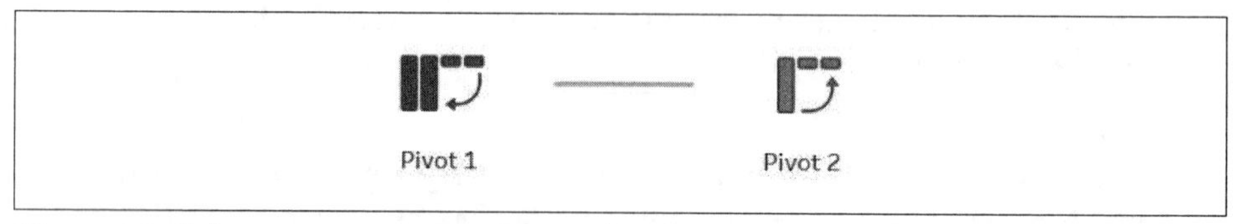

Figure 4-3. Columns-to-rows pivot (Pivot 1) and rows-to-columns pivot (Pivot 2)

Pivot 1, Columns to rows
: Converting from multiple columns of data to additional rows of data. The column header is converted into a new dimensional column that will contain all other column headers that are involved in the pivot (Figure 4-4).

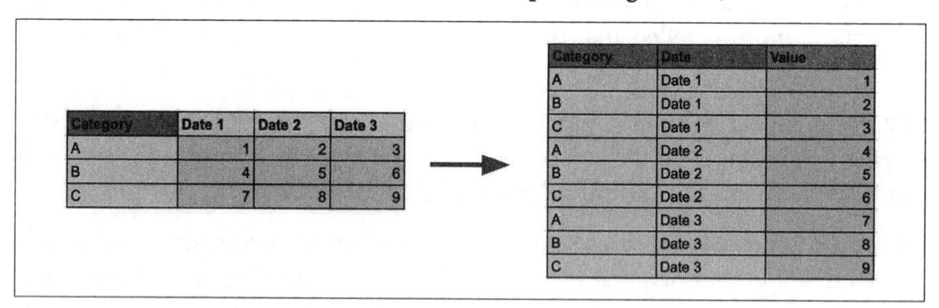

Category	Date 1	Date 2	Date 3
A	1	2	3
B	4	5	6
C	7	8	9

Category	Date	Value
A	Date 1	1
B	Date 1	2
C	Date 1	3
A	Date 2	4
B	Date 2	5
C	Date 2	6
A	Date 3	7
B	Date 3	8
C	Date 3	9

Figure 4-4. The columns-to-rows pivot

Pivot 2, Rows to columns
: This is the reverse of the columns-to-rows pivot. In this pivot, rows of data are converted into additional columns within the data set (Figure 4-5). This requires you to select both the column that will become the headers of the new data fields, as well as the data field that will act as the values for the new data fields. If multiple values are forced into the same cell, you'll need to choose and apply a form of aggregation to them.

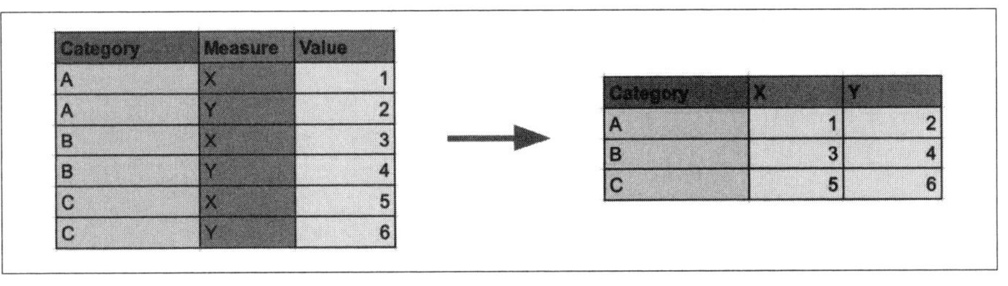

Category	Measure	Value
A	X	1
A	Y	2
B	X	3
B	Y	4
C	X	5
C	Y	6

Category	X	Y
A	1	2
B	3	4
C	5	6

Figure 4-5. The rows-to-columns pivot

Pivots are covered in more detail in Chapters 13 and 14.

Aggregate

The Aggregate step is where values are summed or averaged (Figure 4-6). Aggregations change not only the number of rows, but also potentially the structure of the data.

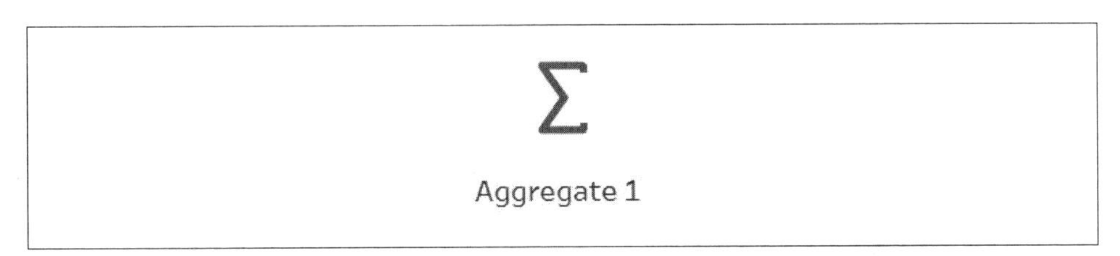

Figure 4-6. Aggregation icon

The only data fields that continue on in the data flow from this step are those included as a Group By field or an aggregation. In this example, Category is the field we are grouping by and whose values are summed (Figure 4-7).

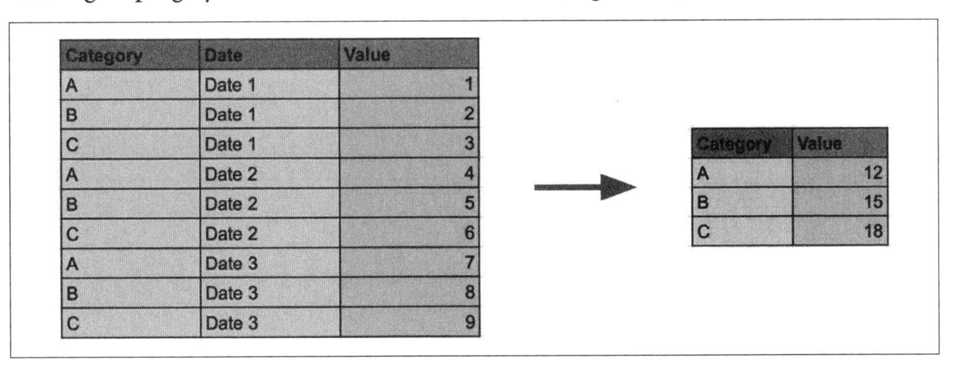

Category	Date	Value
A	Date 1	1
B	Date 1	2
C	Date 1	3
A	Date 2	4
B	Date 2	5
C	Date 2	6
A	Date 3	7
B	Date 3	8
C	Date 3	9

Category	Value
A	12
B	15
C	18

Figure 4-7. An aggregation

Aggregations are covered in detail in Chapter 15.

Join

The Join step (Figure 4-8) adds columns to the original data set from additional data source(s).

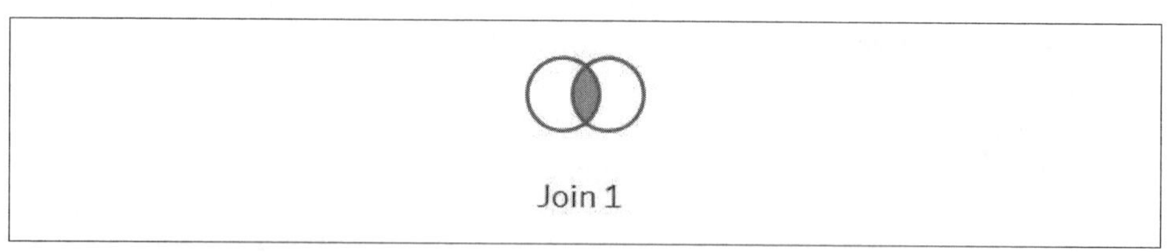

Figure 4-8. The Join icon showing an inner join

Depending on the join type and join conditions you set, the resulting data set will differ in terms of the number of data fields, as well as the number of rows if the granularity of the two data sources is different. In Figure 4-9, an inner join joins the rows from two tables based on their common category to form one table.

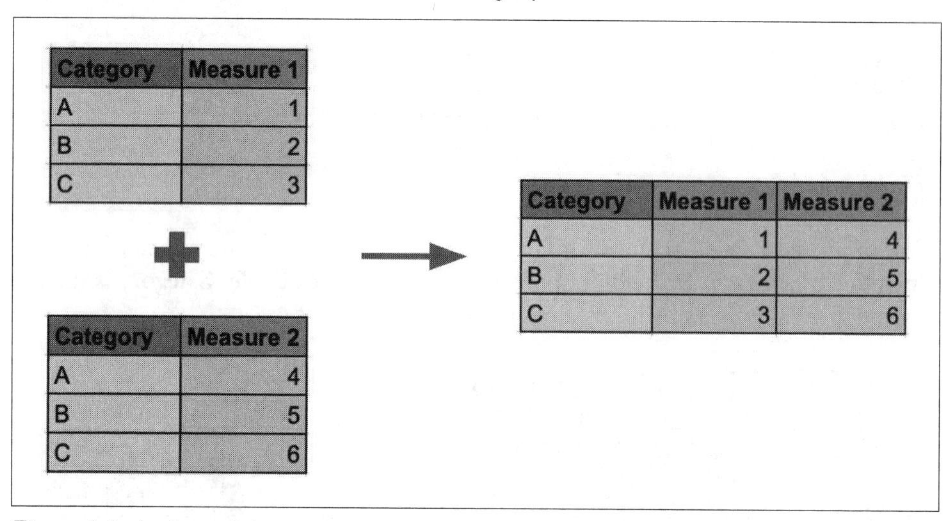

Figure 4-9. An inner join

Joins are covered in more detail in Chapter 16.

Union

The Union step (Figure 4-10) is used to stack data sets on top of each other.

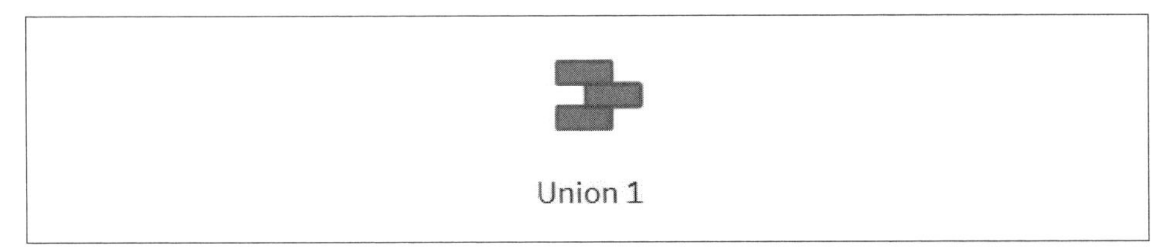

Figure 4-10. The union icon

If the data sets have the same structure with matching data fields, the resulting data set will stay the same width and just include more rows (Figure 4-11).

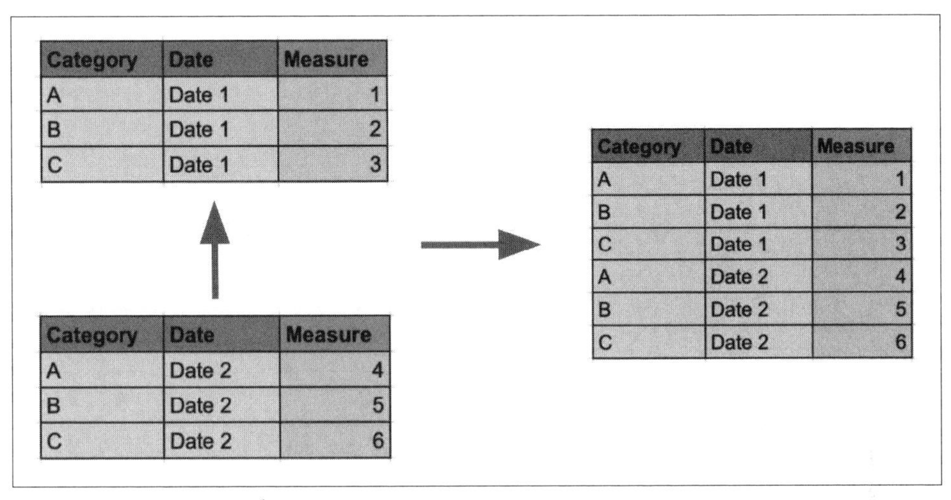

Category	Date	Measure
A	Date 1	1
B	Date 1	2
C	Date 1	3

Category	Date	Measure
A	Date 2	4
B	Date 2	5
C	Date 2	6

Category	Date	Measure
A	Date 1	1
B	Date 1	2
C	Date 1	3
A	Date 2	4
B	Date 2	5
C	Date 2	6

Figure 4-11. Unioning data sets

This step can create a different data structure, as unioning mismatched column headers will create a wider data set. If you don't merge the mismatched fields, any fields that are not contained in both data sets will have null values.

Unions are covered in more detail in Chapter 17.

Applying Restructuring Techniques to the Ice Cream Example

Let's apply these data restructuring techniques to the ice cream scents example.

Step 1: Pivot Columns to Rows

Creating a single column to contain dates will allow us to use the data within Desktop more easily, as the date field can be placed on the Columns shelf of Desktop to create an x-axis covering all the dates in the data set. This makes it easy to perform a time-based analysis. This pivot step creates a column labeled Pivot1 Names to hold the former column headers (Figure 4-12). The Pivot1 Names header should be renamed Date for clarity.

Shelves are where you place data in Desktop to analyze it.

Category	Measure	Pivot1 Names	Pivot1 Values
Mint Choc Chip	Sales	Jan-2019	6933
Mint Choc Chip	Profit	Jan-2019	965
Strawberry	Sales	Jan-2019	3832
Strawberry	Profit	Jan-2019	574
Vanilla	Sales	Jan-2019	5206
Vanilla	Profit	Jan-2019	174
Mint Choc Chip	Sales	Feb-2019	9895
Mint Choc Chip	Profit	Feb-2019	353
Strawberry	Sales	Feb-2019	2512

Figure 4-12. The result of the columns-to-rows pivot

Step 2: Pivot Rows to Columns

To create a column for each measure, we need to convert the Measure column values into headers for the new columns containing the relevant sales and profit data (Figure 4-8). The values in these columns will come from the Pivot1 Values column.

Category	Date	Sales	Profit
Mint Choc Chip	Jan-2019	6933	965
Strawberry	Jan-2019	3832	574
Vanilla	Jan-2019	5206	174
Mint Choc Chip	Feb-2019	9895	353
Strawberry	Feb-2019	2512	775
Vanilla	Feb-2019	1440	921
Mint Choc Chip	Mar-2019	1871	357

Figure 4-13. The result of the rows-to-columns pivot

Depending on the desired analysis and the tool in which it will be conducted, you might want to aggregate the data at this point up to the category level for each measure. Tableau Desktop can easily aggregate a small data set like this. If no additional data is required from any other tables, no further preparation steps are required for this example.

Summary

As you've seen, restructuring data is a key skill to master and practice, as it has made our sample data set much easier to analyze. When manipulating the shape of the data, you must take care to ensure that you haven't inadvertently duplicated records or changed a measure's total. Also keep in mind that a longer and thinner data set is not always the intention. A couple of key elements to aim for are:

- A single data field for each dimension, ideally containing a single data type (like a string or date)
- One column per measure

The more closely your data matches this structure, the easier it will be to analyze flexibly, no matter what the subject of the data is. As always, planning your steps before reshaping your data will make the process much easier. As you progress through the following chapters, you will learn more about the importance of reshaping data sets as well as techniques and tactics for doing so that will make your analysis much easier.

CHAPTER 5
Connecting to Data in Files

Files Upon Files Upon Files

One of the first steps you will take in any data preparation, visualization, or analytics project is to input data. In this chapter, you will learn how and why you should connect data files to Prep Builder, where to find those files, and what challenges data files may pose for preparing data. Data can come in many different forms, but the most common input for Prep Builder is Excel, CSV, or plain-text files (Figure 5-1).

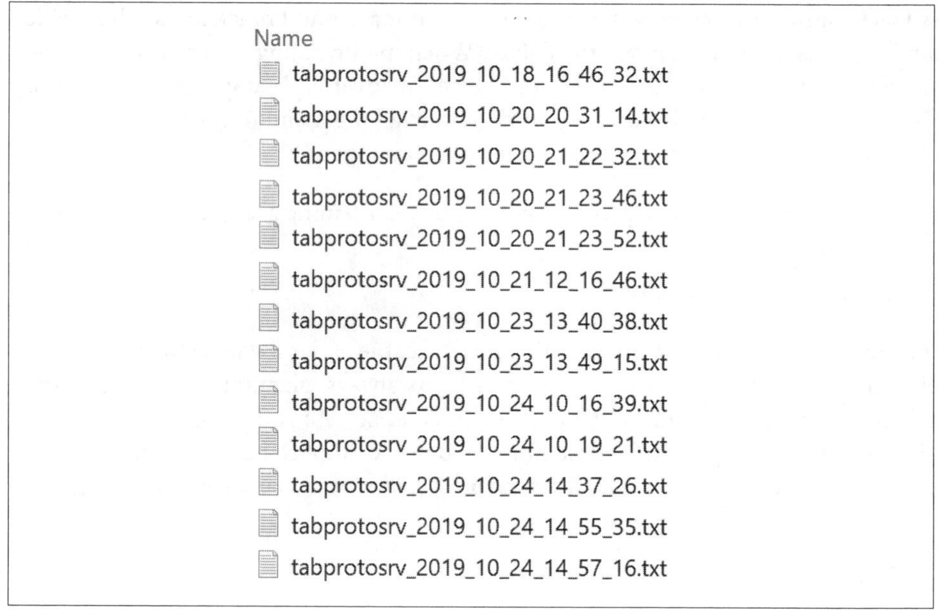

Figure 5-1. System files generated by software

Spreadsheets

For many of us, our first experiences of working with data involved a piece of ubiquitous software: Microsoft Excel. The spreadsheet has become the jack-of-all-trades for data storage, manipulation, and basic reporting. Every organization across the world has tens, if not hundreds, of spreadsheets saved on every computer. All of them contain useful information, raw data, or reference tables that can help answer the questions organizations are posing.

However, spreadsheets raise some potential issues (some of which are not unique to spreadsheets) when it comes to data preparation and analysis:

Accuracy
: The data may have changed since being added to the spreadsheet.

Timelines
: The data may have become outdated since being added.

Optimization
: Any manipulation or analysis is often manual and nonrepeatable.

Size
: With limits on rows and columns, sometimes the data grows beyond spreadsheets.

Sharing
: Multiple people using multiple sheets can create confusion and errors.

Single-point solutions
: Calculations and analysis are limited to the single workbook in which they're conducted.

Organizations are complex. Mergers, acquisitions, and splits can create a complex history and network to unpack, and the data often reflects this. Trying to maintain the integrity of the data and your understanding of it is a challenge in itself.

Other File Types

Data files come in many formats besides Excel spreadsheets—CSV, text, statistical, and PDF files are all likely to house data on your computer that could help you answer the questions you have. All these file types are created and used for multiple reasons. Often, they are default outputs from different programs where you're working with operational or analytical systems in your organization. Let's take a closer look at these files and where they come from:

CSV (comma-separated value)
These files can be output as a data storage file. Whether you are sharing data for programs that are not in the Microsoft suite or simply moving data into databases, CSVs are very flexible file types. Columns of data are separated by commas (hence the name), with new rows of data listed on separate lines within the file. CSVs are actually a text file but with a fixed delimiter.

Text
Though these files are suitable for holding data in a very simple format, they have the potential to be problematic to work with due to their ability to contain differing structures and data formats. Often, text files are delimited in some way, but this format isn't always consistent or as easy to use as a CSV file.

Statistical
R and Python are increasingly being taught in universities. Therefore, these file types are being demanded more in the workplace, especially by data scientists and others working with the statistics resulting from R and Python models and packages.

Portable Document Format (PDF)
These commonly used files often contain useful data sets. How a PDF has been created and formatted makes a massive difference as to how easy it is to extract its data using a data preparation flow. For example, if a table is created as an image, its cells can't be read as data fields and values, but if it's created as a grid of columns and rows, then they can be.

Where to Find Your Data Files

These file types can help hold a range of data in lots of different formats. That's one of the many reasons they are so commonly used, but it creates some issues for us as data preppers. Each file will have its own structure, messiness, and set of challenges that we'll have to overcome before we can use the data for analysis or augment it with other data sets to add extra value.

Not only that, but the data files could be anywhere and everywhere. It's key to know the right people in your organization who may have collaborated with others or created the files themselves. Here are some possible sources and some considerations around each:

On your computer created by you
Building your own data sets is a common task in the workplace. You might be capturing numbers from reports you read, collecting records, or aggregating other data sets that have been sent to you.

On your computer created by others
: Emailed database extracts, survey results, or market research are also common sources of data. The greatest challenge with this data is making sure you're getting the frequent updates you need to keep your analysis fresh. The hard part isn't knowing who to ask for current information and insight; it's waiting for them to reply.

Cloud storage
: Team drives held on cloud computers and servers present more of a challenge, as multiple people will be adding files to shared drives or editing existing files, making it a lot tougher to track down the sources for the information. For example, it can be more difficult to learn how data points have been defined, what aggregations have been applied, or what data has been removed. Even the task of refreshing the data set may be challenging due to the difficulty of tracing the origin of the file(s).

On the internet
: Many files are now composed of downloads from web pages. The type of file downloaded will affect how well structured it is and hence what challenges you might have to overcome when preparing the data for analysis.

This flexibility means your files will contain many useful bits of data. Whether it is projected budgets, targets, or the latest reorganization structure to be applied to your earnings, you will need to be comfortable preparing these data sets for analysis. Now, though, rather than using manual manipulation, you will be able to use Prep Builder to separate the useful from the dross and structure the data in preparation for your analysis.

How to Connect to Files in Prep

Tableau Prep Builder's Home screen is very similar to that of Tableau Desktop. To make any data connection, click the plus sign next to the word Connections (Figure 5-2).

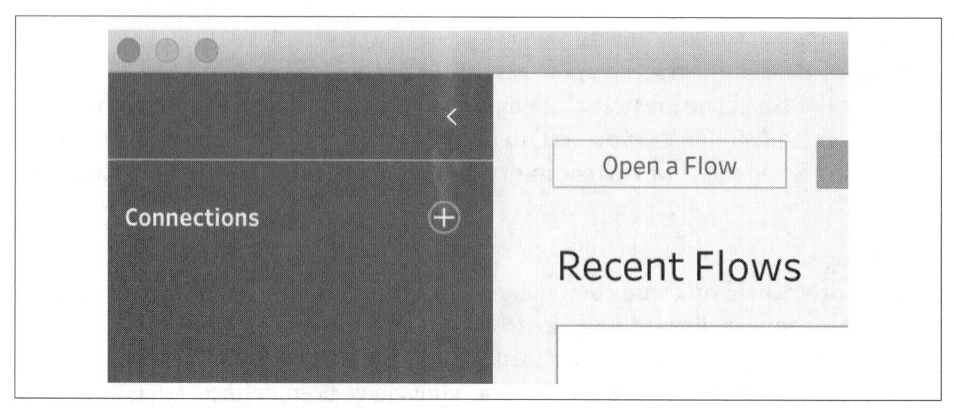

Figure 5-2. Default Connections pane in Prep Builder

As shown in Figure 5-3, this expands the Connections pane to display all the connection types you can make in Prep Builder, including the file types just discussed (except spatial files, as of version 2020.1).

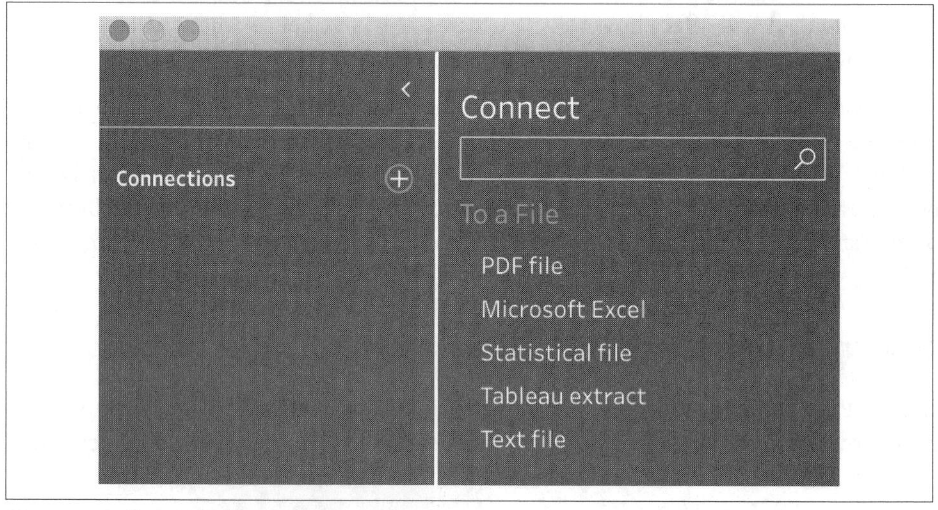

Figure 5-3. Expanded Connections pane

Click the relevant file type to open a File Explorer (Windows) or Finder (Mac) to choose the location of your file. Once you find the file, click Open to connect to that file, after which you're ready to prepare your data for your purposes.

Considerations for Saving Flows with File Inputs

When saving your Prep Builder flows, you have two main options. First, you can create a Tableau Flow File (*.tfl*), which just saves the logic of the flow. This means the input file's location needs to match the location where the current flow is connected. Otherwise, if the flow is rerun then it will throw an error because it can't find the input data. The second option is to create a Packaged Tableau Flow File (*.tflx*), which saves both the flow logic and the input data in the same file. This means you can move the input files and the flow will still run. However, if the original input files are updated, this flow will not process the new files; it runs only the file saved within the packaged flow file.

Summary

Big data is not always the biggest challenge within the data world. Small data files can be hugely beneficial in terms of containing useful data, but that data is not always easy to extract. Using tools like Prep Builder makes preparing these files much easier and repeatable, unlike manual workarounds to achieve the same result.

CHAPTER 6

Connecting to a Database

Using data files will be a significant part of most users' data preparation, but connecting to databases is another option. Most organizations have built up significant data assets, the majority of which are held within databases. In this chapter we will look at what a database is, how to securely connect Prep Builder to one, and when to avoid connecting to a database.

What Is a Database?

Databases, data warehouses, and data lakes are terms that will be familiar to most people using data in their jobs every day, but it's important to distinguish between them for everyone else:

Database
: A piece of software that resides on a computer (often a server) that specializes in ingesting, storing, and providing data to other tools. The database is likely to be split up into different objects, namely tables and views; therefore, the data needs to be well structured.

Data warehouse
: A collection of databases or a particularly large database. Because it stores multiple servers together, a data warehouse allows for sharing resources like memory between them.

Data lake
: A newer concept that is becoming more common and allows for more flexible storage of all data types and files. Data is often held in data lakes while awaiting processing and restructuring for storage in a database.

I will use *database* as a catch-all term for all three storage forms unless otherwise specified.

Databases hold the majority of the data in tables. The main type of storage, a relational database, optimizes storage space and enables easier updating by assigning data numeric IDs that you can look up in reference or dimension tables. Piecing together the data that you require adds another hurdle to working with data: linking multiple data tables using joins (see Chapter 16 for more on joins). Figure 6-1 shows a simple set of dimension tables and a fact table.

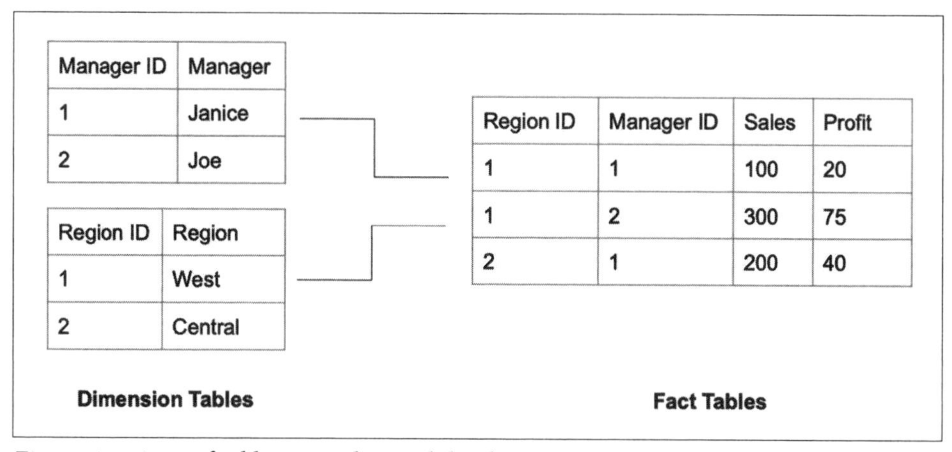

Manager ID	Manager
1	Janice
2	Joe

Region ID	Region
1	West
2	Central

Dimension Tables

Region ID	Manager ID	Sales	Profit
1	1	100	20
1	2	300	75
2	1	200	40

Fact Tables

Figure 6-1. A set of tables in a relational database

Connecting all of the tables together each time you want to use those data sets would be laborious and error-prone. To mitigate this, the database creates *views*, which are accessed very similarly to tables. In fact, only in the database is a view different from a table. The table holds the data, whereas the view is the logic of how tables connect, including the filtering, calculating, and renaming of data fields. The view resulting from the connections made in Figure 6-1 is shown in Figure 6-2.

Region	Manager	Sales	Profit
West	Janice	100	20
West	Joe	300	75
Central	Janice	200	40

Figure 6-2. View resulting from the dimension and fact tables in Figure 6-1

How to Connect to a Database Within Prep Builder

Connecting to a database is a different experience from connecting to a file, but is relatively similar across the different database connections available within Prep Builder. Let's walk through one of the most common databases in the world: Microsoft SQL Server (Figure 6-3).

Figure 6-3. SQL Server connection setup

To connect to SQL Server, follow these steps:

1. Enter the IP address or URL for your server.
2. Pick the type of authentication that the setup requires. You'll need to work with your database administrator to determine this.
3. Enter your username and password.

The username and password credentials are the same as those set up for you to access the database itself. That is, any access set up for people who normally use the database will be available for them within Tableau Prep Builder and Desktop. The same is also true for tables and views that the users *don't* have access to; if they don't have access to the table or view itself, they won't have access in Tableau either.

After clicking Sign In, you'll be taken to the normal Prep Builder interface but with a different Connections pane than the one you see when working with files (Figure 6-4).

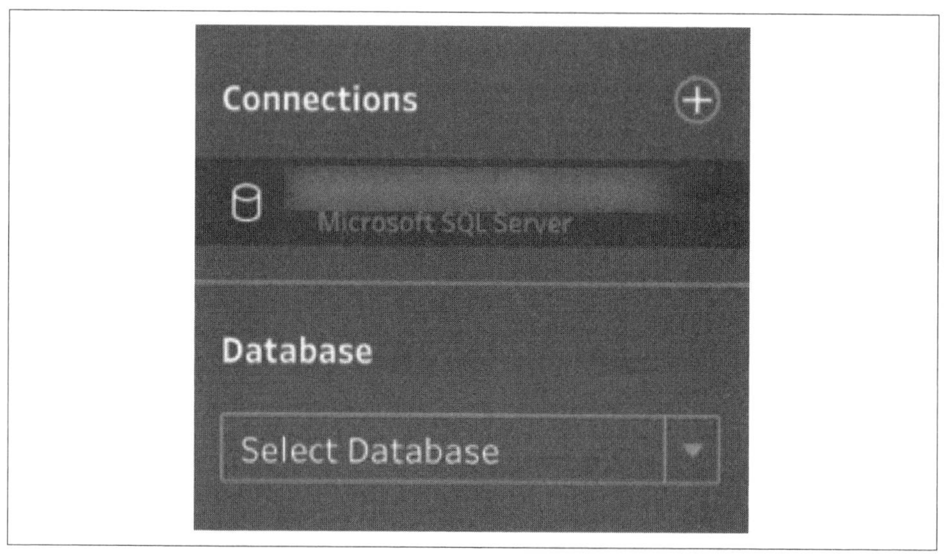

Figure 6-4. Connections pane with drop-down list of available databases

You'll see a drop-down list of databases found at the server address you entered earlier. After selecting the database required, you'll be presented with a list of tables to select where the data resides (Figure 6-5).

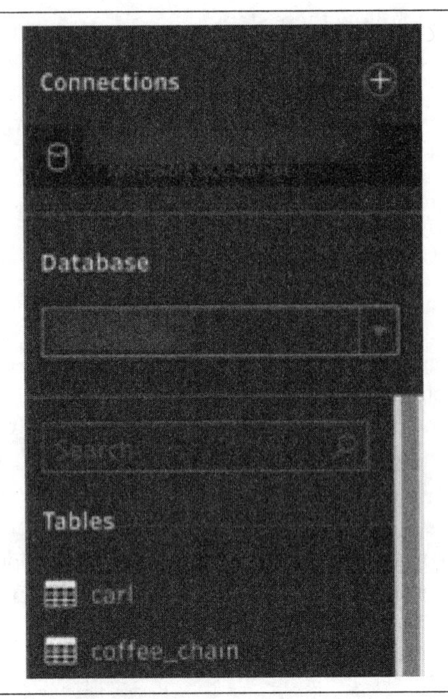

Figure 6-5. View of tables available after selecting a database

Danger!! What Damage Can You Do?

If you're worried about accidentally overwriting database data, don't be. Permission to write to a database is often controlled by database administrators, so you would not have access to perform any tasks you are not trained to do. In most situations, data preparation will involve writing a new output file rather than overwriting existing data.

The only damage you can do is making decisions based on incorrectly formed data sets. Because joins are much more common when working with data from databases, there is greater potential for poorly formed data sets. Prep Builder has a great answer to this, though: the Profile pane. The Profile pane allows you to see the data, have rogue values highlighted through data roles, spot gaps in the distribution of data values, and identify where joins have formed incorrectly. For example, in Figure 6-6, it's easy to see that the value "Sarah" has mistakenly popped into the Category list.

Overall, the benefits of working with a database far outweigh the complexities of connecting to and joining tables, thanks to the database's vast storage and processing capabilities compared to directly using files. It takes some time to learn how to get the most out of a database, but using Prep Builder is a great introduction, as it simplifies a lot of the aspects of using one.

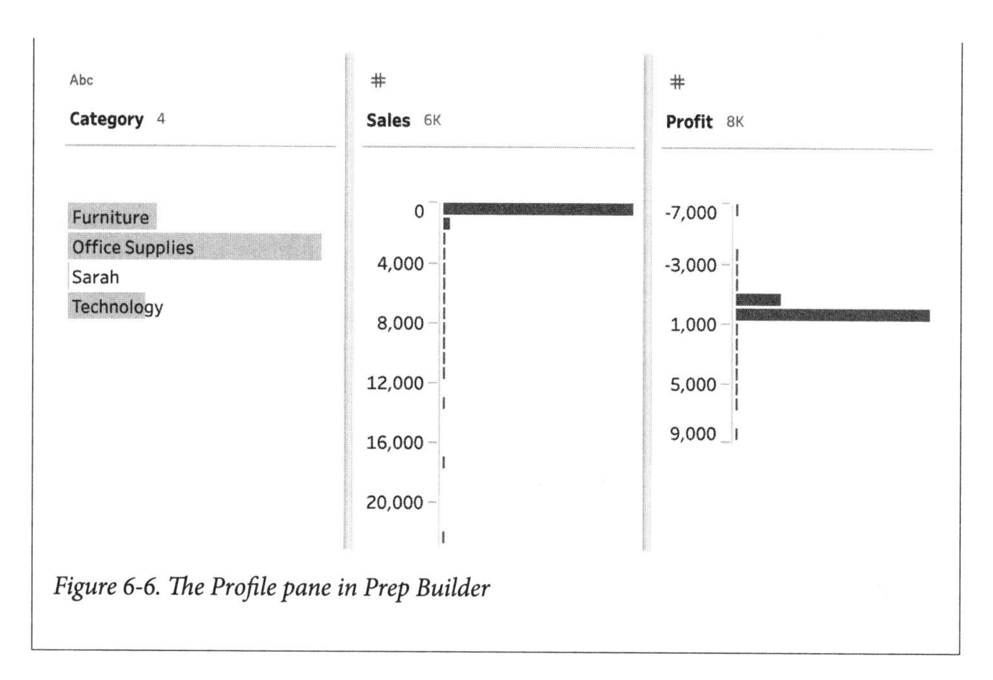

Figure 6-6. The Profile pane in Prep Builder

When to Avoid Connecting to a Database

Despite their power, databases are not the solution for every data preparation task. Databases largely require structured, clean data. If you have files of data that contain useful information, it's worthwhile to access these directly rather than wait to have them loaded into a database first. Obviously, the files can be added to the database if they become a data source for a productionalized report or analysis. *Productionalizing* a flow or report means setting it up to be more restrictive to prevent accidental edits and allow for automated updates if possible.

Gaining permissions to a database can be a blocker to quick, agile work. If the data resides only in a database on a table that you don't have permission to access, you have no choice but to wait. However, if you can source the information from a file more quickly, it might be worth it to get started and you can switch to the official source once permission is granted.

Databases are heavily used in large organizations. In fact, they can actually be slower to work with than extracted data sets held on Tableau Server or elsewhere due to the number of data refreshes and queries, and the sheer volume of data stored across your organization's network. As an alternative, Tableau has its own file type, called Hyper, for extracting data sets. As Hyper files are constructed to be optimized for use within the Tableau product range, they offer very good performance even on large data sets. If Prep Builder begins to take a long time to process anything you do within the tool while using a database, work with your data team to see if there are other options to host the data outside of the database.

Summary

Having to use coding to access a database may be a barrier preventing many potential users from gaining the benefits of the data it contains. Prep Builder removes this barrier by allowing you to perform most actions through drag-and-drop menus and clicks in the configuration panes. Once you gain experience connecting to databases and you know your tables inside and out, there is very little to stop you from creating a strong data analysis. Data stored in databases often requires less cleaning than data stored in files, and the time you spend setting up joins and applying filtering logic is an investment that yields powerful gains. Working with databases requires more care, though, for reasons from permissions hurdles to performance speed. However, the time spent practicing and learning how to use databases will reap a lot of rewards.

PART II

Data Types

CHAPTER 7
Dealing with Numbers

When many people hear "data," they immediately picture large tables full of numbers like the excerpt shown in Figure 7-1. Numerical data sits at the heart of most analysis; therefore, being confident and comfortable with numbers is key to successfully preparing data for analysis and sharing. In this chapter we will cover considerations for using numerical data.

Returned Orders	Total Orders	% Returned	Type
760	1,477	51.5	Bar
750	1,519	49.4	Massage Bar
788	1,527	51.6	Bath Bomb
742	1,477	50.2	Liquid
3,040	6,000	50.7	All
1,272	2,531	50.3	Fruit
1,768	3,469	51	Non-Fruit
502	1,000	50.2	2019-05-01
502	1,000	50.2	2019-04-01
511	1,000	51.1	2019-03-01
478	1,000	47.8	2019-02-01
512	1,000	51.2	2019-01-01

Figure 7-1. Normal table of data

What Do We Mean by Numbers?

If only we could just think about 0, 1, 2, 3, 4, 5, 6, 7, 8, 9. These numerals form the basis for using numbers in data analysis, but it's much more complex than that. Your analyses will often focus on numerical questions, such as:

- What is my total sales?
- How many staff work in our organization?
- How many customers have we sold to?
- What percent of suppliers have billed us?

As discussed in earlier chapters, we commonly call these numbers the *measures* of our data. Analyses will typically compare these measures or break them down by categories to give them more context.

Types of Numbers

There are two types of numbers:

Integer
: Whole numbers (i.e., those without decimal places) such as 4 and 16,874

Decimal/float
: Numbers with decimal places such as 3.1415 and 0.31

Category or Measure?

Numerical data is often used as a category, or dimension, instead of a measure. Customer IDs (identifiers) are a classic example of numerical data that you would not want to simply aggregate and treat as a measure. The average of a Customer ID field wouldn't make much sense, after all.

In some cases, numerical data can be used as either a category or a measure within Tableau Desktop. Age, for example, makes sense as both:

Age as a measure
: Calculating average age within an analysis can be useful for tailoring products, services, or marketing channels to a certain age demographic.

Age as a category
: Looking at the number of purchases or average sales by the age group of your customers can also be useful for understanding customers' life cycles.

Aggregation

After determining if the numerical data field is a measure, the next question an analyst will often ask is: So how should we aggregate the values? You might remember from previous chapters that *aggregation* means summing, averaging, or finding the minimum or maximum value. Even in a simple 100-row data set, just knowing each individual sales figure doesn't really help us analyze the numbers. Therefore, calculating the total, average, maximum, or standard deviation is key in finding the data set's underlying story.

You should start considering this question even before you're using Business Intelligence (BI) tools like Tableau Desktop; you should assess it during the data preparation stage of your work. If you can aggregate your data to a higher level of granularity, it can help speed up your analysis, as it will reduce the number of rows that the BI or visual analytics tool needs to process. The data preparation stage is the perfect point to do this aggregation—you can process the data once to the desired granularity rather than requiring each user to process the aggregations.

Formatting Numbers

During the data preparation stage, you don't need to be too precise about the final formatting of the values in your data set. In Prep Builder, there aren't many options to format the values that will be carried over into Tableau Desktop. However, there are a few exceptions where preformatting data will save you from writing excess calculations in the future:

Currencies
: Rounding values to two decimal places will produce values that are easy and sensible to use in your analysis. However, avoid adding currency symbols to your numbers; doing so will convert the number into a string, preventing data aggregation. If necessary, specify the symbols in your column headers (though it is often better to use currency abbreviations, such as GBP or USD). Also avoid rounding too early within the flow so you don't lose precision in your calculations. Some organizations have regulatory restrictions on where the rounding should and shouldn't occur.

Percentages
: Depending on your analysis, you may or may not want to multiply your percentages by 100. When using software like Tableau Desktop, if you set the formatting as a percentage, then you will want your 31% to actually be stored as 0.31 in the data set. If your percentage is stored as the full value, 31 in our example, you can format the percentage correctly in Tableau Desktop using the custom number formatting of 0\%. As with currencies, naming your column headers logically will help clarify to users how they should interpret and use the column.

Functions for Mastering Numerical Data

Calculated Fields create a new column of data in your data set (unless you give the Calculated Field the same name as an existing field, in which case it overwrites the original content). You can create a calculation from the ellipsis menu at the top of a data field in the Profile pane of the Clean step (among other places). The blue text below the Calculated Field is the function of the calculation, and it determines what happens to the data field. In Figure 7-2, the `round()` function is rounding the Profit Ratio data field to one decimal place.

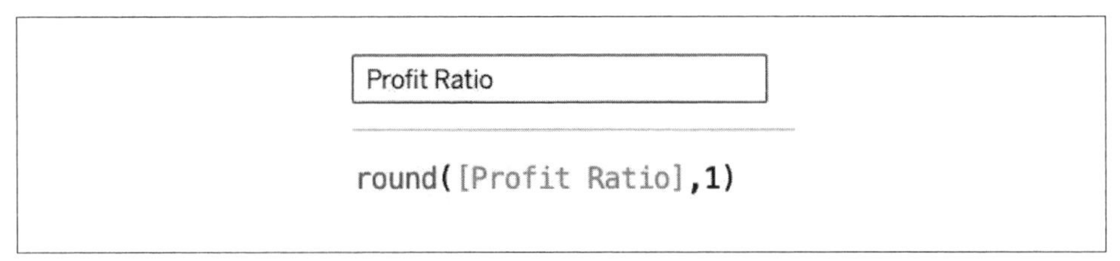

Figure 7-2. Rounding the Profit Ratio field to one decimal place

I use both uppercase and lowercase function names throughout this book, as Tableau Prep accepts both.

When creating Calculated Fields, be sure to pay attention to these elements to avoid returning an error:

Parentheses ( )
: After the function, the text contained in the parentheses indicates what the function acts upon.

Square brackets [ ]
: These are used where the calculation has a space in the name (such as "Profit Ratio"). They surround the name of the data field, turning it orange.

Speech or quotation marks ' ' " "
: These indicate a string field.

Numbers
: When numbers are entered into a Calculated Field without any other punctuation, they are constants.

Double forward slash //
: Using a double forward slash allows the user to leave comments—text that the processor will ignore—within Calculated Fields.

Here are some other useful numeric functions that will help you prepare your data:

`round()`
: Rounds values to a set number of decimal places.

`ceiling()` *or* `floor()`
: Rounds numbers up or down to the nearest integer.

`abs()`
: Short for "absolute," this function returns only positive values; for example, `abs(-7)` becomes 7.

`zn()`
: Short for "zero if null," this function returns a value of `0` if the data field has a `null` response.

`sign()`
: Returns `-1` if the value is below `0`, returns `0` if the value is `0`, and returns `1` if the value is positive.

Each function is described on the right-hand side of the Calculation Editor (Figure 7-3).

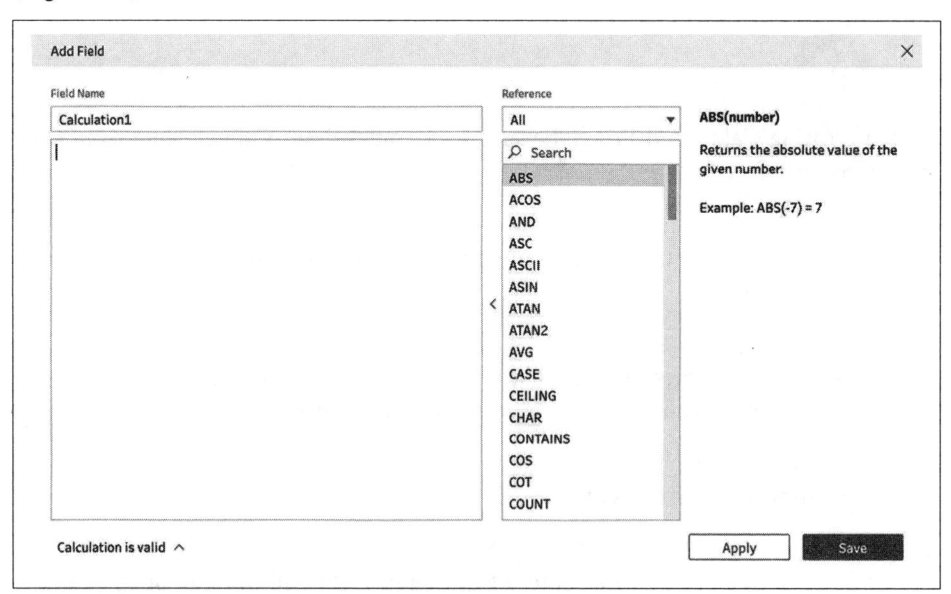

Figure 7-3. The Calculation Editor in Prep Builder

Chapter 18 looks more at how to build calculations from scratch.

Summary

Numbers are probably what most people associate with data, and understanding basic mathematical equations lies at the heart of data preparation and analysis. Knowing when to aggregate numbers and what functions to use to create the metrics you want is a key skill to develop.

CHAPTER 8
Dealing with Dates

Sorry to anyone looking for dating advice—this chapter isn't for you. For those who are battling dates in their data sets, it *is* for you! In this chapter we will explore why dates are important in the world of data, how to break up dates, what different types of dates you might come across and need to prepare, and how to use date functions in Prep Builder to clean up your dates so they're ready for analysis.

Why Are Dates Important?

A basic date field can start to tell us so much:

- How many sales were made on Saturday?
- How many students joined the program this term?
- How many games does my team play this month?

We can begin to answer all of these by counting the number of rows and breaking them down by the different parts of the date.

Parts of a Date

When most of us think of dates, we think of days, months, and years. Which side of the Atlantic Ocean we are on generally determines whether the months or days are listed first. As I am British, the rest of this chapter (and book) will unapologetically place days in front of months.

The date format we commonly use is dd/MM/yyyy (e.g., 25/09/2019). The `d` represents the day of the month; we use `dd` as there are potentially two digits for the day. `M` stands for month, and again `MM` is used because there are potentially two digits in the month. The `y` is for year, and you'll use either two digits for the last two digits of the year, represented by `yy` (e.g., `19` for the date in Figure 8-1), or four digits for the full year, represented by *yyyy*.

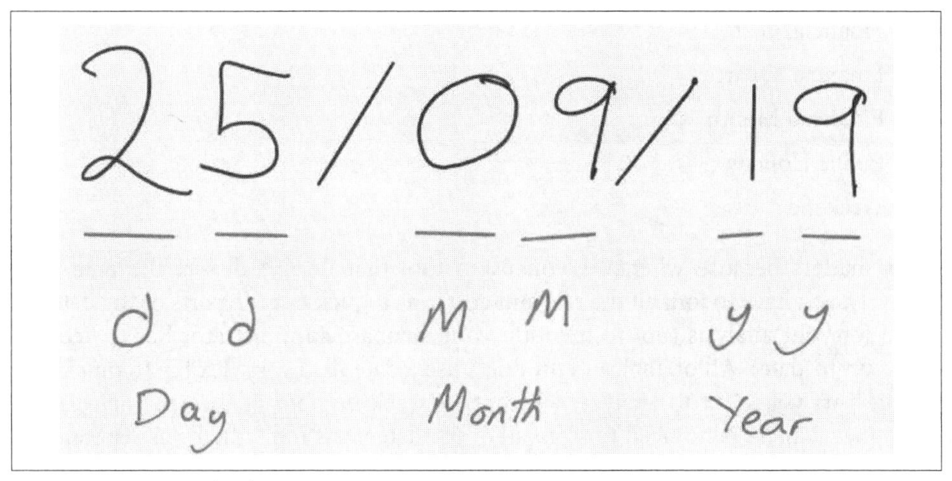

Figure 8-1. Parts of a date

The date parts are case-sensitive. We use a lowercase `d` and `y` for days and years, but an uppercase `M` for months. This is because the abbreviation for minutes (for when we have the time as well as the date) is a lowercase `m`, and we need to ensure clarity between the different elements we're representing.

The date can tell us more than just what day it is, in what month, in what year. Weekday, week, and quarter can all be determined from this basic date format. For example, 25/09/2019 is a Wednesday, in the 39th week, in the third quarter of 2019. The newest BI tools let you dissect the date fields in all of these ways, but that hasn't always been the case.

Date Lookup Tables

In most database setups, you traditionally have access to a reference table that will dissect the same date in different ways. For example, you will often find some of the following columns in the date reference table:

- Date
- Year
- Quarter
- Month
- Week

- Day
- Day of Year
- Weekday
- Financial Year
- Financial Quarter
- Financial Month
- Public Holiday
- Weekend

This matters because whenever you use a tool that doesn't dissect the date more natively, you have to join all the relevant columns to pick out the parts of the date you need for your analysis (not to mention your database administrator has to keep the table up-to-date). All of Tableau's products use a date field very flexibly to quickly get to the part you want to analyze, whether it's week number of the year or weekday. This means there is no need to store all of the date parts individually as separate columns. If you structure your date as a single column, the analysis in Desktop is much easier because no matter what date part you want to use, you are adding the same field into your view. This allows you to flexibly analyze data by dissecting measures by different date parts rather than having to continually swap data fields in and out of your analysis.

Epoch Dates

Unix epoch dates are a date format that is commonly found in data sets but makes little sense to anyone at first glance. As integers (whole numbers) are easier to store in databases than long strings, the epoch date reflects the number of seconds that have passed since 1st January 1970 00:00:00 (UTC/GMT), not counting leap seconds. For example, midnight for our example date 25/09/2019 would result in a value of `1569369600`.

The challenge for many data preppers is to convert this format for use in modern BI tools. In Tableau, you can convert this number back to a date relatively easily using the `dateadd()` function (shown in Figure 8-2), which adds the epoch time in seconds to the base epoch date of 1st January 1970.

Function names in Tableau are not case-sensitive, so you'll see them formatted in both uppercase and lowercase throughout the book.

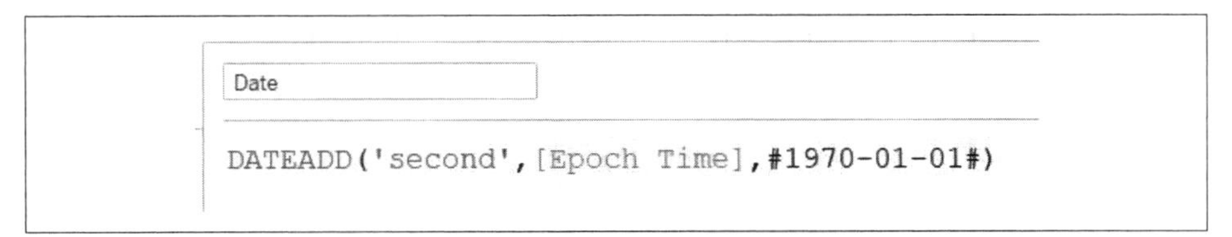

Figure 8-2. Converting an epoch date with the **dateadd()** *function*

If you want to check whether your Epoch conversion is correct, I recommend the Epoch Unix Time Stamp Converter (*https://oreil.ly/WV6fZ*).

Excel Serial Number

If you have worked with data, you may have come across a value that is similar to the Unix epoch date but much smaller. Excel's serial number, as discovered by Hall of Fame Zen Master Jonathan Drummey, is the number of days since 00/01/1900 (a date that has never existed), so 1 would represent 01/01/1900, but, bizarrely, the first leap year is skipped and ignored. The Excel serial number equivalent to our example date is 43731. Using the `dateadd()` function, we can convert this number to a date for analysis much like we did in the epoch solution, but this time we will use increments of days and a start date of 1900-01-01 (Figure 8-3).

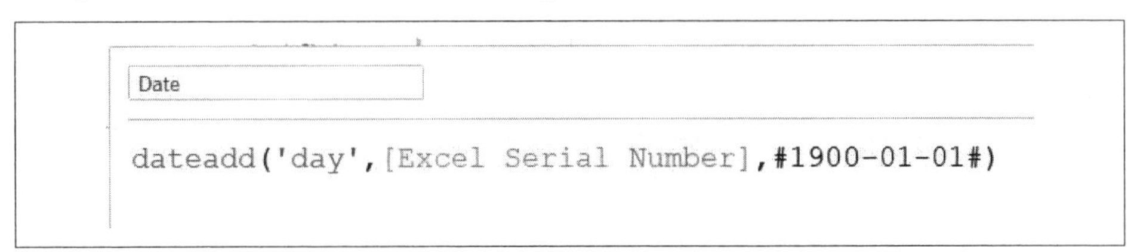

Figure 8-3. Converting an Excel serial date with the **dateadd()** *function*

Entering Dates

You may have noticed in Figure 8-2 and Figure 8-3 that the dates of 01/01/1970 and 01/01/1900 were surrounded by hash marks. This is a common syntax found in many of the calculations and tools you use. There are more common ways to convert dates too. In Tableau, there are a few date-specific techniques that will help make your analysis very easy.

The makedate() Function

If you have each part of the date as numbers in separate fields, you might want to consider using the `makedate()` function, shown in Figure 8-4.

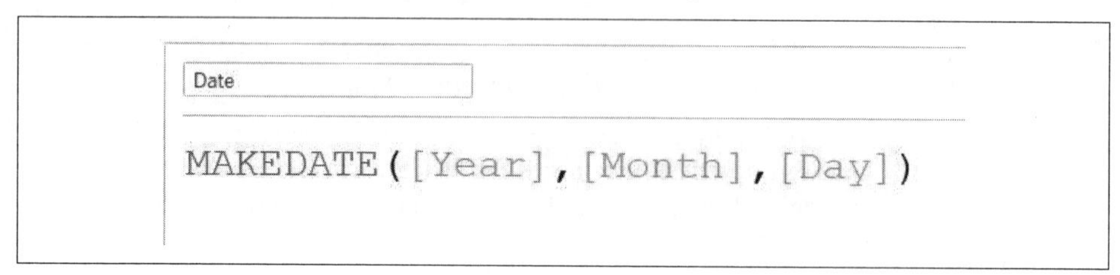

Figure 8-4. The `makedate()` function

This function enables you to piece together the individual components of the date to be analyzed however you want.

The dateparse() Function

Strings are very common for working with dates. Here are some of the ways you might find "September" written:

- S
- Sep
- Sept
- September

Coupled with the different ways to separate the different parts of the dates (spaces, commas, slashes, dashes, etc.), string data for dates can come in many different forms, such as:

- 25 Sep 19
- Sept.25th.2019
- 25-9-19

The `dateparse()` function can take any string and separate out the parts of the date as long as they are consistently formatted throughout the date field. The key to the `dateparse()` function, then, is properly calling the date parts of the string field you are referencing. Table 8-1 shows a sampling of formats for the date and time parts you might call for 25th September 2019 07:19:19.1203.

Table 8-1. Formats for calling date and time parts with the `dateparse()` function

Date part	Abbreviation	Example	Format
Year	Y	2019, 19, 9	YYYY, YY, Y
Month	M	September, Sep, 9	MMMM, MMM, M
Week of year	w (1–52)	39	ww
Day of month	d	25	dd
Day of year	D (1–365)	268	DDD
Hour	h (1–12), H (0–24)	7, 19	h, HH
Minute	m	19, 9	mm, m
Second, millisecond	s, A	19, 9, 1203	ss, s, AAAA
Period	a	PM, pm	aa
Era	G	Anno Domini	GGGG

Figure 8-5 shows how to convert the date format 25.September.2019 to a date with the `dateparse()` function.

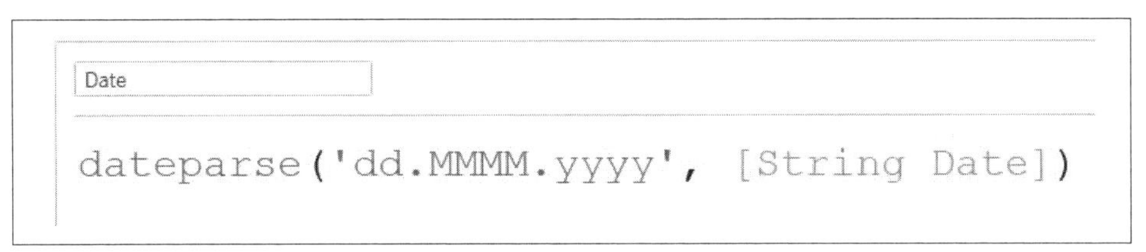

Figure 8-5. The `dateparse()` function

Summary

Dates are fundamental to your analysis, so it's critical to get comfortable using them as part of your data preparation toolkit. Tableau has many specific functions to allow you to flexibly work with dates. These functions can take a while to get used to, but eventually they'll allow you to work quickly to conduct strong time-based analyses. Preparing dates in a data set can save those data users significant amounts of time and avoid misunderstandings about date formats. There's more calculations and use cases than those covered here, however.

CHAPTER 9

Dealing with String Data

When I think about preparing data, I instantly think about battling fields that contain string data. Customer names, product descriptions, messages, and geographic names are all strings that you will likely need to manipulate when you prepare your data for analysis. This chapter will cover what a string is and how it differs from numbers. Having that background will allow you to further explore working with string data fields through both calculations and Prep Builder's built-in functionality.

What Do We Mean by Strings?

When most people think of data, they think about the measures we are summing, averaging, or finding the minimum and maximum of. But those measures don't tell us much until we start analyzing things at different levels—for example, average sales *by region* or total sales *per company*.

Strings are often those dimensions by which we categorize data in tools like Tableau Desktop. Tableau indicates strings with the Abc icon (Figure 9-1).

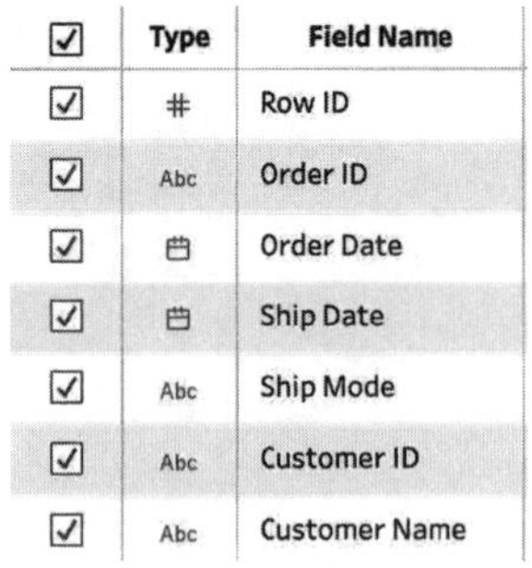

☑	Type	Field Name
☑	#	Row ID
☑	Abc	Order ID
☑	📅	Order Date
☑	📅	Ship Date
☑	Abc	Ship Mode
☑	Abc	Customer ID
☑	Abc	Customer Name

Figure 9-1. String fields in Tableau are indicated by the Abc icon

String data fields are the most flexible in terms of the values they can hold. Typically, they can contain everything from A through Z and 0 through 9 as well as punctuation and most other characters. With that flexibility, though, also comes pain, as the variety of values can create headaches for you as you prepare your data.

How String Data Is Different

We approach cleaning string fields very differently than cleaning numeric fields. This section looks at some of their unique properties.

Character Order

When we analyze string data fields, the order of the characters is very important; we look at which character occupies which position in the string. Figure 9-2 shows an example with the string `Preppin' Data`.

Figure 9-2. String with position values shown

Counting from the left, you can see the first letter in the string is a capital `P`. Let's look at some other facets of this string:

- The last letter is in the 13th position from the left and is a lowercase `a`.
- The apostrophe (') occupies its own position in the string.
- The space between the two words also occupies a position in the string.

Formatting Considerations

There are also some unique formatting challenges to consider when you are working with string data.

Names

Splitting
: Sometimes names are in one column (field) of data and you need to split them into two separate columns—that is, for first name and last name. This can be

challenging, as names are often different lengths. Some people have two, three, or four names, so it can be difficult to designate a surname/family name.

Initials
: If you have a full first name in your data set, you may want to use only the first initial instead.

Case sensitivity

UPPER/lower
: Many data tools are case-sensitive, meaning "THIS" and "this" and "This" are all different values to the software. Standardizing all the string fields to the same case is important when grouping your data or joining different data sets.

Title case
: This is tricky, as not every piece of software has a simple title case function whereby the first letter in each word is capitalized and every subsequent letter is lowercased.

Addresses

Geographic names
: Like people's names, city and country names can often evolve in response to geopolitical changes over time. For example, with the fall of communism in the Soviet Union came name changes for many Eastern European countries. Be sure you are comparing like to like when analyzing data over many decades.

Address lines
: When filling out my address on a form, I find that I enter the details slightly differently every time. Whether a form has a company name line or one or two address lines has a significant impact on my data entry. Processing this data consistently can be a challenge.

Postcodes in the UK
: Postal codes are one area where the US definitely beats the UK. The UK's postcodes come in multiple forms like AA1 1AA, or A1 1AA, or A1A 1AA. Trying to piece these together and also preserve the space can be a significant task.

Spaces

When matching strings, you need to consider not only visible characters but also characters that you can't even see on the screen. Hidden spaces often lurk at the start or end of words, the result of splitting strings or the data entry method. These spaces in particular are very difficult to check for, but fortunately, data software developers have given us a handy function to help with this, `trim()`. If you need to remove spaces within strings, you can write custom calculations to do so. Finally, if you need to remove all spaces, you'll find that option within the ellipsis menu of Prep's Clean step, as shown in Figure 9-3.

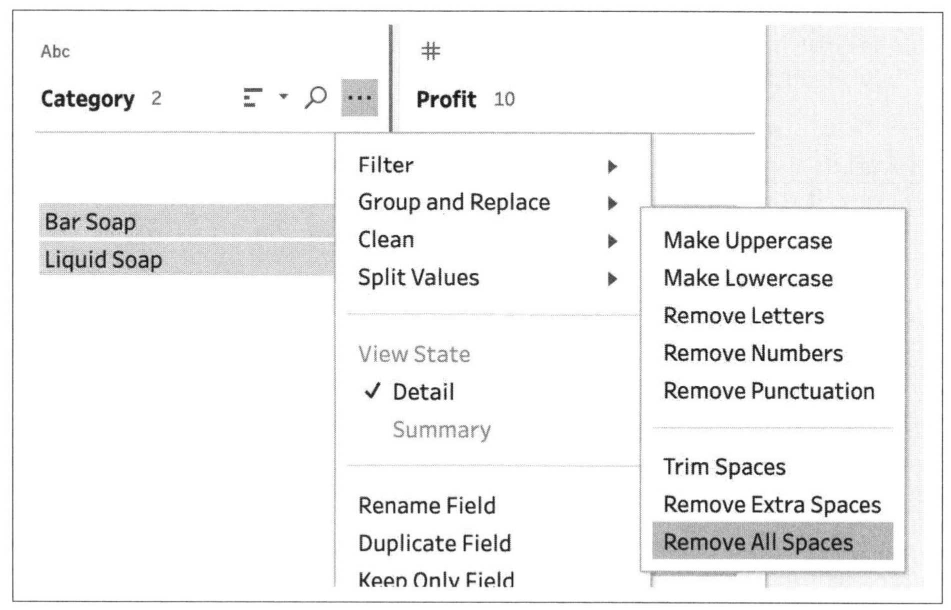

Figure 9-3. Remove All Spaces option within the Clean step

Poor or inconsistent spelling

Subpar spelling is often a result of poor planning during IT system setup or not thinking about the data analyst who will inevitably be tasked with analyzing the data a system spits out. Processing this data can be simple or migraine inducing.

Common Functions for Preparing String Data

These are the functions you'll turn to most frequently when preparing string data:

`split()`
: Breaks up longer strings of data into shorter strings based on a specified character. Often you can export data fields from operational systems merged together with symbols. This function allows you to break apart these strings into separate data fields.

`trim()`
: Removes the leading or trailing spaces from any given field. BI software isn't intelligent enough to ignore a leading or trailing space by default and see that string as the same as one without the space.

`upper()` *and* `lower()`
: Converts all alphabetic characters into uppercase or lowercase. This enables you to join together data sets from different inputs if the cases match. If the cases don't match, the join will not consider the strings the same even if they are otherwise identical.

`left()` *and* `right()`
: Extracts a specified number of characters from either the left or right side of the string. If you want to just take the initial from a customer's or employee's name, for example, you can use a `left()` function to return the first character of the name.

`mid()`
: Extracts a specific number of characters starting from a specified position within the string. Coupled with `find()` or `findnth()`, this is a powerful function that allows you to break down more complicated strings and return substrings in the middle of a longer string.

`find()`
: Locates the first occurrence of a specified character or substring within a string and returns its position within the string. This can help you identify where a substring starts if it's not in a consistent position within a larger string.

`findnth()`
: Finds the position of the *n*th instance of a specified character or substring within a string. This function is particularly useful when you don't want to find the position of the first instance of a character, so `find()` will not suffice.

`len()`
: Returns the number of characters in a string. This function can help you find leading or trailing spaces and perform other error checking.

`replace()`
: Searches within a string data field for a specified character or set of consecutive characters and replaces them with a specified different character or set of characters. This function can also replace the existing character or set of characters with nothing if you enter two consecutive quotation marks as the replacement value.

Grouping and Replace Options for Working with String Data

In Prep Builder, the grouping functionality goes beyond the Desktop version of Tableau and can help you clean up messy data sets much more quickly than with Desktop's manual selection approach. Prep Builder provides three alternative grouping and replace options for working with strings (Figure 9-4):

- *Pronunciation* groups strings based on how syllables sound when put together.
- *Common Characters* groups strings that have similar letters.
- *Spelling* groups together similar words with either a missing or different letter compared to the correct spelling.

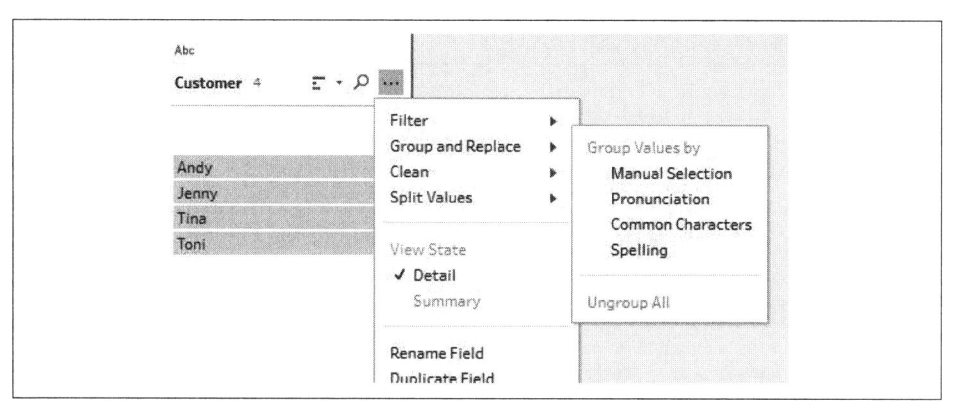

Figure 9-4. Group and Replace options available in Prep Builder

We go into grouping strings and values in a little more detail in Chapter 26 and applying the functions mentioned earlier in Chapter 18.

Summary

Data preppers can manipulate string data using either calculations or Prep Builder's built-in functionality. By understanding the importance of the position of characters in string data, you'll have a lot of options for cleaning structured columns. More advanced string cleaning will be discussed in Chapter 31's coverage of regular expressions.

CHAPTER 10
Dealing with Boolean Data

My favorite data joke is, "What's a ghost's favorite data type?"

"BOO-lean."

Now that that's out the way, let's talk about potentially the simplest data type but one that sits at the heart of a lot of what we do in data analytics and therefore is an important part of data preparation: Boolean data. In this chapter, we'll cover what it is and why it is useful when analyzing data.

What Is Boolean Data?

The word *Boolean* comes from the mathematician George Boole. He was Cork University's first mathematics professor, and his theorems were eventually applied to computing. Boolean data is simply a `True` or `False` response to a conditional statement or test.

Why Is It So Useful in Data Analysis?

The response of `True` or `False` is often encoded as `1` or `0` behind the scenes in the software we use. Therefore, the performance of calculations that use Boolean data is very quick. Computing is based on `1`s and `0`s, so Boolean data is easily processed by a computer.

A simple column of `1` or `0` responses is actually amazingly useful in data analysis for many reasons beyond just performance:

Indicators
: Here, an indicator refers to a field or set of fields that indicate whether or not each record fits some criteria. These can be analyzed very simply in most data tools. For example, if you want to count how many customers have a certain

product type, it's very simple to sum a Boolean indicator of `1`s and `0`s in most tools (Figure 10-1).

Customer ID	Current Account Ind	Savings Account Ind	Credit Card Ind
11135754	0	1	1
11510245	1	0	1
2140216	1	0	1
2101259	0	1	0
11489581	1	1	0
4755569	1	1	1
9515504	1	1	0
9937959	1	0	1
1130351	0	1	1

Figure 10-1. Indicators demonstrated as `1` for yes and `0` for no

In Prep Builder or Desktop, you could easily aggregate these values using either a `SUM()` or `AVG()` function to form simple summary statistics about the data set without having to build complex calculations or functions.

Logical tests to filter

Determining whether or not something meets a condition is a key part of data analytics. As mentioned, Boolean tests simply result in `True/False` responses, allowing you to decide how to proceed. These tests can be useful in cleaning data as well as in developing more complex logic for solving specific difficult problems. For example, using the banking data set in Figure 10-1, if we are analyzing only customers with credit cards, we can easily extract the relevant data by checking if a customer has a `1` value in the Credit Card Ind column (Figure 10-2).

Customer ID	Current Account Ind	Savings Account Ind	Credit Card Ind
11135754	0	1	1
11510245	1	0	1
2140216	1	0	1
4755569	1	1	1
9937959	1	0	1
1130351	0	1	1

Figure 10-2. Customers with Credit Cards data set

One aspect of using Boolean data fields to be mindful of is losing the meaning of the data points. Getting a response of `True/False` or `1/0` might be useful to the analyst but may not be as meaningful to the user of the resulting analysis. Therefore, it's a good idea to use an alias to describe what the Boolean data type refers to (Figure 10-3).

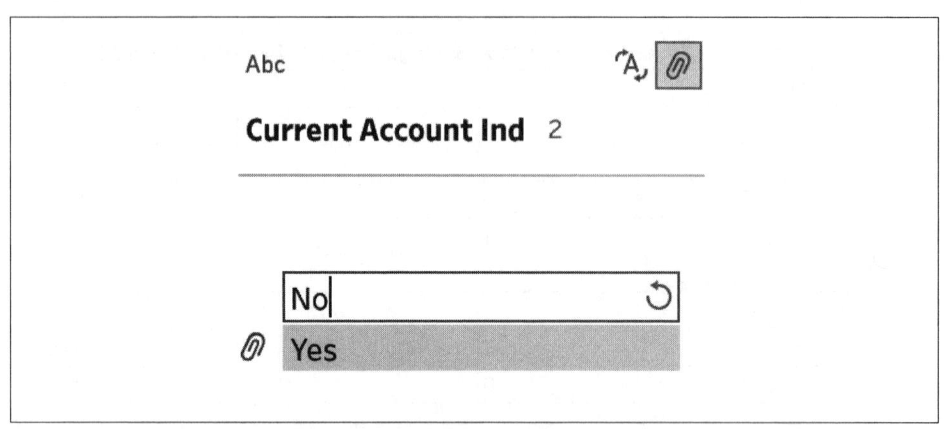

Figure 10-3. Aliasing Boolean values in Prep

In Prep Builder, the Profile pane is the perfect place to change your values, but doing so would negate some of the performance benefits of Boolean values. Therefore, it is often more beneficial to save these changes for later in the flow. Simply double-clicking a value (in this case, 1) allows you to replace it with an alias, such as "Yes." Remember to change your numeric data type from an integer to a string first, or Prep Builder will not accept the change.

Functions Featuring Boolean Logic

Many calculation functions feature Boolean logic—that is, they assess whether a value in a data field meets a condition to be output as either `True` or `False`. The following are some of the most useful.

IIF()

This is short for an "Immediate IF" function. `IIF()` functions are the equivalent of choosing what to output instead of just `True` or `False` when a condition is or isn't met.

Here's how to set up this calculation:

```
IIF(logical test, if True return this, if False return this)
```

The function returns the `True` result (second part of the function) if the logical test is met, and the `False` result (third part of the function) if the logical test is not met. Using an `IIF()` function can save a lot of typing of traditional `IF` statements and thus prevent keying errors. For example, an `IIF()` function can be very effective to assess whether sales targets have been met. Here's how you would use it to return a value of `'Met'` if sales are over 100 and `'Missed'` if they are not:

```
IIF( [Sales] > 100 , 'Met ', 'Missed' )
```

contains()

This is a great function to test whether certain words, terms, or characters are present in a string. This prevents you from having to split fields up into individual words or characters in order to check whether they contain the search term. It also avoids data being converted into massively long data sets and values being duplicated.

For example, if you're searching for a single word within a tweet, the `contains()` function can tell you if that term exists without splitting each word out into an individual row and having the values associated with that tweet replicated within the data set (e.g., number of retweets or likes).

```
contains([tweet],'keyword')
```

Figure 10-4 shows the `contains()` function applied to the Preppin' Data challenge from 2019: Week 9 (*https://oreil.ly/pM3ED*).

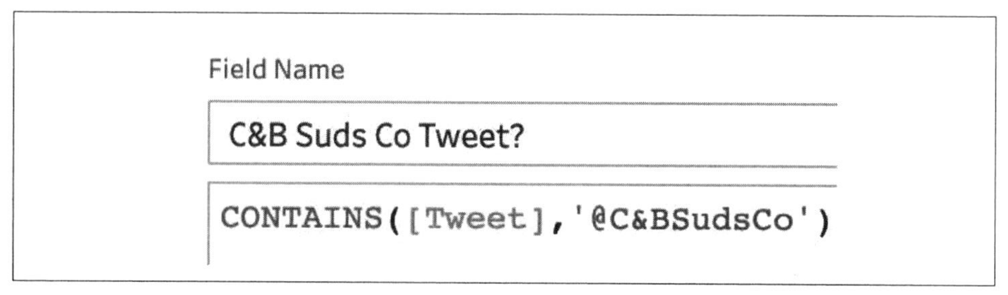

Figure 10-4. Using `contains()` to find the substring `'@C&BSudsCo'`

For each row of the data set, the function returns `True` if it finds the keyword within the given field, and `False` if it doesn't. Figure 10-5 shows the resulting data set using the example from the Preppin' Data challenge.

C&B Suds Co Tweet?	Tweet
True	My wife has accused me of having an affair you morons. You've over perfumed your Soap Bar @C&BSudsCo
True	I just wanted a bar of soap, not to smell like a brothel?! Do you even smell your own products @C&BSudsCo
False	Your soap has made my beard itchy, what the hell do you put in it?
True	@C&BSudsCo when r u coming to Paris?

Figure 10-5. Resulting data set for the `contains()` calculation in Figure 10-4

IsDate()

This function tests whether a column is a valid date. When you are working with manual date recording (for example, capturing records of events in Excel), it is very useful to be able to identify invalid dates so you can clean the inaccurate records to ensure only valid data is captured for analysis.

```
IsDate([Date])
```

The function returns `True` if the Date field (shown in square brackets) is in the format yyyy-MM-dd. This means the year is two to four digits, month one or two digits, and day one or two digits. The date parts need to be valid; that is, the day value has to be less than or equal to 31 and the month values have to be below 13. The Date field must separate the date parts with hyphens.

In this example, let's use the string `'2020-02-29'` to test whether Prep Builder recognizes this date as valid. As you can see in Figure 10-6, Prep Builder validates that 2020 did have a leap year.

Valid Date?	Date
True	2020-02-29
True	2020-02-29
True	2020-02-29

Figure 10-6. Testing the `IsDate()` function

As Figure 10-7 shows, Prep Builder returns `False` for dates that are not possible, like the 30th of February.

Valid Date?	Date
False	2020-02-30
False	2020-02-30
False	2020-02-30

Figure 10-7. `IsDate()` correctly recognizing an invalid date

IsNull()

Nulls in data can often be valid, but finding them in fields where you don't expect them can be a sign of issues that need to be resolved. Branching off records that have unexpected nulls so you can either alert an analyst or prepare that data differently can prevent misanalysis.

```
IsNull([Field])
```

If the field isn't null—that is, it contains any value or string (including a space character)—the function will return `False`. If the field being assessed returns `True`, then the field will be a null.

IF/THEN

`IF` statements are logical tests that allow you to specify what value should be returned when the conditions set within the statement are met. When an `IF` statement is called, Prep Builder will assess each condition in turn. Once a condition is met, the resulting `THEN` clause sets the specified value into the calculation's resulting data field. If the calculation's first condition isn't met, the next condition will be tested. This process continues until either:

- All records have met a condition.
- No condition is met, but the `IF` statement contains an `ELSE` statement to force a value to be entered.
- No condition is met and there is no `ELSE` condition, so the calculation's resulting data field will contain a null for that record.

Using the Tweets data set from the Preppin' Data 2019: Week 9 challenge (*https://oreil.ly/pM3ED*), the `IF` statement shown in Figure 10-8 tests if the tweet contains the word "soap."

Field Name

Soap?

```
if contains([Tweet],'soap') then TRUE
Else FALSE
end
```

Figure 10-8. `IF` statement assessing whether `Tweet` contains the word "soap"

The resulting column would be Boolean, as the values returned are `True` (where the word "soap" is found in the tweet) and `False` if not. Notice in Figure 10-9 how Prep Builder changes the case of the `True` or `False` return value to match its normal output for a Boolean field.

Soap?	Tweet
True	Hey @C&BSudsCo you suds are soap...I expected beer!
True	WTF?! You're Soap has just filled my bathroom full of bubbles!! Way to bubbly for me @C&BSudsCo
False	What kind of moron name is @C&BSudsCo?
True	No where near enough bubbles from your Soap Bar @C&BSudsCo. I wanted a bar of soap, not a chocolate bar

Figure 10-9. Resulting data set from the `IF` statement calculation in Figure 10-8

The `IF` statement in Figure 10-8 yields the same result we would get if we used a `contains()` function. So why use it? The benefit of using an `IF` statement is twofold:

- You can test more than two conditions. An `IF` statement can have as many conditions as you like. My personal record is nearly 80. The order of the conditions is important because as soon a record meets a condition, it is no longer assessed for any of the following conditions.
- You can specify different return values. The resulting values of the `IF` statement needn't be Boolean; you can have it return dates, values, or strings as well.

Using the Tweets data set again, Figure 10-10 shows an `IF` statement that identifies whether a specific type of soap was mentioned in the tweet.

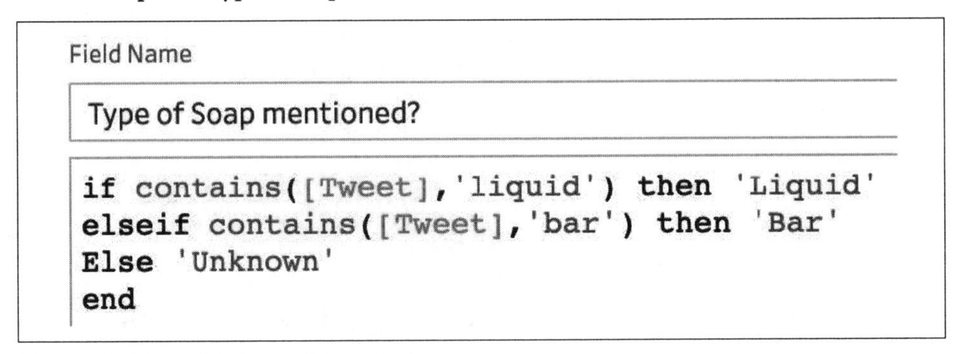
Field Name
Type of Soap mentioned?
if contains([Tweet],'liquid') then 'Liquid'
elseif contains([Tweet],'bar') then 'Bar'
Else 'Unknown'
end

Figure 10-10. Multiple condition `IF` statement

We add another condition by using the term `elseif`. So, for every subsequent condition you want to test, you would just add another `elseif`. The results of the statement in Figure 10-10 are shown in Figure 10-11. Notice that the last tweet in the data pane contains the word "liquid" as well as "bar," but because the "liquid" condition is assessed first, the resulting value is `Liquid` rather than `Bar`.

Type of Soap mentioned?	Tweet
Bar	My wife has accused me of having an affair you morons. You've over perfumed your Soap Bar @C&BSudsCo
Bar	I just wanted a bar of soap, not to smell like a brothel?! Do you even smell your own products @C&BSudsCo
Unknown	Your soap has made my beard itchy, what the hell do you put in it?
Unknown	@C&BSudsCo when r u coming to Paris?
Unknown	I just bought my first @C&BSudsCo soap! Because I'm worth it! Actually I'm worth a lot more!
Unknown	@C&BSudsCo Who thought glitter in a beard shampoo was a good idea???
Unknown	@C&BSudsCo OMG my beard fell off!!
Liquid	@C&BSudsCo I hate liquid soap, I only like Bars

Figure 10-11. Resulting data set for the `IF` statement in Figure 10-10

`IF` statements can contain quite complex conditions. Understanding the impact of using AND or OR conditions, as well as the mathematical order of operations, is important. To return a `True` value, AND conditions require every condition specified to be met, whereas OR conditions require only one to be met. Familiarize yourself with the mathematical order of operations, often nicknamed BODMAS (UK) or PEMDAS (US), to ensure your `IF` statement returns the values you expect:

1. Brackets or Parentheses
2. Of (powers of) or Exponents
3. Division or Multiplication
4. Multiplication or Division
5. Addition
6. Subtraction

Multiplication and division have the same precedence (they are always done left to right), which is why they're transposed in the BODMAS and PEMDAS acronyms.

Because of their flexibility, you will find `IF` statements very useful in data preparation, whether for returning a simple Boolean value or a field with many differing values based on multiple conditions.

CASE

Whereas `IF` statements test *conditions* in Tableau, `CASE` statements test *values*. During data preparation, you can use `CASE` statements to change the output value of a field as either a conditional test or simply to rename it to something easier for users to understand. To demonstrate this, let's use a simple data set on the number of corners found in different shapes (Figure 10-12).

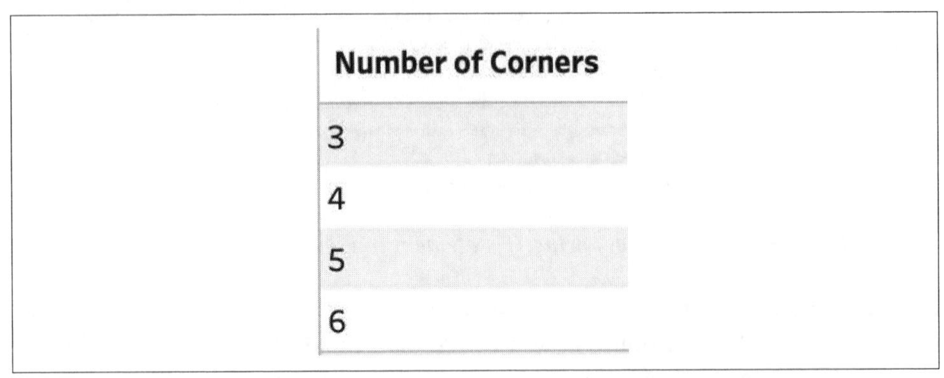

Number of Corners
3
4
5
6

Figure 10-12. Number of Corners data set

The CASE statement, with its simple syntax (see Figure 10-13), saves a lot of coding compared to the equivalent IF statement.

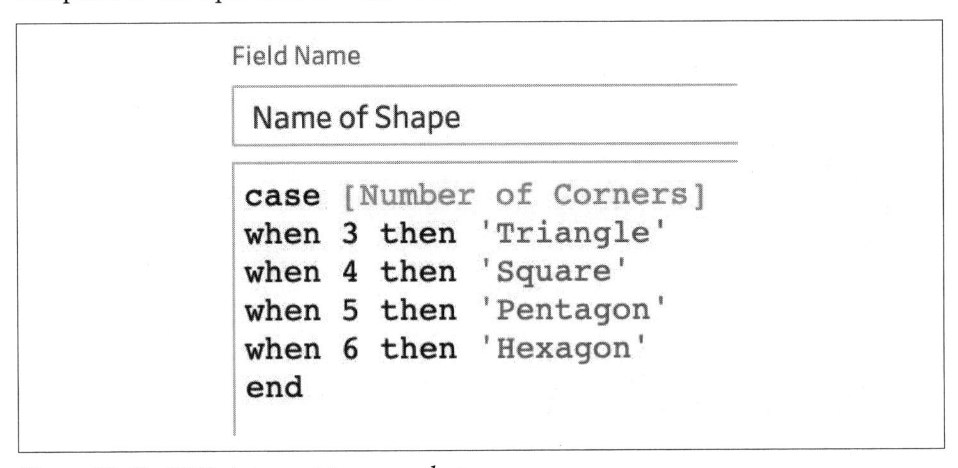

Figure 10-13. CASE statement to name shapes

Unlike the SQL language, Tableau does not allow CASE statements to perform the Boolean algebra required to test conditions.

Summary

As you have seen in this chapter, Boolean data type fields can be very powerful. Practicing their use is key to becoming skilled at data preparation. The speed with which computers can process Boolean fields makes this data type especially effective in large data sets.

PART III

The Shape of Data

CHAPTER 11

Profiling Data

The art of data preparation is understanding the data set in order to determine what you might need to do to prepare it for analysis. Understanding the *profile* of the data is key to forming a full view of the data. Without profiling the data, you can easily miss an obvious preparation step or add in unnecessary work. This chapter will explore what profiling is, why profiling data is important, and how Prep profiles data.

What Is a Profile?

By *profile*, I mean the characteristics of the data set. As discussed in earlier chapters, understanding the types of data you have in the data set is essential to your analysis. Equally important is understanding the number and variance of the categorical data fields of the data set. Determining the data set's level of granularity will help you to identify how many unique records there are, or whether there are duplicate records that you need to remove in the data preparation process. All of these factors form the foundation of the data set profile, which comprises these factors:

- Minimum, maximum, and range of values: Does the range between the minimum and maximum values make sense?
- Data outside of limits: Are there natural limits in the data, like 100%, or current dates that cannot be exceeded but have been?
- Outliers: Do the values lie inside a certain range except for one or a few that sit outside of it?
- Irregular number of records: Is there a consistent number of rows for certain dimensions, and does this number suddenly change? For example, do you expect a set number of records for each date within the data set?

- Irregular spelling: Can you identify the correct spelling for names and words in the data set?
- Duplicate records: Were duplicates created prior to data preparation, or during the previous steps of this process?
- Missing data: Are there certain values that aren't present in the data set but should be? Are there nulls where you would expect a value?

Checking for all these factors can be quite time-consuming if you have a lot of columns, but there are ways to make this task easier and more intuitive.

Why Visualizing the Data Set Is Important

One of the most important strategies for profiling your data sets is visualization.

Anscombe's Quartet

If you have read any books on data visualization, then you have likely come across Anscombe's Quartet, the very best argument for why descriptive statistics (minimum, maximum, average, etc.) isn't enough to understand what is truly going on in a data set. In 1973, Francis Anscombe constructed four data sets comprising pairs of x and y values (Figure 11-1).

Observation	x1	y1	x2	y2	x3	y3	x4	y4
1	10	8.04	10	9.14	10	7.46	8	6.58
2	8	6.95	8	8.14	8	6.77	8	5.76
3	13	7.58	13	8.74	13	12.74	8	7.71
4	9	8.81	9	8.77	9	7.11	8	8.84
5	11	8.33	11	9.26	11	7.81	8	8.47
6	14	9.96	14	8.1	14	8.84	8	7.04
7	6	7.24	6	6.13	6	6.08	8	5.25
8	4	4.26	4	3.1	4	5.39	19	12.5
9	12	10.84	12	9.13	12	8.15	8	5.56
10	7	4.82	7	7.26	7	6.42	8	7.91
11	5	5.68	5	4.74	5	5.73	8	6.89

Figure 11-1. Anscombe's data set

As you can see in the figure, there is a significant range of values, but all four sets have largely the same descriptive statistics, namely:

- Means: $x = 9, y = 7.5$
- Sample variance: $x = 11, y = 4.5$
- Correlation, linear regression, and coefficient of determination are all similar to two or three decimal places.

However, when the individual data points for each set are visualized, we can see the data sets are actually very different (Figure 11-2).

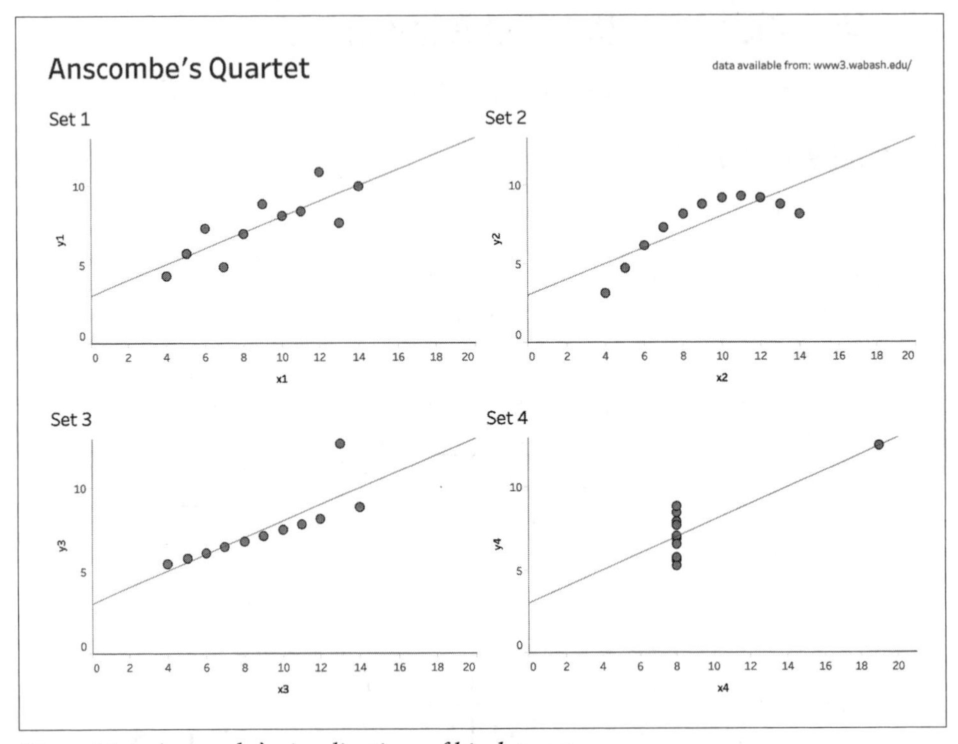

Figure 11-2. Anscombe's visualizations of his data set

Therefore, to adequately prepare the data, you must visualize it in some basic ways—not necessarily to form an analysis yet but to understand whether the data is as expected. If it isn't, don't simply remove those rogue data points, but also make an effort to understand why they differ from what you expected.

Visualizations Versus Data Tables

When viewing the underlying data for Anscombe's Quartet, maybe you spotted some variables that were not as expected, like the outlier on the x- and y-axis in Set 4. But it's tough to make those kinds of assessments in a table, especially in a data set larger than Anscombe's. For example, try finding all of the outliers in Figure 11-3 at a glance. Your eyes are a fantastic tool at spotting patterns, so let's make life easier for ourselves when it comes to finding oddities within our data by using data visualization techniques.

Show	Timing	Date	Seats	Value	% Attended
Les Miserables	Afternoon	01/01/2020	142	18132	85
Les Miserables	Evening	01/01/2020	175	17049	105
Les Miserables	Afternoon	02/01/2020	183	24950	68
Les Miserables	Evening	02/01/2020	153	22022	91
Les Miserables	Afternoon	03/01/2020	162	22304	96
Les Miserables	Evening	03/01/2020	188	20371	97
Les Miserables	Afternoon	04/01/2020	134	23330	84
Les Miserables	Evening	04/01/2020	130	15726	60
Les Miserables	Afternoon	05/01/2020	120	17336	0
Les Miserables	Evening	05/01/2020	136	21781	68
Les Miserables	Afternoon	06/01/2020		24505	
Les Miserables	Evening	06/01/2020	182	23277	88
Les Miserables	Afternoon	07/01/2020	141	17938	61
Les Miserables	Evening	07/01/2020	149	23218	17
The Lion King	Evening	01/01/2020	154	17924	91
The Lion King	Evening	02/01/2020	115	16452	89
The Lion King	Evening	03/01/2020	162	19356	86
The Lion King	Afternoon	04/01/2020	136	27094	79
The Lion King	Evening	04/01/2020	118	19422	68
The Lion King	Afternoon	05/01/2020	108	13023	82
The Lion King	Evening	05/01/2020	161	16390	99
The Lion King	Evening	06/01/2020	117	12230	72
The Lion King	Evening	07/01/2020	130	12457	87

Figure 11-3. Can you quickly find the trends in this table of data?

To that end, let's use Prep Builder's built-in Profile pane to help us spot things to investigate in the London Theatre Shows data set shown in Figure 11-3.

How Prep Builder Profiles Data

When you load the London Theatre Shows data set into Prep Builder and add a Clean step, Prep Builder instantly visualizes the data profile in the Profile pane (Figure 11-4).

Prep automatically samples data when you input larger data sources to help maintain its response time. The algorithm used by Prep aims to maintain the data set's profile. Sampling is covered further in Chapter 12.

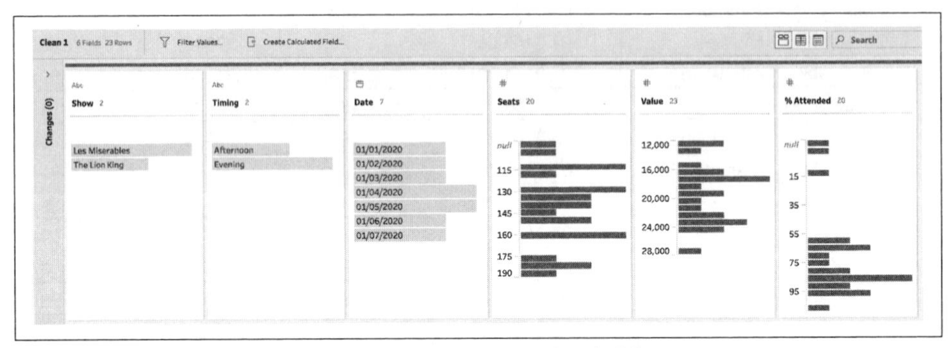

Figure 11-4. Profile pane of Prep Builder with data loaded from Figure 11-3

Though it looks like a relatively straightforward data set in the table format, the Profile pane demonstrates that actually it contains a lot of variations. Before assessing this particular data set, let's dig in further to what happens in the Profile pane in Prep Builder.

Generating Histograms and Mini-Histograms

Before Prep Builder, a lot of my investigations of new data sets involved building histograms. Often, I was looking at the number of records found for each bin (range of values) or the date to determine the data profile and understand its completeness.

Thankfully, now Prep Builder does this automatically when data is loaded into a Clean step. Time previously wasted building these charts can now instead be spent cleaning, investigating more data sets, and getting to the analysis sooner. This is the genius of the Profile pane: you can see the profile of your data instantly. Each histogram helps you spot trends and gaps quickly. Each bar length, regardless of whether it represents a bin or specific value, indicates the number of rows containing those values (Figure 11-5).

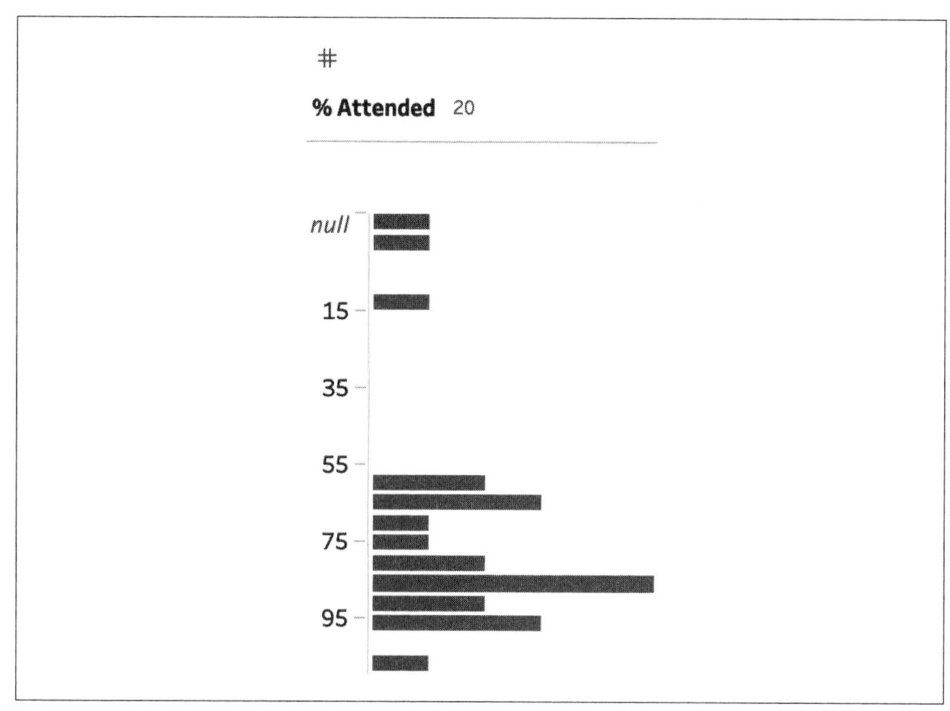

Figure 11-5. Histogram built in the Profile pane showing the % Attended data field

Prep Builder has summarized the data into bins of similar ranges of data, indicated by the dark blue histogram bars. Therefore, you will see gaps in your data where no records exist within that range. You can very quickly see the most common values too, as they will be the longest bars (Figure 11-6).

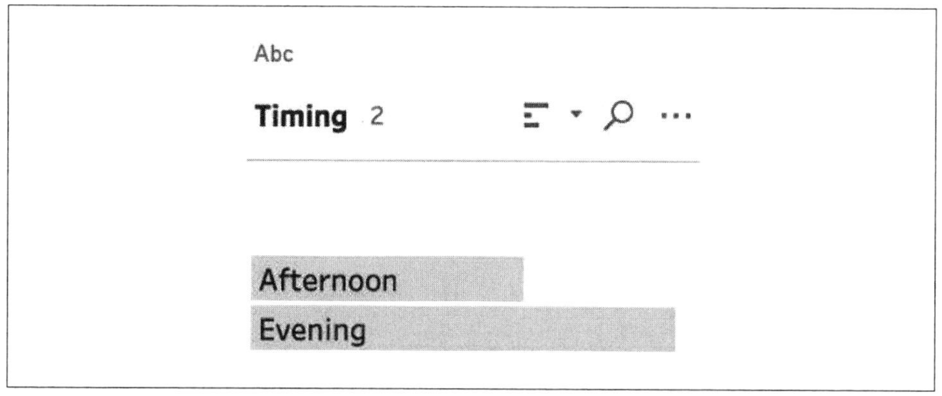

Figure 11-6. Profile pane histogram showing most common values

The gray bars are used when either there are not enough values to form bins of similar values or you are analyzing a dimensional value instead of a numerical one.

Because the gray bars are used only for values actually found in the data set, you will not spot missing values as easily as you can with the dark blue bars; that is, there will be no gaps like you have with bins. If the data has a logical order and one of those values is missing, however, you might be able to identify the missing value.

When Prep Builder cannot fit the whole histogram into the Profile pane space, it will create a scrolling histogram of all the values or bins within that data field, as well as a mini-histogram in the top-right corner of the data fields Profile pane (Figure 11-7).

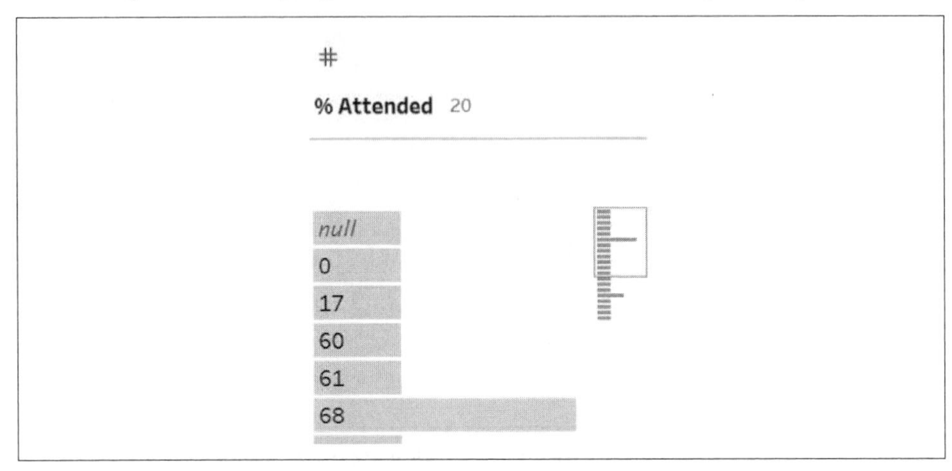

Figure 11-7. Mini-histogram

Although this mini-histogram looks like an icon simply indicating there is more data, this isn't the case. The mini-histogram is a representation of the actual data. The box around the mini-histogram highlights what part of the data field is being displayed in the data field's Profile pane.

Selecting Summary Versus Detail Views

Prep Builder will automatically create a summarized view of the data where there are lots of different values. This is very common for measures, but it also happens for dates. If you want to check exact values rather than the summary shown in histogram bins, you can change the View State option in the data field's menu in the Profile pane (Figure 11-8).

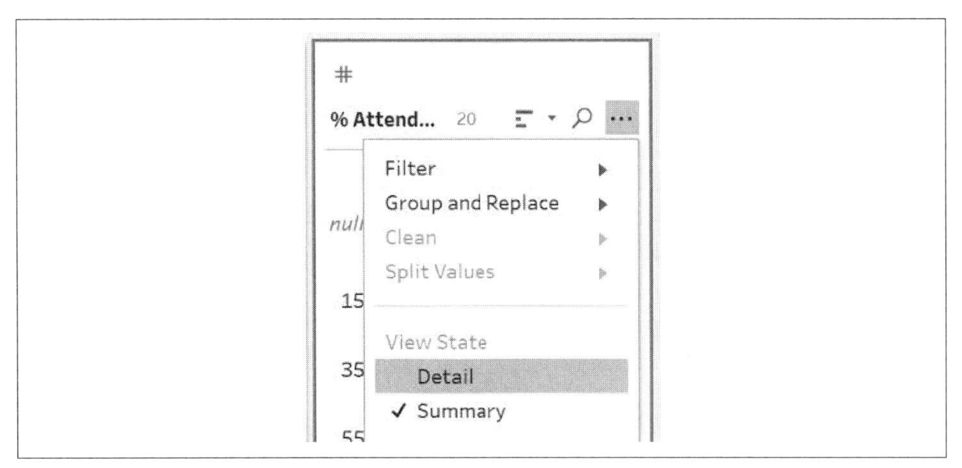

Figure 11-8. Changing the histogram View State

The greater precision of the Detail view can help you investigate patterns within the histogram that you didn't expect to find.

Highlighting Values

Clicking on a value in the Profile pane will highlight all the values in other columns that relate to (exist in a row with) the selected value. This feature is another way to profile and gain insight into the data (Figure 11-9).

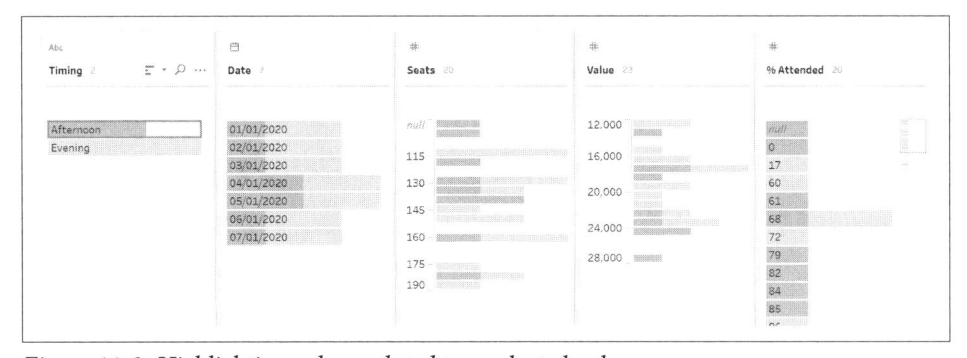

Figure 11-9. Highlighting values related to a selected value

The highlighted area in each value forms a new blue histogram within the existing gray one, allowing you to profile the data for the selected value. You can then narrow down this data further by selecting additional values in other columns, or expand it by holding down the Ctrl key while clicking additional values within the original column.

This highlighting functionality allows you to dig deeper into the data set to ask the follow-up questions that inevitably arise as you investigate your initial question.

This technique is particularly useful when you are assessing the records that have null values in certain columns (Figure 11-10). By selecting the null values in one field, you can see the dates when those records originated or other bits of key information to help you determine whether the nulls should be replaced by other values or removed altogether.

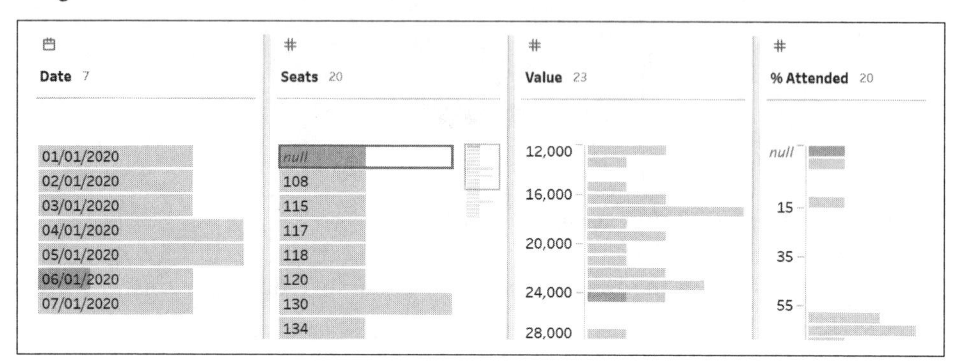

Figure 11-10. Selecting a null value

Viewing Dimension Counts

The final way that Prep Builder profiles the data is by displaying how many members (unique values) there are in a dimension. This count is shown at the top of a data field within the Profile pane (Figure 11-11).

Figure 11-11. Count on different dates for a data field

In this case, there are seven different dates within the data set. When this count is more or less than you expect, it can prompt you to investigate where an issue might lie.

Sorting

Within each data field in the Profile pane, you have the option to sort the values in either alphabetical or numerical order (Figure 11-12). The numerical sort is based on the number of rows in the data set Prep is using (remember this may be a sample of the full data set) and can be set to either Ascending or Descending. The alphabetical sorting order can be set to either A to Z or Z to A.

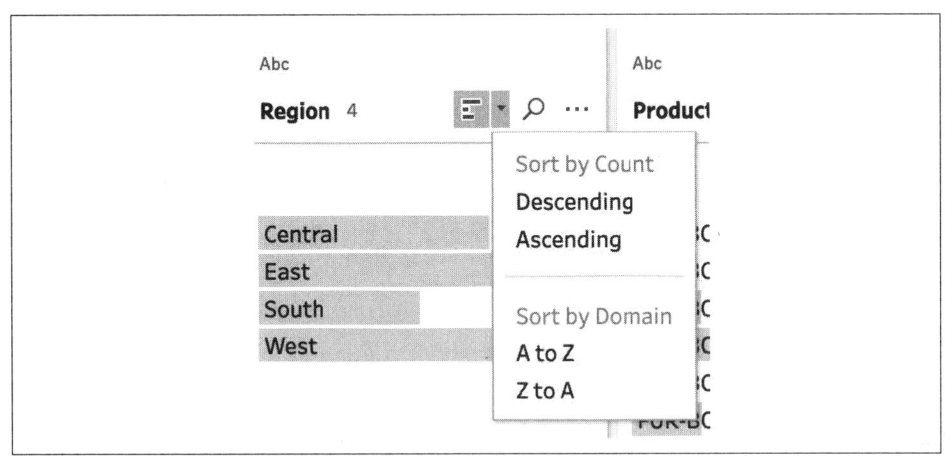

Figure 11-12. Sorting options in the Profile pane

To apply a sort, click the Sort icon and then click the drop-down arrow to specify which sort to use. Each click on the Sort icon changes the sort order as follows:

1. Sorts from highest to lowest number of records
2. Sorts from lowest to highest number of records
3. Reverts to the original alphabetic or numeric sort order

Summary

This chapter has emphasized the importance of profiling your data through visualizations. Prep automatically visualizes the data through the Profile pane in the Clean step, saving you a lot of work. Tableau Prep makes it much easier to profile your data set, which means you'll be able to prepare your data more thoroughly and with fewer iterations than in the traditional, manual data profiling process.

CHAPTER 12

Sampling Data Sets

To sample or not to sample? That is the question. In a world where data volumes are growing, storage solutions are getting cheaper, and data creation is easier than ever, data preppers must decide whether to use a sample subset and understand the implications of doing so. This chapter will look at why sampling should be used with caution, when you might need to sample, and what techniques you can use to sample data in Prep Builder.

One Simple Rule: Use It All If Possible

The reason we use data is to find the story, trends, and outliers that will help us make better decisions in our everyday and working lives. So why not aim to use all the data and information you can?

Using the full data set is not always possible, though, frequently due to the size of the data set. The reason Preppin' Data exists is because data often needs to be prepared for analysis. To do that, we need to know what is possible to clean completely and what is not. If it isn't possible to clean data sets completely, then it makes sense to remove sections that can't be cleaned. This is not what is meant by sampling, though. Sampling means using a subset of the full data set—not because the data can't be cleaned but for lots of other reasons.

Sampling to Work Around Technical Limitations

A sample allows you to take the data you need to clean and freeze it in time to deal with the two main technical challenges of data prep:

Volume of data
: A sample lets you set up your analysis and then run the full data set against that logic.

Velocity of data
: A sample limits the amount of continual change, allowing you to set up the logic before permitting more frequent updates.

Let's look at both in more detail.

Volume of Data

The world is swimming in data and there is no sign of stopping. Many data sets have become too big to store in files, so databases are being increasingly used to support the volume of data. The good news is that if your data is in a database, someone has at least architected the data to be stored that way. The likely data prep challenges here are:

- Joining other data sets together. Database tables are rarely in the perfect form for your analysis. When working out what is useful for your analysis and what isn't, it can be beneficial to take small samples of the data tables to assess how they can be joined or what data fields you need from each. Samples also allow you to work faster by avoiding the long processing times of joining large data sets together, especially if you don't get the right join condition(s).
- Determining your ideal structure for analysis. Database tables are designed for storing data, not optimizing your analysis. Pivoting data, removing unnecessary fields like database keys/IDs, and filtering to relevant time periods are common techniques that you can apply more quickly to samples than the full data set. Related considerations are:
  - What columns are there? Do you need to add more (calculations) or remove some?
  - How clean are those data fields? Check whether the data set has concatenated fields that you need to break apart, clean up strings of text to turn them into meaningful categories, and make sure there are no foreign characters slipping into your measures. By taking a sample, you can break down the overall cleaning step into more manageable sections. After you've added this cleaning logic back to the original data set, you can tweak the logic to cover other challenges in the data set.

Velocity of Data

Media streaming services and the Internet of Things are two areas currently facing the daily battle of data velocity. Long gone are the days where an overnight batch run to update key data sources was sufficient. The speed at which data is created by users of modern media services and digital platforms means that data preparation can be a constantly moving target. By using samples of this data, you can avoid many of the pitfalls of trying to use all the data, all the time. You create logic for a sample, and as more data floods in, you can apply that same logic to the live stream by simply removing the restriction you used for the sample.

Other Reasons for Sampling

There are other aspects of data preparation where sampling can prove beneficial.

Reduce Build Times

Like many others, I spent my early career working in large corporate companies. While these experiences can present fantastic opportunities, they can also produce a serious amount of frustration with slow computers, servers, and connections between them. In one institution, I used two computers so one could run queries while I was building the next set of queries on the other. My love of coffee comes from having to fill my time when running queries on both machines at the same time. Using samples to set up the data structure and analysis was key to keeping my pace of delivery high and caffeine levels lower.

When I had the queries structured, joins checked, and relevant filters in place, I still had to wait for the full data set to run, but I did so with confidence that I had done everything I could to just run it once.

Determine What You Need

Sampling also comes in handy when you don't actually know what you need. Chapter 2 advocated sketching out what you need to complete your analysis. However, sometimes the only way to iterate that need is to try to start forming that analysis. Multiple iterations of data preparation are needed just as much as multiple iterations of data analysis as people learn and ask follow-up questions. Samples of data can give you a feel for additional changes you might want to make. Writing out an output that takes a long time to form is only an issue if you have to do it time and time again.

Sampling data not only speeds up the preparation process and makes it easier to understand your data but also allows you to more easily communicate the challenges you are facing to others.

Sampling Techniques

By default, Prep Builder uses samples whenever you connect a data set to it (Figure 12-1).

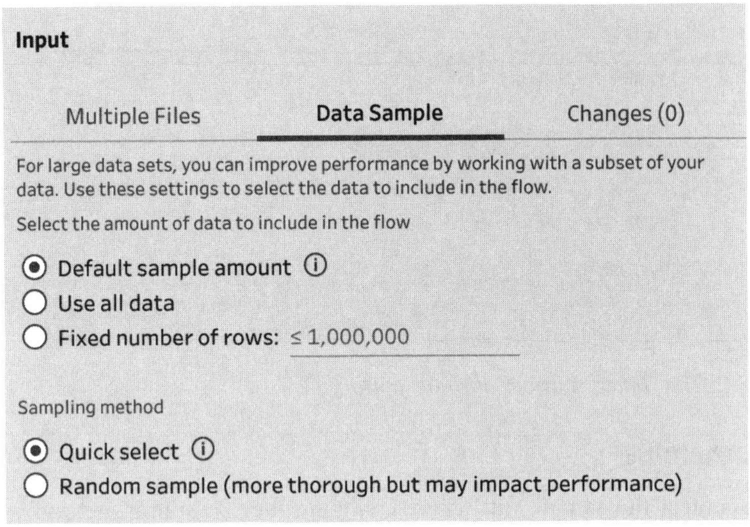

Figure 12-1. Tableau's default input sampling

Through the clever use of algorithms, Prep Builder automatically allows you to begin to see the full shape of your data, categories, and distribution of values without having to run the full data set through each step until you are ready to generate the final output. (At which point Tableau will use the full input data set unless instructed otherwise.) If you have a small data set, Tableau will use the full data set, but if you have wider data sets (more columns), Tableau's default sampling will return fewer rows and records of data than if you have thinner data sets (fewer columns).

Even though Prep Builder is sampling the data, you might want to control the sample yourself. There are two basic controls for this within the Input step.

Fixed Number of Rows

The fixed number of rows will be taken from the first rows in the data source. These will not necessarily be in any prescribed order, as data sets hold data in different ways based on how they load and store data. Figure 12-2 shows where you can set the number of rows you want to return.

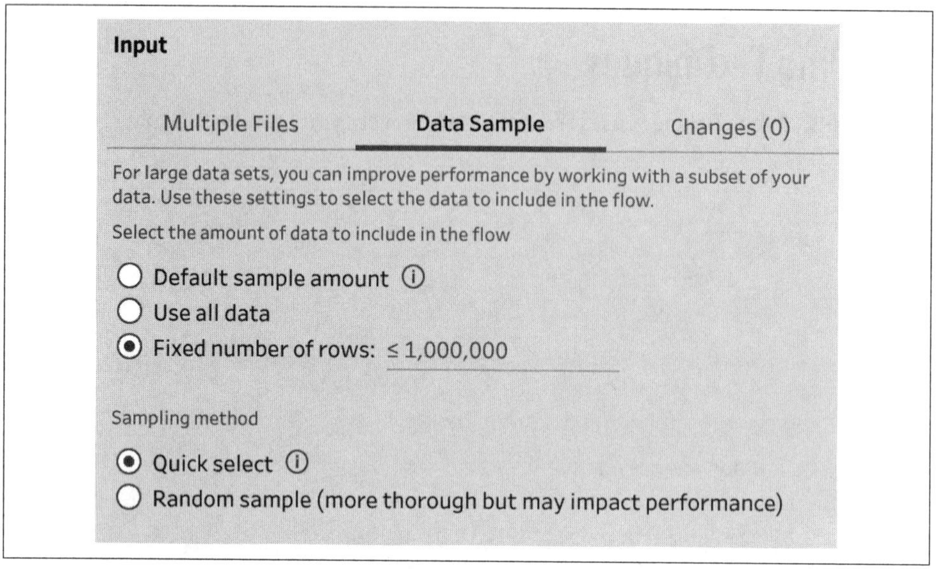

Input

Multiple Files | Data Sample | Changes (0)

For large data sets, you can improve performance by working with a subset of your data. Use these settings to select the data to include in the flow.

Select the amount of data to include in the flow

- ○ Default sample amount ⓘ
- ○ Use all data
- ◉ Fixed number of rows: ≤ 1,000,000

Sampling method

- ◉ Quick select ⓘ
- ○ Random sample (more thorough but may impact performance)

Figure 12-2. The "Fixed number of rows" option

Random Sample

You can control the sample further by changing the "Sampling method" option to "Random sample" (Figure 12-3). Unlike the "Quick select" method, with this option Prep Builder will select random rows of the data set. To make this selection, Prep Builder has to work through the data set and return various rows.

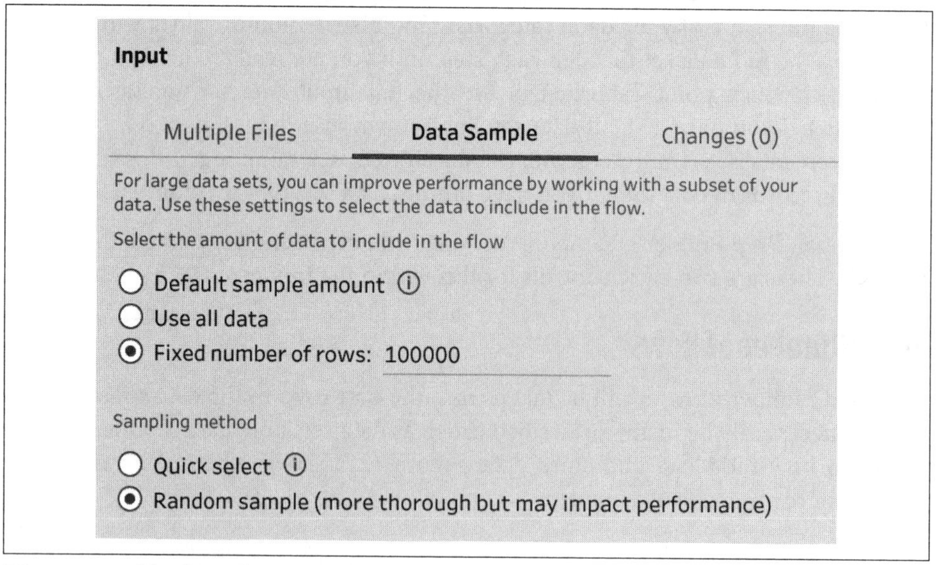

Input

Multiple Files | Data Sample | Changes (0)

For large data sets, you can improve performance by working with a subset of your data. Use these settings to select the data to include in the flow.

Select the amount of data to include in the flow

- ○ Default sample amount ⓘ
- ○ Use all data
- ◉ Fixed number of rows: 100000

Sampling method

- ○ Quick select ⓘ
- ◉ Random sample (more thorough but may impact performance)

Figure 12-3. The "Random sample" option

You can always set your own sample using a bespoke filter in Prep Builder so you can control what the smaller set of data contains (Figure 12-4). Just remember, by setting this yourself you are biasing the data set, so be careful with any analytical conclusions you draw from this approach.

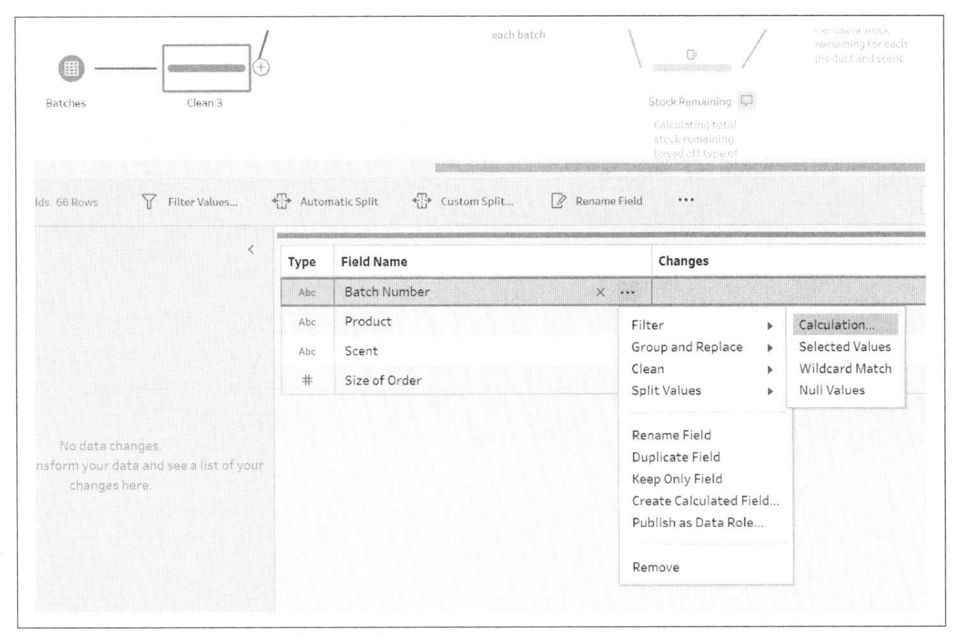

Figure 12-4. Setting up a filter in the Input step

When Not to Sample

Sampling can speed up the data prep process, but what if you want to use Prep for the analytical process too? With the Profile pane showing the distribution of values and highlighting records of data, it's possible to answer your stakeholders' questions within Prep itself. In this case, you would want to avoid using a sample to prevent issues when you're asked questions like:

- How many of these are there?
- When was the first sale we made?
- Can you confirm this has never happened?

If the input data set were sampled, it would be difficult to give a confident answer to these questions. So, in this situation, you'd want to simply change the input to use all the data in Prep. This might increase the processing times, but you gain the assurance that all the data is present. When you run your flows by writing outputs, you write the whole data set regardless of whether you have a sample on the input, so you only need to switch to the whole data set on the Input step when answering questions in Prep.

Summary

Sampling data can be a useful technique for increasing the speed and agility of your data preparation. But any sample by definition is not the whole data set. To produce reliable analytical findings, you really need to use the full data set where possible and make it available to your users. This way, outliers and patterns can be discovered that a sample might not have revealed.

CHAPTER 13
Pivoting Columns to Rows

I grew up in the '90s, so when I hear the word "pivot," the image that instantly pops into my brain is the *Friends* characters trying to get Ross's sofa up the winding apartment staircase (*https://youtu.be/n67RYI_0sc0*).

We first looked at pivoting data (not sofas in stairwells) in Chapter 4. This chapter will build on that discussion, looking at when and how to pivot columns into rows of data in Prep Builder. This type of data transition is also referred to as *transposing*.

When to Pivot in Tableau Prep Builder

Tableau Desktop needs the data to be in structured columns. Each column can hold one type of data and can be used on either Desktop *shelves* or *cards* to alter the view onscreen. Accordingly, a common use case for pivoting is when you are adding a new column for each new date in a data set (Figure 13-1).

Type	Jan-19	Feb-19	Mar-19
A	1	4	7
B	2	5	8
C	3	6	9

Figure 13-1. Adding a new column per month to the data set

The data structure in Figure 13-2 isn't great for two reasons. First, if you want the dates displayed on an x-axis timeline, you can't do that with this structure, because you need one column to hold all the different dates and another column to hold all the relevant values for each of those dates. Second, if the file is updated, Tableau Desktop won't automatically add the new column into your analysis. If the new data appeared as additional rows instead, it would be included in the view (as long as

Desktop is able to read it). Figure 13-2 shows the view in Tableau Desktop of the data from Figure 13-1 in its current structure.

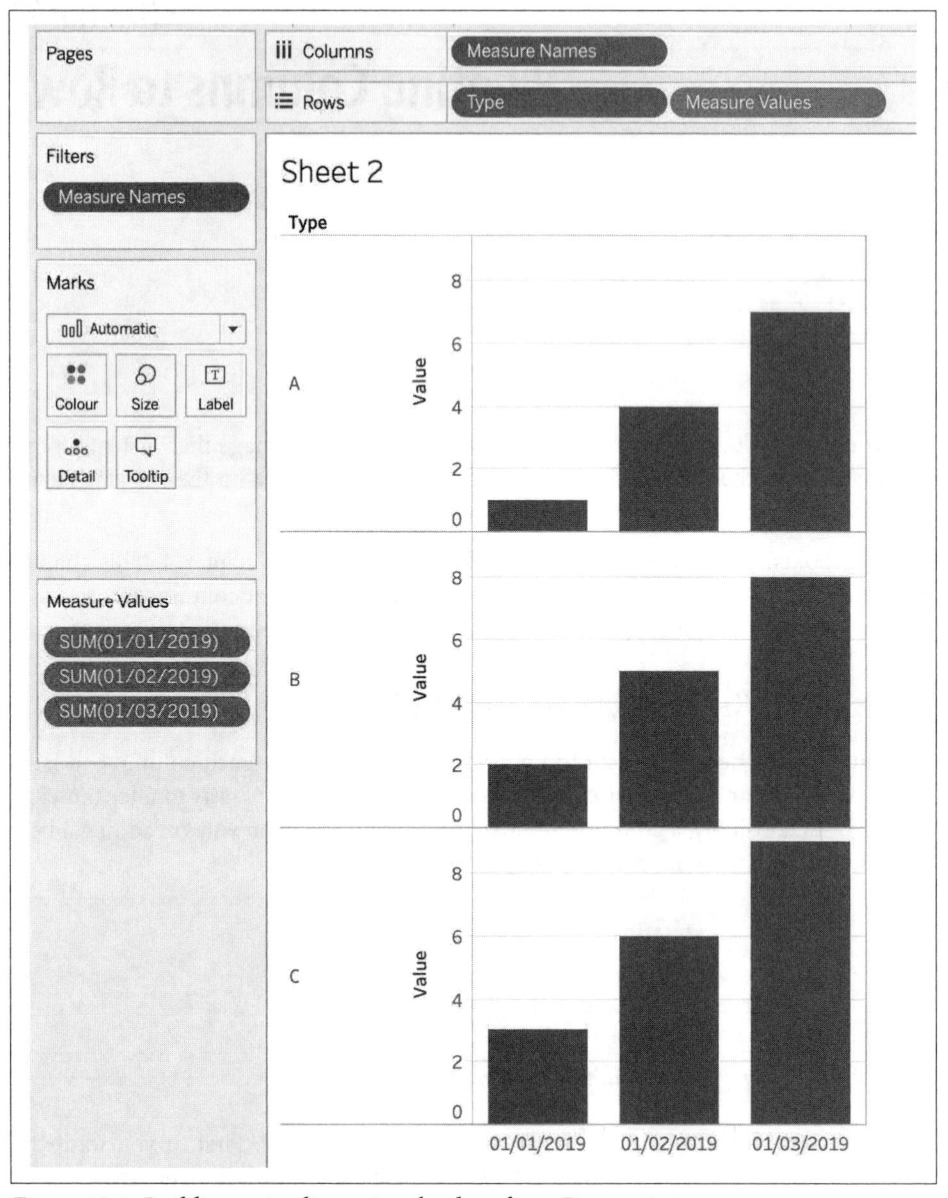

Figure 13-2. Building a timeline using the data from Figure 13-1

Using Measure Names and Measure Values is more complicated than just dragging a single date field to the Columns shelf in Desktop. We need to pivot this data.

There are two types of pivot in Tableau to select from:

- Columns to rows
- Rows to columns (discussed in Chapter 14)

How to Pivot Columns to Rows

To resolve the date situation, we need to change the columns into rows of data instead. After adding a Pivot step, select the Columns to Rows pivot option as shown in Figure 13-3.

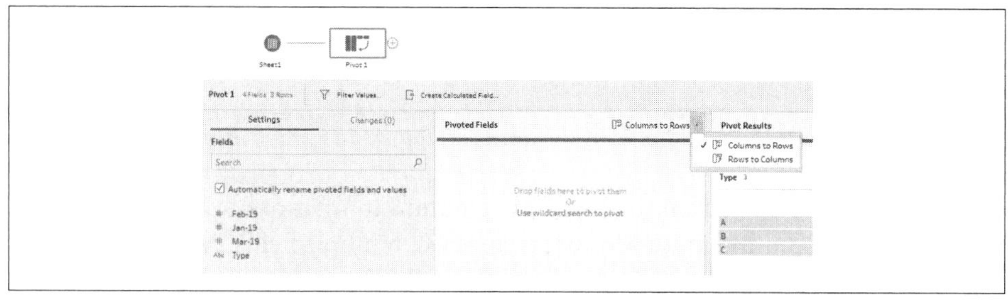

Figure 13-3. Pivot options

The equivalent in Excel is to transform Table A into Table B (Figure 13-4).

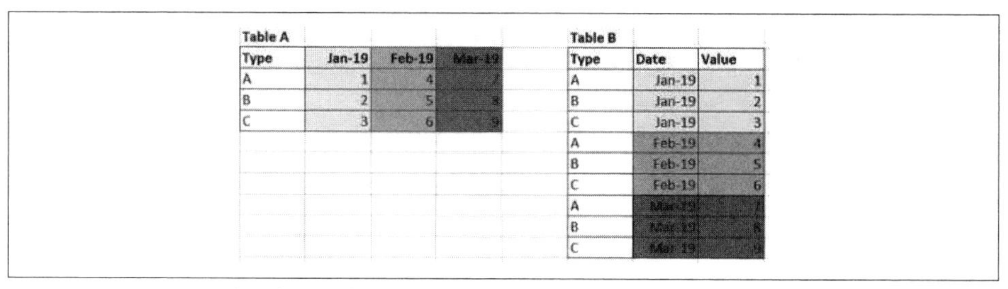

Table A

Type	Jan-19	Feb-19	Mar-19
A	1	4	7
B	2	5	8
C	3	6	9

Table B

Type	Date	Value
A	Jan-19	1
B	Jan-19	2
C	Jan-19	3
A	Feb-19	4
B	Feb-19	5
C	Feb-19	6
A	Mar-19	7
B	Mar-19	8
C	Mar-19	9

Figure 13-4. Example of transformation of Table A to Table B

Let's look at how to achieve this result in Prep with the Pivot step.

1. Add the Pivot step. All the fields (columns) in your data set will be listed in the left-hand pane (Figure 13-5).

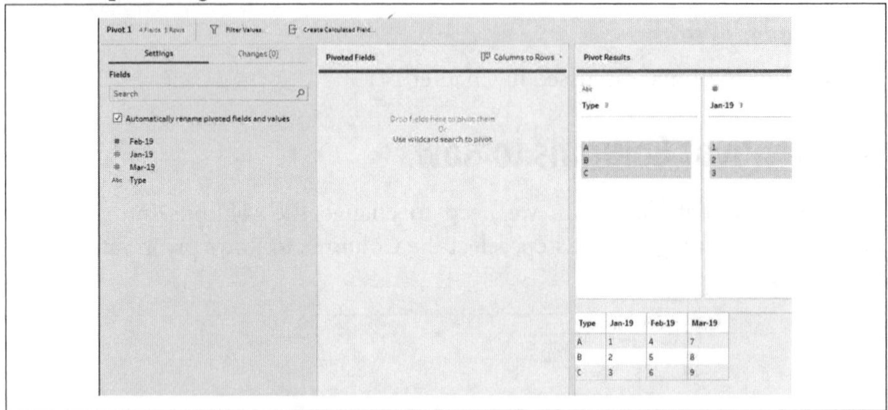

Figure 13-5. All the data fields in the data set appear in the left-hand pane

2. Select the fields you want to pivot from columns to rows of data. You can use the Ctrl key (or Command for Mac users) to select multiple fields. After selecting the fields, drag them to the middle pane as shown in Figure 13-6.

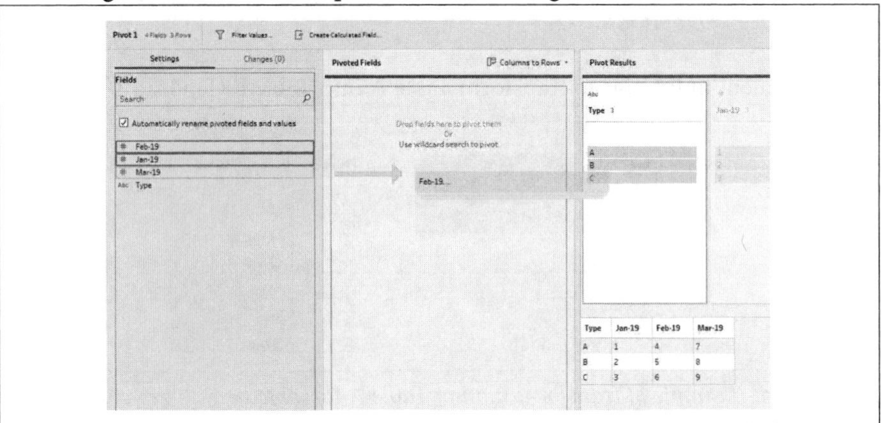

Figure 13-6. Dragging the date fields to be pivoted

3. You should see your pivoted data in the bottom-right-hand corner of the screen (Figure 13-7).

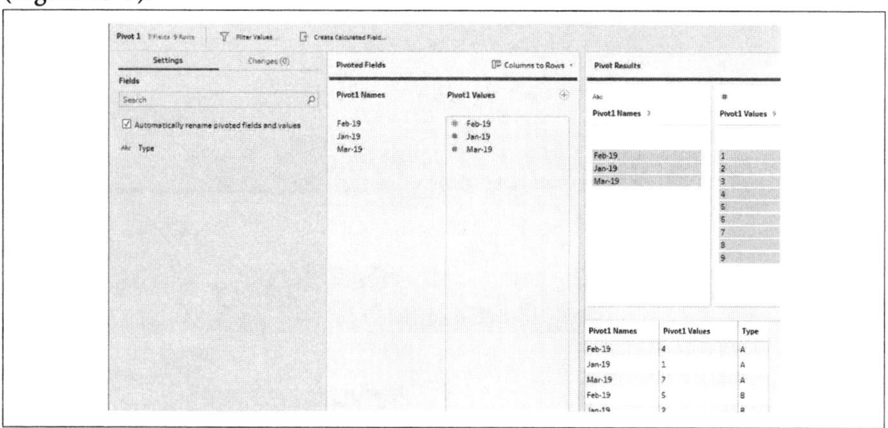

Figure 13-7. Viewing the pivoted date fields

4. Rename your columns to match your data by double-clicking the field name (Figure 13-8).

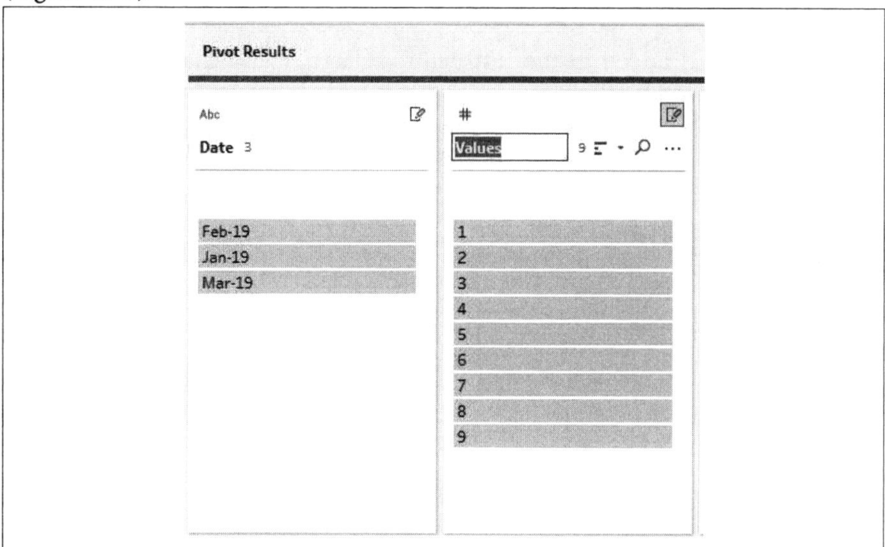

Figure 13-8. Renaming a field in the Profile pane

Now that the data is reshaped, it's much easier to visualize the data. As you can see in Figure 13-9, there are a lot fewer data fields in this view than in Figure 13-2 and no need to use Measure Names or Measure Values in Tableau Desktop.

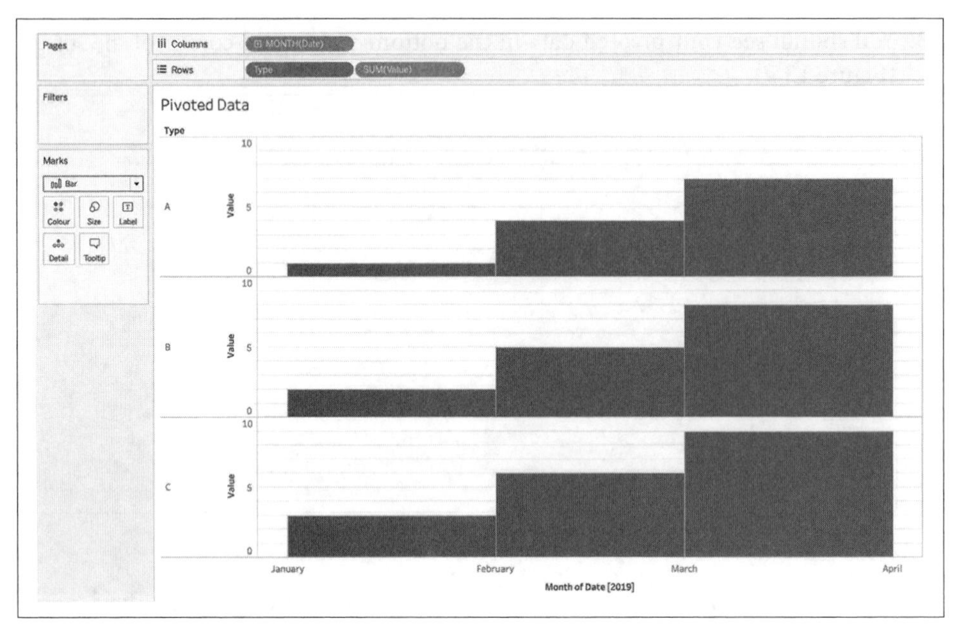

Figure 13-9. Visualizing the pivoted data

When new data is added to the restructured data set, the chart will automatically update once you run the Prep flow again. Not only that, but all the rich date-related functionality in Tableau Desktop is now available to you too (Figure 13-10). This allows you to conduct a deeper and richer analysis.

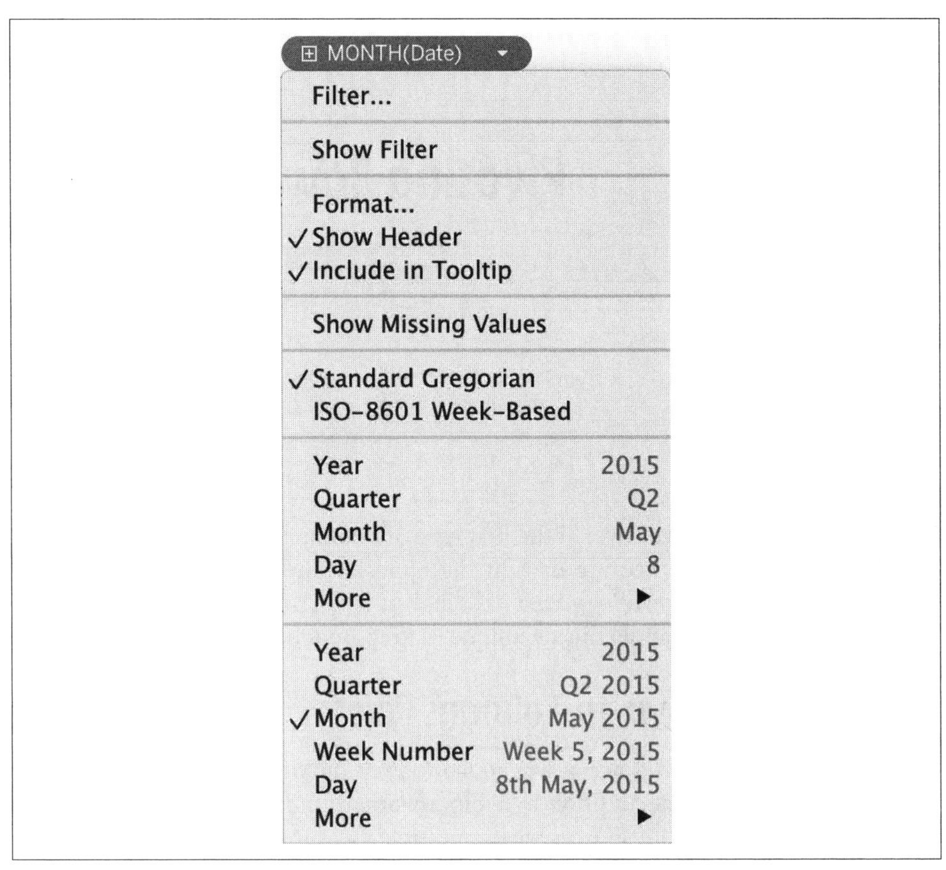

Figure 13-10. Date options for a Date data field in Tableau Desktop

Summary

Pivoting columns to rows allows you to automate updates to your data set. Without this technique, you would have to manually update your analysis each time the data is added as a new column. Pivoting enables you to spend more time analyzing your data rather than prepping it.

CHAPTER 14

Pivoting Rows to Columns

In the previous chapter, we covered the first type of pivoting: columns to rows. In this chapter, we will look at pivoting data in the opposite direction—rows to columns (sometimes known as *unpivoting*). First we'll look at when to pivot rows into columns and then how to actually apply this technique in Prep Builder.

When to Use a Rows-to-Columns Pivot

The Rows to Columns pivot option was added to Prep Builder in a later version of the tool, and it was a welcome addition to avoid a workaround for a common use case. Data frequently comes from sources where multiple measures or dimensions are held in a single column. To make analysis easier in Tableau Desktop, you need to separate these out into an individual column per measure or dimension. If you don't separate them, it is technically still possible to analyze the data in Tableau Desktop (Figure 14-1).

However, if you want to create a scatterplot comparing two metrics, you can't do so without creating multiple calculations, or duplicating data fields and filtering them. Remember, the aim of data preparation is to make the analysis easier for end users.

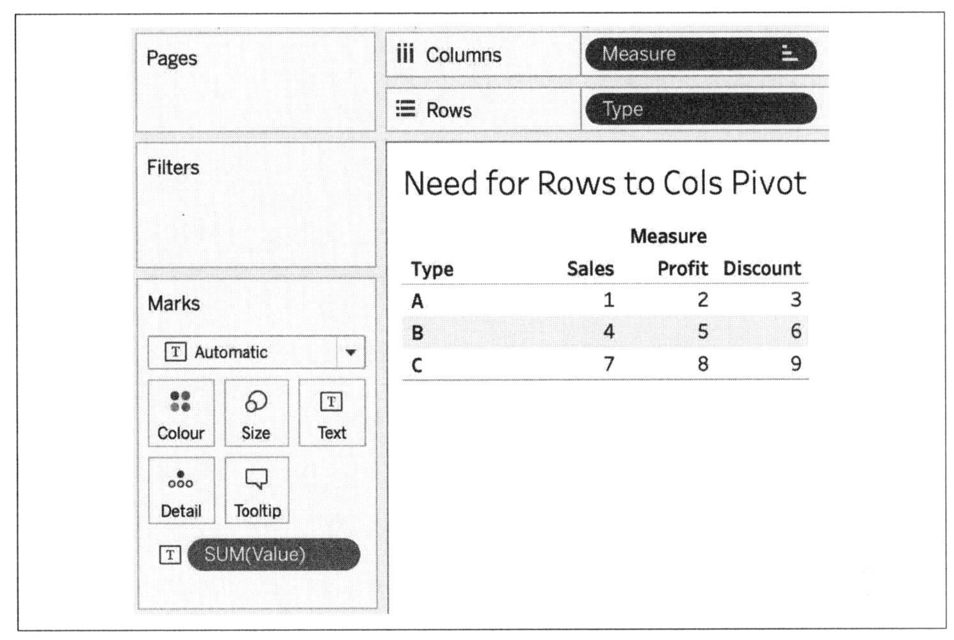

Figure 14-1. Visualizing the single data field of metrics

Figure 14-2 shows the visualization we want to achieve.

Type	Measure	Value
A	Sales	1
A	Profit	2
A	Discount	3
B	Sales	4
B	Profit	5
B	Discount	6
C	Sales	7
C	Profit	8
C	Discount	9

Type	Sales	Profit	Discount
A	1	2	3
B	4	5	6
C	7	8	9

Figure 14-2. The data visualization after pivoting from rows to columns

How to Pivot Rows to Columns

Let's take a look at how to pivot rows to columns.

1. In a Pivot step, use the drop-down menu at the top right of the Pivoted Fields pane to select the Rows to Columns pivot option (Figure 14-3).

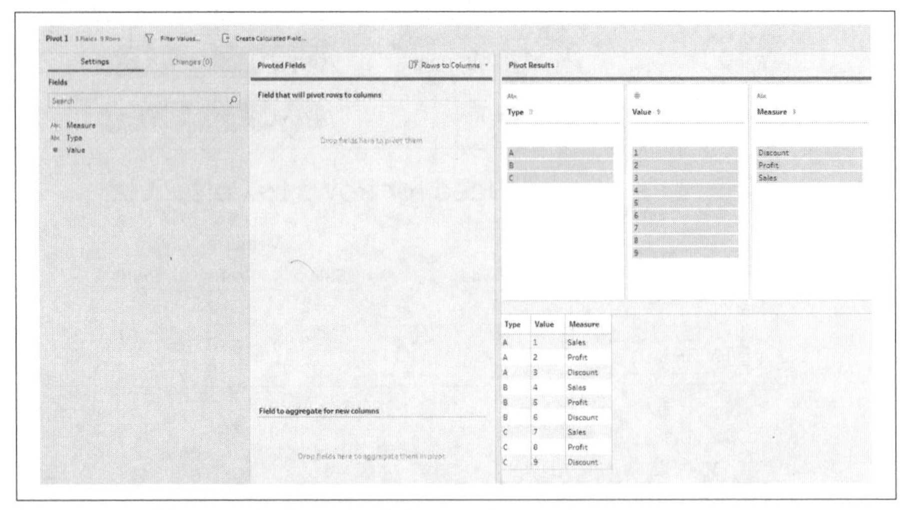

Figure 14-3. Selecting the Rows to Columns pivot option

2. Adding the Measures field to the Pivoted Fields pane will display each category in the data (Figure 14-4) and create a new column in the field.

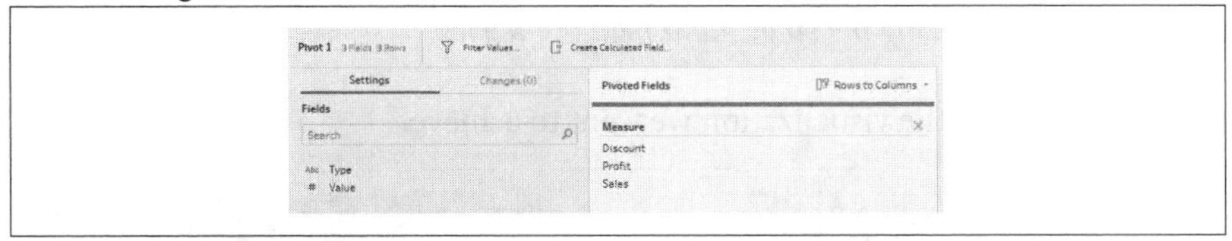

Figure 14-4. Setting the field to create new column headers

3. Add the field that contains the values for your new column headers into the lower part of the Pivoted Fields pane, under the "Field to aggregate for new columns" header (Figure 14-5).

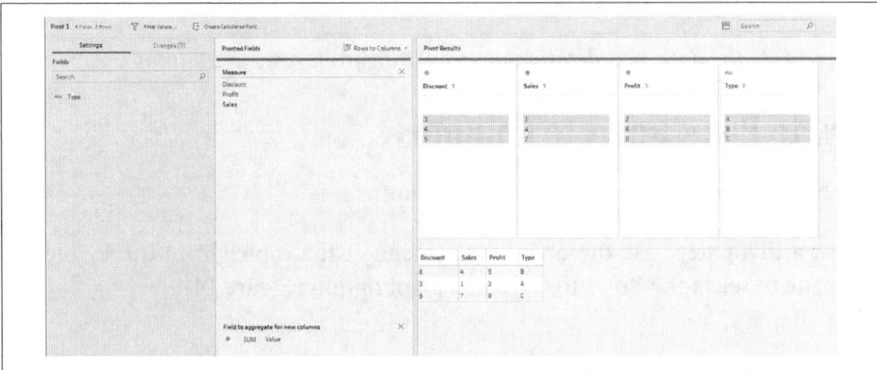

Figure 14-5. Setting the values for the new headers

You will need to choose how multiple rows will be aggregated. The aggregation will happen at the level of whatever remains in the left-hand pane of the step. This is the equivalent of a Group By (see Chapter 15) for the fields in the left-hand pane.

Now we have the data in an easier form to analyze. Figure 14-6 shows the resulting scatterplot.

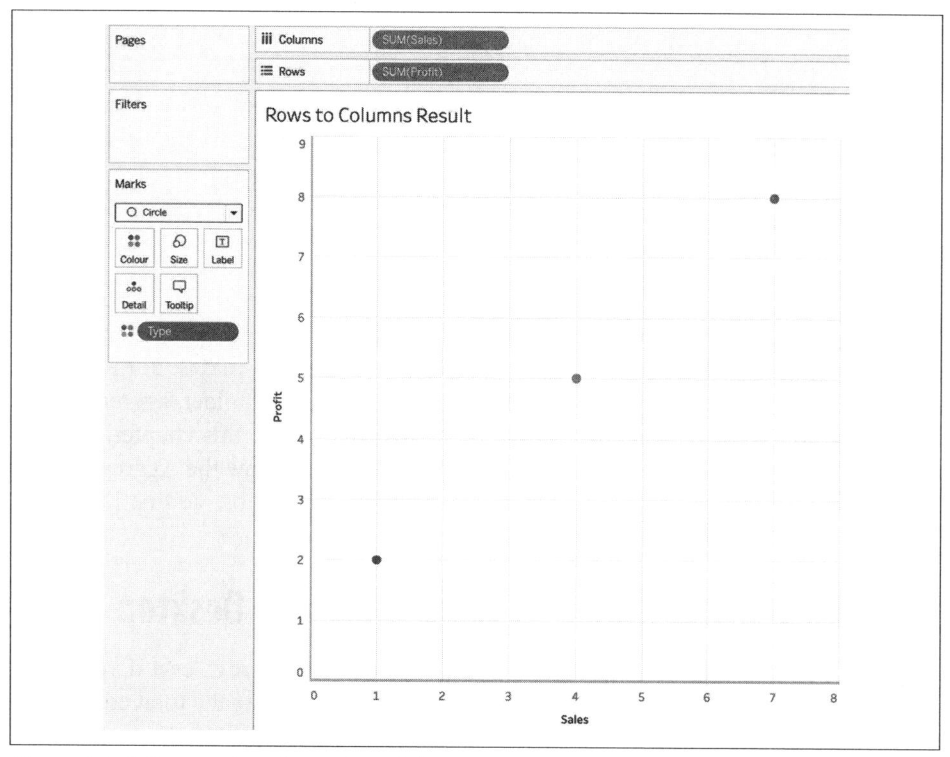

Figure 14-6. Scatterplot of the pivoted data

Summary

Getting used to pivoting data takes some time, but practice makes perfect! If you are using data that has been sourced from an Excel spreadsheet and has been held in a pivot table, the Rows to Columns pivot option will enable you to convert the data into the form that works best with Desktop.

CHAPTER 15

Aggregating in Prep Builder

Thanks to the two tools' similar look and feel, Tableau users can prepare their data in Prep Builder for visual analysis in Desktop quickly and easily—that is, until some of those familiar features start to differ. When I'm teaching Prep Builder, aggregation is a common source of pain for my students for this reason. In this chapter, we will cover how aggregation in Prep Builder differs from Desktop, how the Aggregate step works, and how to overcome the biggest challenge of aggregation: adding back into Desktop any column that you may want for further analysis.

Comparing Calculations in Prep Builder and Desktop

Like Desktop, Prep Builder will work out each calculation at a row level if you don't use any form of aggregation. For example, if I wanted to calculate the total cost in the simple Bathroom Renovation data set shown in Figure 15-1, I would simply add one cost value to the other, one row at a time.

Type of Item	Item	Price	Manufacturing Cost	Distribution Cost
Furniture	Bath	250	105	30
Furniture	Shower	150	70	10
Accessory	Shower Screen	75	35	30
Decoration	Tiles	80	20	20
Decoration	Paint	40	15	10
Furniture	Sink	100	65	20

Figure 15-1. Bathroom Renovation data set

To add these costs together, we would write the same calculation in Prep Builder as we would in Tableau Desktop (Figure 15-2).

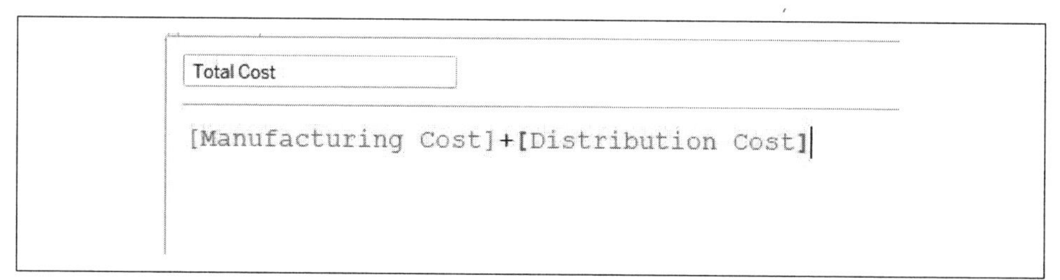

Figure 15-2. Calculation for total cost

This adds the values together for each row, while creating a new column to hold the total (Figure 15-3).

Type of Item	Item	Price	Manufacturing Cost	Distribution Cost	Total Cost
Furniture	Bath	250	105	30	135
Furniture	Shower	150	70	10	80
Accessory	Shower Screen	75	35	30	65

Figure 15-3. Result of total cost calculation

The total cost calculation and the additional column to hold it are identical between the two tools. For heavy Excel users who haven't used a tool that works with columns, the approach is the same in Excel: to calculate this value, you simply select which cells to add together rather than selecting the column names as you do in Tableau (Figure 15-4).

A	B	C	D	E	F
Type of Item	Item	Price	Manufacturing Cost	Distribution Cost	Total Cost
Furniture	Bath	250	105	30	=D2+E2
Furniture	Shower	150	70	10	
Accessory	Shower Screen	75	35	30	
Decoration	Tiles	80	20	20	
Decoration	Paint	40	15	10	
Furniture	Sink	100	65	20	

Figure 15-4. Excel equivalent of the calculation in Figure 15-2

Which Calculations in Prep Builder Differ?

So if our row-based calculations are the same in both Prep Builder and Desktop, which calculations aren't? You've probably guessed the answer: calculations where you are aggregating multiple rows together in the same column. Aggregations in Prep Builder allow you to change the granularity of the data set and create subtotals or totals.

Let's use totals as an example. In Desktop, to add up each item's total cost, we would write the calculation shown in Figure 15-5.

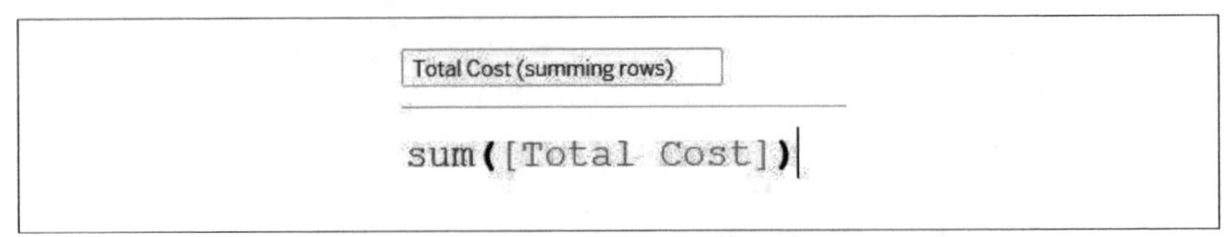

Figure 15-5. Calculating a data set total in Tableau Desktop

If you've used Desktop for a while, you probably know that if you add an aggregation to your calculation, Tableau will add up all those values based on what discrete fields are within your view (Figure 15-6).

Discrete fields are blue and *continuous* fields are green. These fields are sometimes called *pills* because of their distinctive shape.

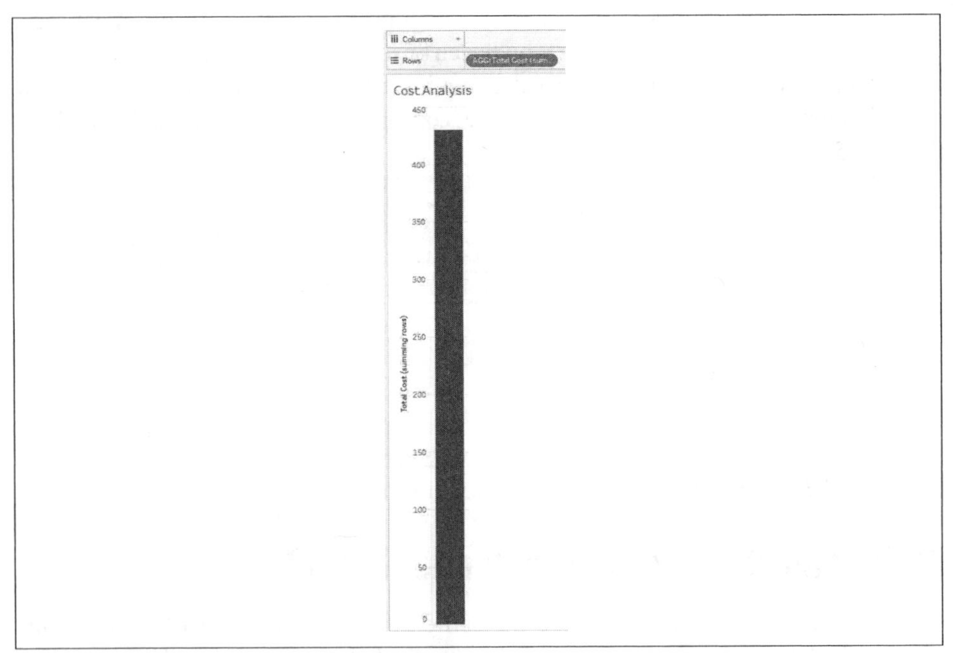

Figure 15-6. Desktop aggregates by the dimensions used on a worksheet

We don't have any discrete fields in our view, so Desktop simply sums up all of our items' total costs. The equivalent in Excel looks like Figure 15-7.

Type of Item	Item	Price	Manufacturing Cost	Distribution Cost	Total Cost
Furniture	Bath	250	105	30	135
Furniture	Shower	150	70	10	80
Accessory	Shower Screen	75	35	30	65
Decoration	Tiles	80	20	20	40
Decoration	Paint	40	15	10	25
Furniture	Sink	100	65	20	85
					=SUM(F2:F7)

Figure 15-7. The Excel equivalent of the calculation in Figure 15-5

But in Prep Builder, the `sum()` function doesn't exist within Calculated Fields, so trying to use it throws an error (see Figure 15-8).

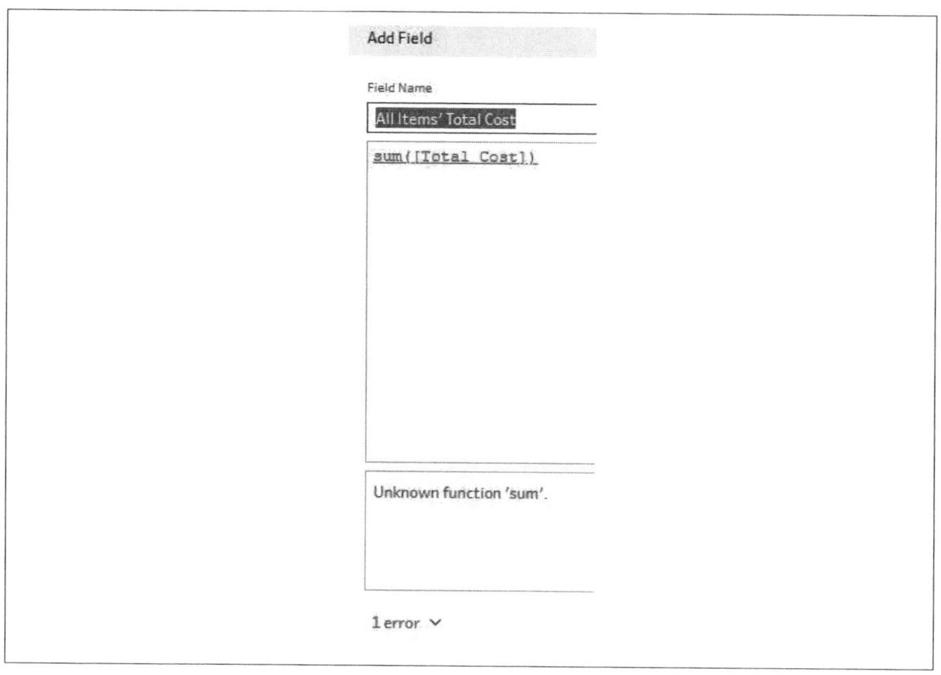

Figure 15-8. Trying to use `sum()` in Prep throws an error

The reason for this error is because we need to tell Prep Builder what to sum up and to what level of granularity. In Desktop, Tableau determines this from the discrete fields in the view (placed on the Columns or Rows shelf or used on the Marks card). In Figure 15-9 we have put Type of Item on the Columns shelf, so Desktop is now showing the total costs for each type of item rather than the sum of all of them together.

The *Marks card* controls how the data point is shown in the visualization. You can control the type, color, size, or even whether the mark has a label attached to it.

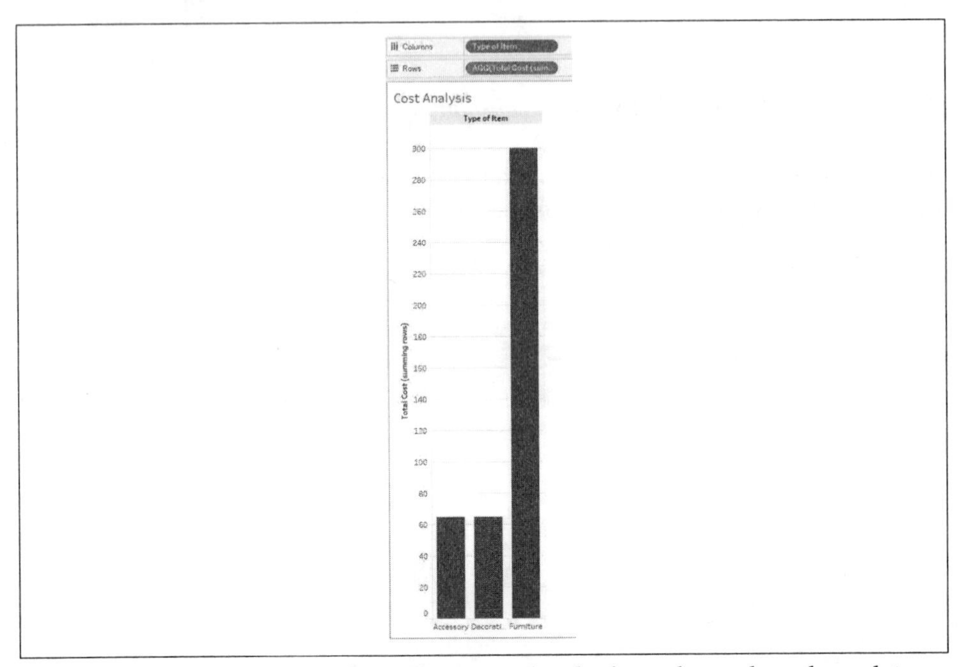

Figure 15-9. Adding the Type of Item dimension breaks down the total cost by each item

As you can see, this gives you a huge amount of flexibility and is why it's so easy to explore data in Desktop. You'd only use Prep Builder for this aggregation when you know what level of granularity you want, so we set this up using the aggregation step rather than a Calculated Field.

Some examples of where you would want to aggregate data in Prep Builder include:

Showing percent of total calculations
: You can use the Aggregate step to calculate totals, or, by adding a column to the "Group by" (which we'll cover shortly) to calculate subtotals. You can then use these totals and subtotals as the denominator for the individual records.

Changing the granularity of the data
: Input data sets are not always at the level of granularity you need for analysis. Reducing the granularity by aggregating the data set can boost performance when you are analyzing the data in Desktop. You might also change the granularity so it matches that of another data set in order to make joining data sets easier.

Analyzing data in Prep Builder
: You can use aggregation when you want to answer questions about the data within Prep Builder instead of having to output the values to Desktop.

Adding the Aggregate Step

Before we dive into the "how" of aggregation, let's go over adding an Aggregate step. First, click the plus sign at the end of your flow (or wherever you want to add it) and select Add Aggregate (Figure 15-10).

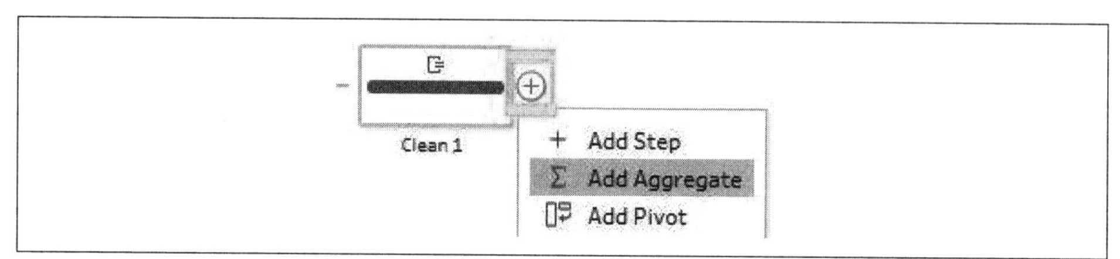

Figure 15-10. Selecting Add Aggregate in Prep

The basic Aggregation pane is shown in Figure 15-11.

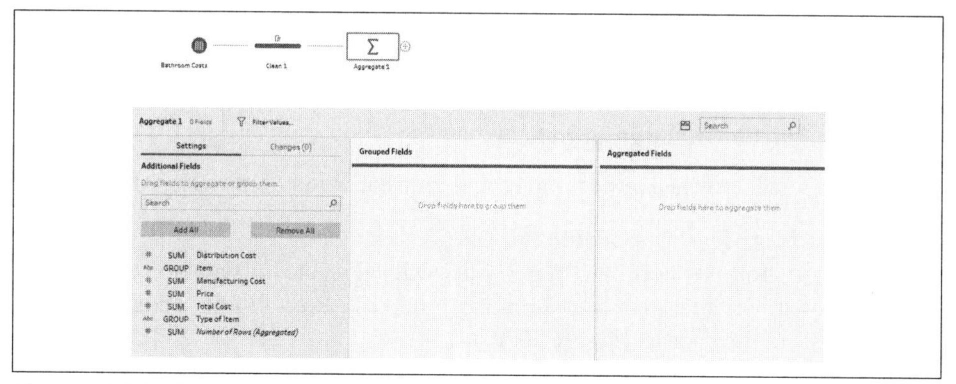

Figure 15-11. The Aggregation pane in Prep

The gray pane on the left shows all the fields in your data set and their default aggregation based on their data type. There are two ways to configure this tool. The first is to drag the fields you want to aggregate to the Aggregated Fields pane on the right and select "Group by." You can change the aggregation by clicking on the current type of aggregation (e.g., Sum) at the top of the column (Figure 15-12). The second option is to click the SUM or GROUP BY text next to each field in the left-hand pane. This opens a drop-down menu of all possible methods of aggregating. Selecting "Group by" will move the field to the Grouped Fields pane at the center of the screen, and selecting a method of aggregation will move the field to the Aggregated Fields pane at the right.

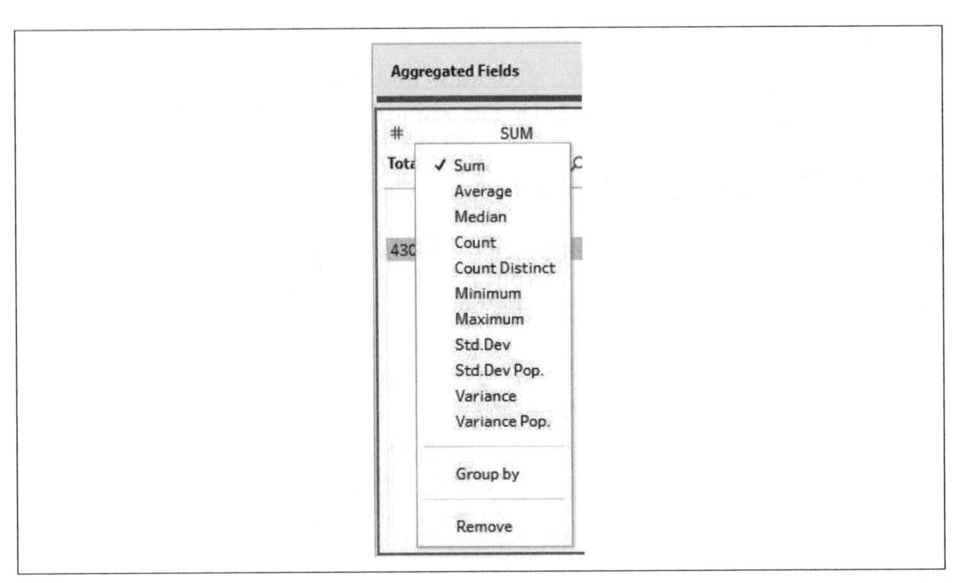

Figure 15-12. Changing the type of aggregation

If you haven't used SQL or other data coding languages, the concept of "Group by" might be a bit strange at first. Basically, adding a field to "Group by" is the equivalent of saying, "For each of these 'things,' I want the aggregated value to be returned." If you have multiple fields in the Group by, then this means you get an aggregated value for each combination of the grouped fields.

To clarify this, let's look at a few examples with our Bathroom Renovation data set:

Group by: Nothing; Aggregate: `sum(Total Cost)`
: When you group by nothing, Prep Builder will aggregate the data however you've asked for it. For this example, summing total cost, Prep Builder adds up all the total costs in the data set (Figure 15-13).

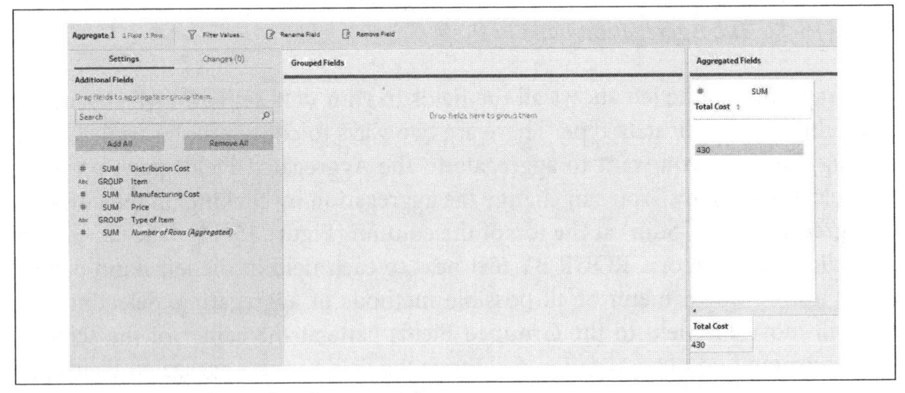

Figure 15-13. Total cost for the entire data set

The equivalent in Excel is summing the whole rightmost column (Figure 15-14).

Type of Item	Item	Price	Manufacturing Cost	Distribution Cost	Total Cost
Furniture	Bath	250	105	30	135
Furniture	Shower	150	70	10	80
Accessory	Shower Screen	75	35	30	65
Decoration	Tiles	80	20	20	40
Decoration	Paint	40	15	10	25
Furniture	Sink	100	65	20	85
					=SUM(F2:F7)

Figure 15-14. Calculating total cost for the data set in Excel

Group by: Type of Item; Aggregate: `sum(Total Cost)`

By adding Type of Item to "Group by," we are asking Prep Builder to break down the total cost by each different type of item in our data set (Figure 15-15).

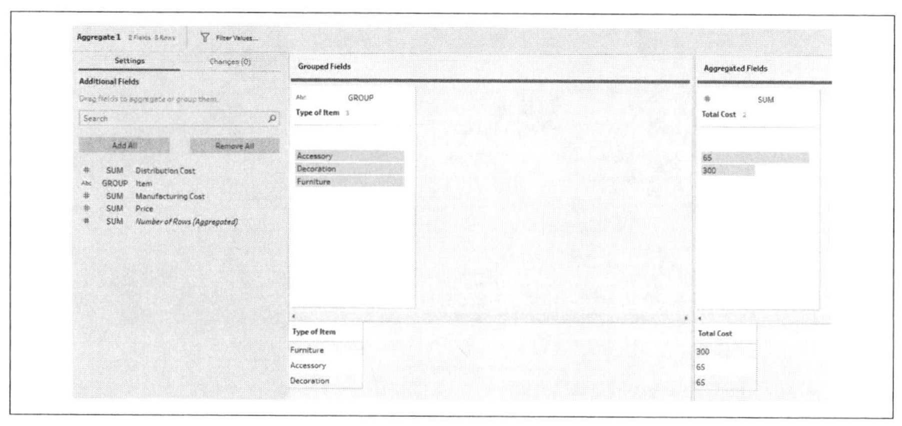

Figure 15-15. Breaking down the total cost by type of item

Notice how the data set has three different types of items, so there are three rows of data shown at the bottom right of the screen. The Profile pane has only two values, as two rows have the same value (65) in our output.

In Excel, we would use the `SUMIF()` function to calculate the three group totals (Figure 15-16).

If a different, new type of item were added to the data set, the next time Prep Builder is run, it would create a fourth row of data.

Type of Item	Item	Price	Manufacturing Cost	Distribution Cost	Total Cost		
Furniture	Bath	250	105	30	135		
Furniture	Shower	150	70	10	80		
Accessory	Shower Screen	75	35	30	65		
Decoration	Tiles	80	20	20	40		
Decoration	Paint	40	15	10	25		
Furniture	Sink	100	65	20	85		
				Totals	300	65	65

Figure 15-16. Breaking down total cost by type of item in Excel

Group by: Type of Item, Item; Aggregate: `sum(Total Cost)`

If we group by each categorical data field in our view (i.e., Type of Item and Item), the aggregated values look the same as they did before the aggregation step (Figure 15-17).

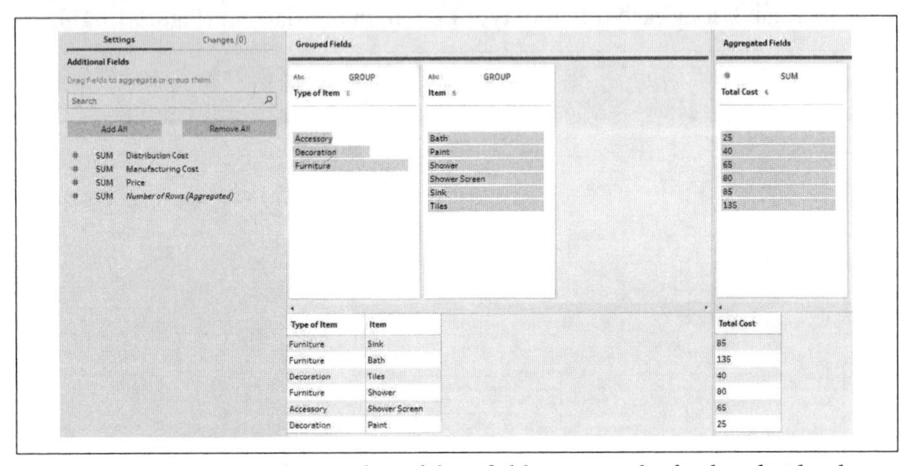

Figure 15-17. Increasing the number of data fields to group by further divides the total cost

In this example, this aggregation isn't actually aggregating anything, as each row of data was already a unique combination of the Type of Item and Item dimensions. Therefore, we have as many rows flowing in to the step as flowing out (so it's a pointless aggregation for this particular data set!).

Where's the Rest of My Data?

Following your aggregation, you might have additional preparation steps to take. The challenge with the Aggregate step is that you output only the fields in your Group by and Aggregation setup.

Let's go back to the second Group by example, where we are grouping only by Type of Item, and see what flows into the next Clean step (Figure 15-18).

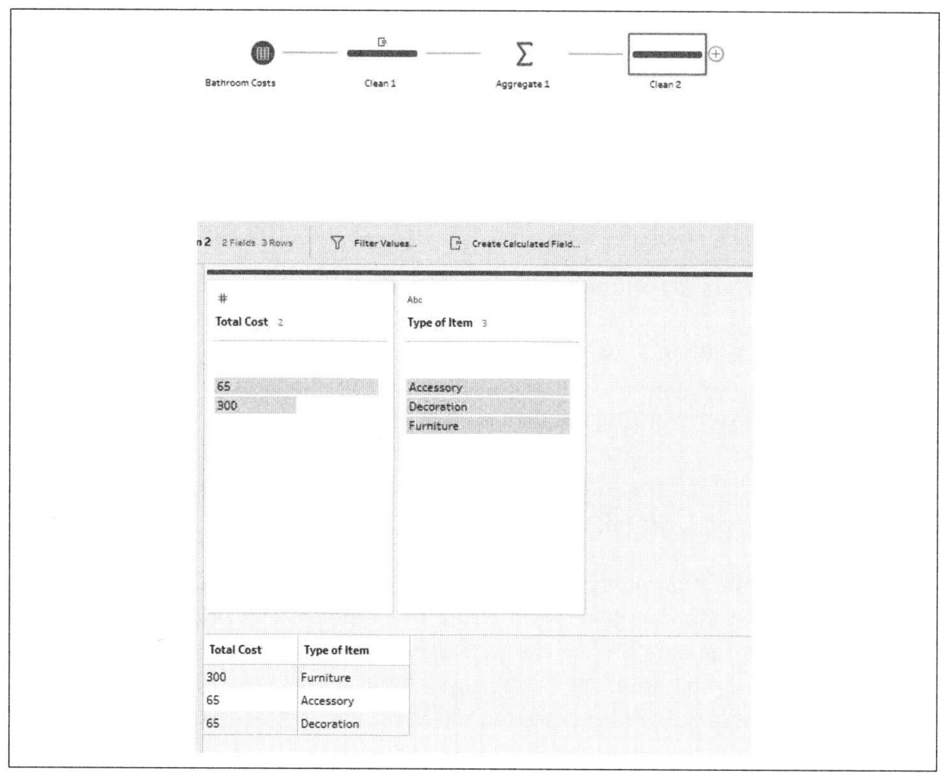

Figure 15-18. The Aggregate step outputs only grouped by or aggregated data

We have three different types of items, so we have three rows of data, but notice that we have left behind all the other columns of data. To get these back, you need to join the total you have just added back to the original data set. Where you draw the data from before the Aggregate step depends on what changes have been made to that data. Your flow will look something like Figure 15-19.

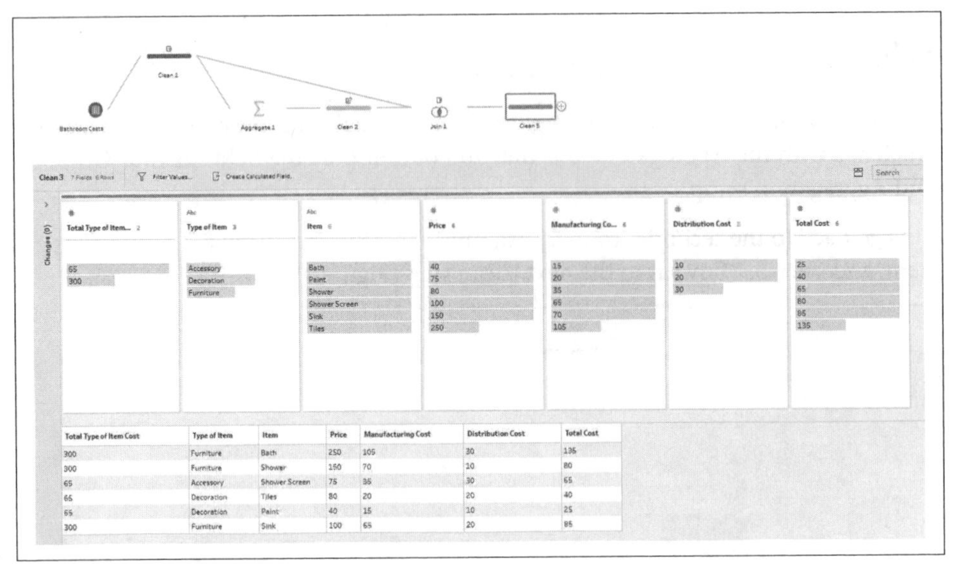

Figure 15-19. Self-join to reintroduce the data lost during an Aggregate step

To add the total value to your existing data, use a join condition(s) for whatever is within the "Group by" part of your aggregation. In this example, the join condition is where Type of Item in our original data set has the same value as Type of Item in our aggregated data set.

Level of Detail Calculation Option

If you want to avoid rejoining your data after the aggregation, consider using *Level of Detail* calculations. As Chapter 33 will cover, this type of calculation doesn't change the granularity of the data set like the Aggregate step does. Instead, you can set how and to what level one measure is aggregated but leave the rest of the data set unchanged. Therefore, there is no need to rejoin the data set.

Summary

The Aggregate step in Prep Builder differs significantly from how Desktop users may think about data aggregation. Changing the granularity of the data is a common aspect of data preparation, as it enables you to join data sets together more easily or reduce the number of records to be processed during analysis. You must be careful not to *overaggregate* the data source if you are visualizing the data in Desktop, as this could negate some of the tool's benefits. Specifically, because Desktop aggregates the data based on the discrete fields within the view—giving you the freedom to analyze what you want, how you want—you should not aggregate your data unless it is necessary, or you risk losing this flexibility.

CHAPTER 16

Joining Data Sets Together

Whether your data comes from files, databases, or both, you'll need to master joins, as most software requires one large table of data to reference for analysis. Your data sets will come from a lot of different sources, so being able to combine them is key. This will allow you to take columns from each data source and use them alongside one another in the output. The rows of data they contain are added to the resulting data set.

Most of the examples in this chapter will use the 2019: Week 29 inputs (*https://oreil.ly/-Dk2g*). We will cover the logic and terminology for working with joins, how to join data sets together in Prep Builder, and which type of join to use in different situations.

How to Join Data Sets in Prep Builder

Joining separate data sources is very easy in Prep Builder. Simply drag one data source toward another, and Prep Builder will give you a number of options (Figure 16-1).

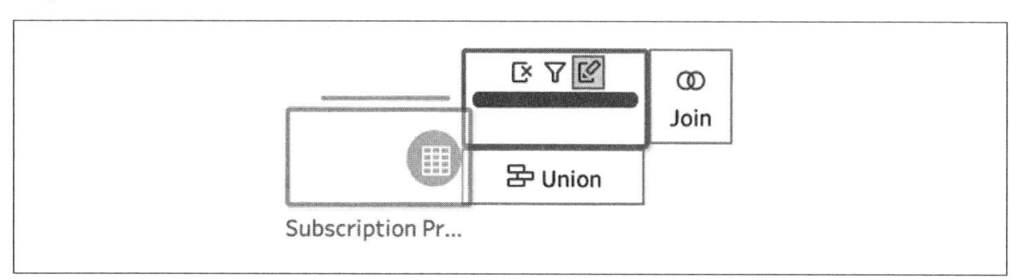

Figure 16-1. Dragging one step toward another triggers the Join/Union options

By dragging the second data set (in this case Subscription Pricing) to the original (blue) data flow (Clean 2), you'll be given the option to union or join the data sets. If you hover over the Join option and then release the mouse button, Prep Builder will automatically link the two flows of data together (Figure 16-2).

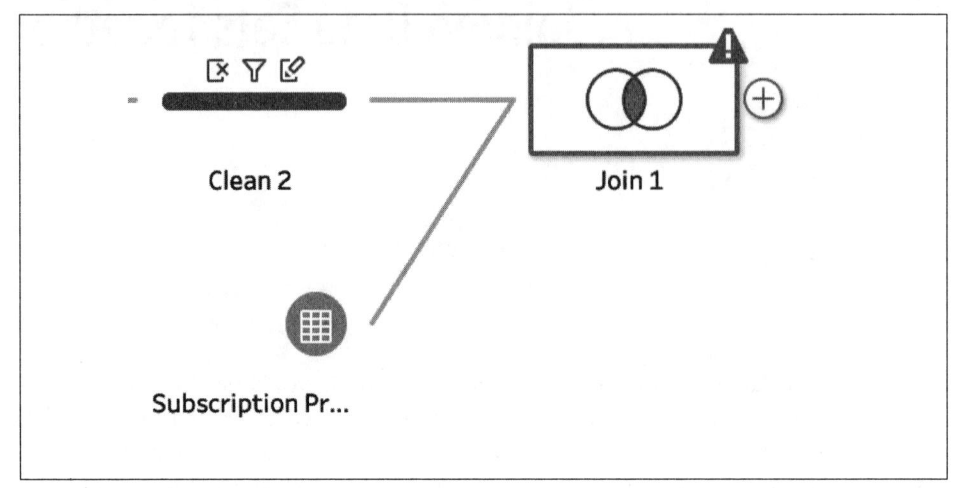

Figure 16-2. The Join step in Prep

Another approach to joining two data sets is to add a Join step, drag the second data set onto it, and then release your mouse button above the Add section (Figure 16-3).

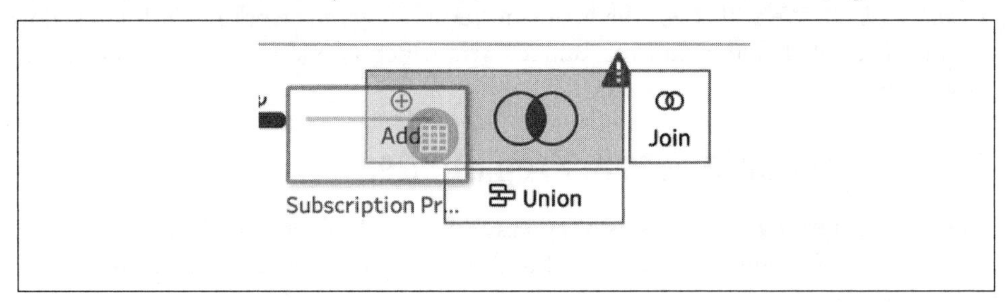

Figure 16-3. Adding a data set to a Join step

You might have noticed that an error is generated in the flow in Figure 16-3; this is because the join condition hasn't automatically been set. This occurs when Prep Builder doesn't find matching data field names in the two separate inputs (Figure 16-4). Be careful to check that the join condition makes sense—just because the column headers match, it doesn't mean the fields have exactly the same values.

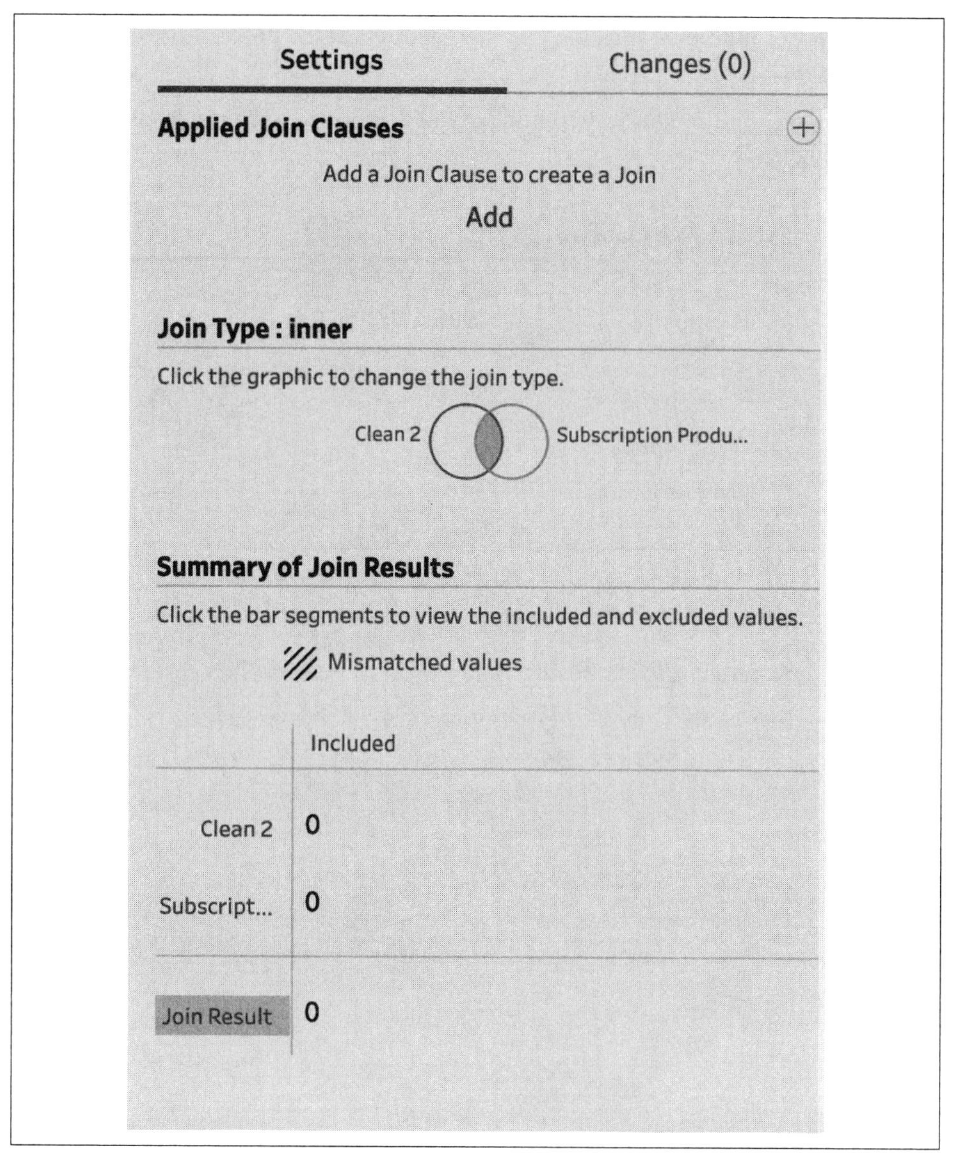

Figure 16-4. Join step configuration pane without a join condition set

You can add the join condition by clicking Add or the circled plus sign in the Applied Join Clauses section of the Join step. To set the condition, select a field to assess from each data source and then choose the type of assessment. Equals is the most common, but you can also set qualifiers like Does Not Equal or Less Than to handle more complex logic and save the data's end users from having to filter the result. If you want to add more join conditions, click the plus sign and repeat the process for each addition join condition.

Once the join condition is set, Prep Builder will demonstrate the results of the join, as shown in Figure 16-5.

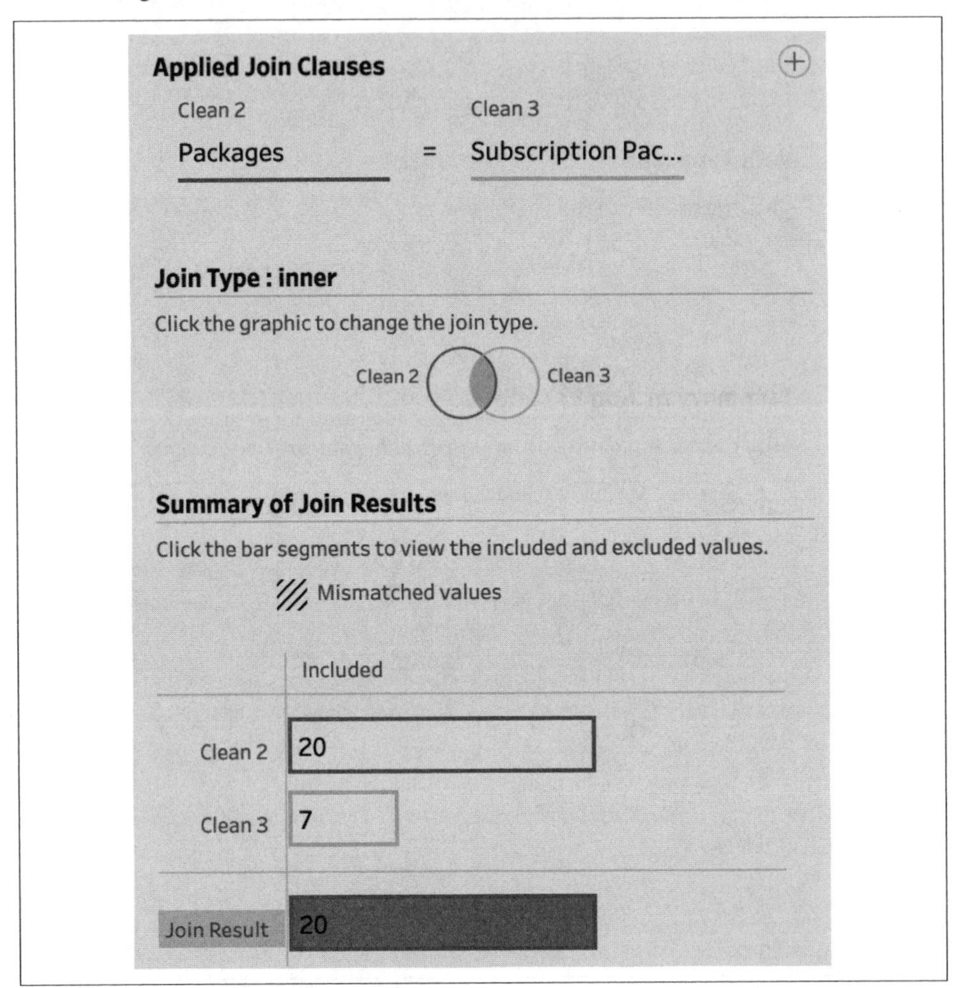

Figure 16-5. Viewing the effects of the join

The Summary of Join Results at the bottom of this view lists the number of rows in the resulting data set (Join Result), as well as how many rows from each original data set are included. If the Join Result is significantly higher than either of the input data sets, you should investigate if the join is creating duplicates or other additional rows. This information can give you insight into the potential impact of different joins as you decide which type to use. To change the join type, click the different sections of the Venn diagram in the central Join Type area. Any areas shaded in dark gray will be included in the resulting output. If you want to prevent a section of the Venn diagram from being returned, click that section again to deselect it.

Join Logic and Terminology

The majority of data software, including Prep, uses similar logic and terminology for joining data sets together. However, joining data sets together in Prep is a lot more user-friendly than in other tools.

Figure 16-6 and Figure 16-7 show two sales-related tables that we might want to join together to analyze revenue. Since the number of subscription packages sold and the package price appear in separate tables, we need to join the tables to find out the revenue generated.

Packages	Name	Frequency
7	Charlie	1
5	Jonathan	2
1	Jamie	2
6	Ellie	1

Figure 16-6. Customer Order Frequency table

Subscription Package	Product	Price
1	Active	5
2	Relax	5
3	Home	10
4	On the Go	10

Figure 16-7. Subscription Pricing table

As noted earlier, there are two aspects to creating this join:

- Join condition, which determines how to link the two tables together
- Join type, which determines what will be returned by the join

In the sales example, the join condition would be where the values match in the data fields Packages and Subscription Package. To conduct the revenue analysis, we'd want to analyze the Price field for each package along with the number sold. Remember, when joining data sets together, you can set multiple join conditions using different fields, so it doesn't have to be simply a case of one value matching another in a different data set. Join conditions need to match not just the values but the data types as well: a string with the value 1 would not match an integer with the value 1.

Joins in many tools, Tableau included, are represented by a Venn diagram. This is because we need to consider what parts of the tables we return, which determines the join type we'll use. The terminology associated with joins is aligned to Venn logic also; you can perform *left joins* and *right joins*. One data set corresponds to the left-hand circle in the Venn diagram, and the other to the right-hand circle. Keep in mind that you can join only two data sets at a time. If you have more than two, simply join two together and add another Join step for each additional data source.

Types of Join in Prep Builder

To explore the impact of using each join type in Prep Builder, we'll use two simple data sets: Left Table (Figure 16-8) and Right Table (Figure 16-9). Each example will demonstrate the output that would result if we set the join condition such that the category from the Left Table is the same as the category from the Right Table.

In a real-world scenario, the output from Prep Builder would contain both tables' Category columns, but these are usually removed by the user, so I've just recorded one here.

Category	Measure 1
A	1
B	2
C	3

Figure 16-8. Left Table data set

Category	Measure 2
A	4
B	5
D	6

Figure 16-9. Right Table data set

Prep Builder gives you the following Join Type options:

inner

The inner join will return only records that meet the join condition(s) (Figure 16-10).

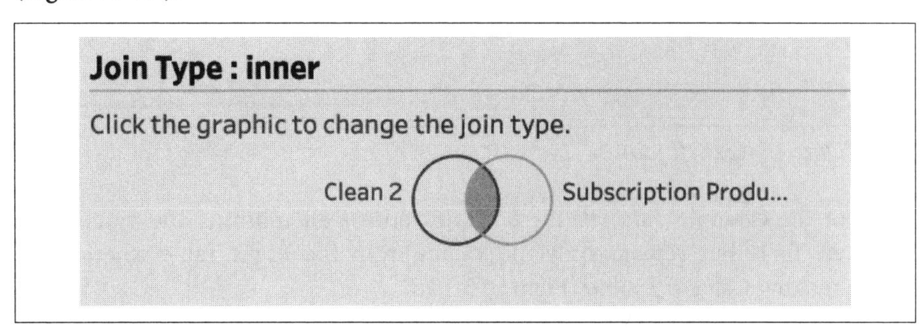

Figure 16-10. Inner join

For the example data set, an inner join would return only Category values `A` and `B` but would have values for both Measure 1 and Measure 2. Figure 16-11 shows the result.

Category	Measure 1	Measure 2
A	1	4
B	2	5

Figure 16-11. Results of the inner join

left

The left join returns all rows from the left table, plus the records from the right table that meet the join condition(s). Any rows from the right table that do not meet the join condition will not be carried through to the next step (Figure 16-12). Records in the left table that do not have a match in the right table will have a null value in the resulting table.

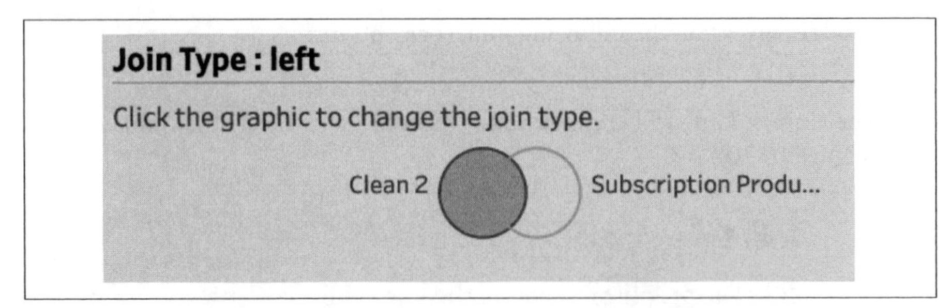

Figure 16-12. Left join

For the example data set, the left join returns all columns and values from the Left Table but returns only the values from the Right Table where there is a matching Category value (Figure 16-13).

Category	Measure 1	Measure 2
A	1	4
B	2	5
C	3	null

Figure 16-13. Results of the left join

leftOnly

The leftOnly join returns records that do not meet the join condition(s) only from the left table (Figure 16-14).

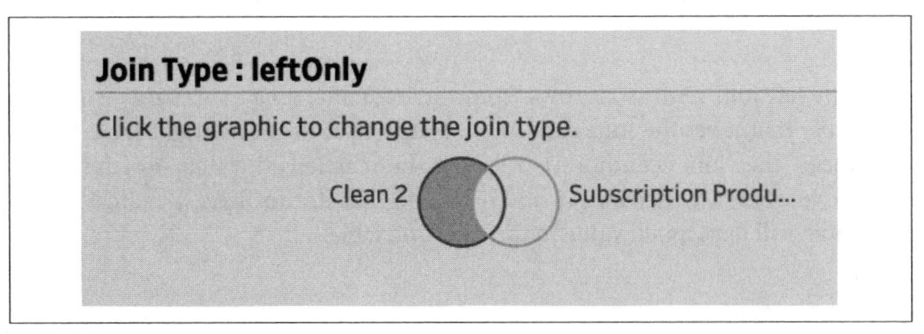

Figure 16-14. leftOnly join

The example data set has only one unmatched Category value and returns only one row of data (Figure 16-15).

Category	Measure 1
C	3

Figure 16-15. Results of the leftOnly join

right

Similar in theory to the left join, the right join returns everything from the right table, plus records from the left table that meet the join condition(s) (Figure 16-16).

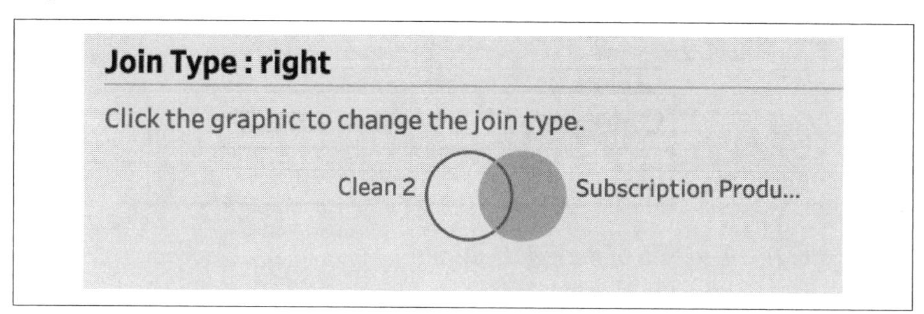

Figure 16-16. Right join

The example data set would return all the data for the right table, but only the two matching Category values from the left table (Figure 16-17).

Category	Measure 1	Measure 2
A	1	4
B	2	5
D	null	6

Figure 16-17. Results of the right join

rightOnly

Similar to leftOnly, the rightOnly join will return records that do not meet the join condition only from the right table (Figure 16-18).

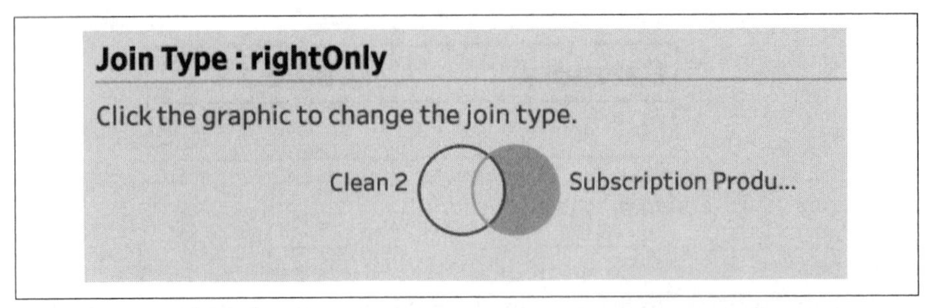

Figure 16-18. rightOnly join

As in the leftOnly example, one value is returned for the rightOnly join (Figure 16-19).

Category	Measure 1
D	6

Figure 16-19. Results of the rightOnly join

full

The full join produces a data set that contains all the values from both tables (Figure 16-20). When a value in either table doesn't have a match in the other table, you'll see a null value in the resulting data set.

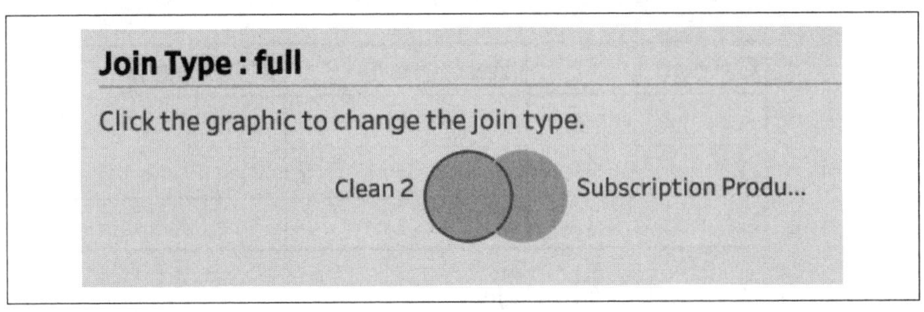

Figure 16-20. Full join

Using the example data set, all the values from both the Left Table and Right Table data sets are returned, but nulls appear where the Category fields don't have a corresponding value in the other table (Figure 16-21). Unlike the other join examples where I've omitted the duplicate Category field, I've left it in here to show the differences in the output. The second Category field will automatically be appended with the value `-1` as each data field in a Prep Builder data set must have a unique name.

Category	Measure 1	Category-1	Measure 2
A	1	A	4
B	2	B	5
C	3	null	null
null	null	D	6

Figure 16-21. Result of the full join

notInner
: The notInner join will return the same data as the full join except it will not return the records that meet the join condition (Figure 16-22).

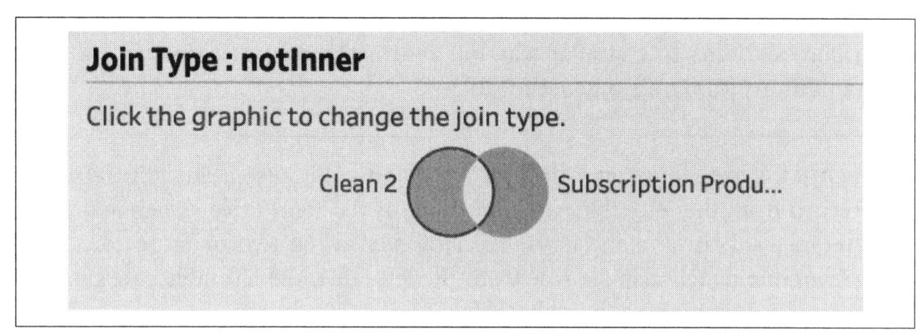

Figure 16-22. notInner join

As with the full join, I've retained the duplicate Category field for the notInner join in Figure 16-23 to highlight the difference in its output compared to the earlier join types.

Category	Measure 1	Category-1	Measure 2
C	3	null	null
null	null	D	6

Figure 16-23. Results of notInner join

When to Use Each Join Type

Just because a join returns the correct number of records, it doesn't mean the join is correct. You must carefully think through what data you'll be including and excluding when choosing a join condition and type of join.

Let's consider a few situations that call for a specific join type:

inner
: The inner join is the workhorse of join techniques. You use it as a way of filtering out missing records that would be returned as a null value in a left or right join, thereby protecting calculations that require only non-null values.

notInner
: This initially seems to be a bizarre join type. Why join two data sets together that fundamentally *won't* join due to not matching the specified join condition? The answer is data quality. By creating a notInner join, you can treat the data that does not meet the join condition as if it were going to be rejoined in the future. This is a good error-checking technique, as it helps you validate what has been returned by an inner join. Depending on the data set, you could use leftOnly or rightOnly joins in a similar way but return the data that doesn't meet the join conditions from only one of the tables.

left
: Left joins are almost as common inner joins. Because a left join returns every record from the left table, you can think of the right table as being appended to the relevant rows in the left table. However, when two or more rows from the right table match a single row from the left table, the left table row gets duplicated for each row from the right table.

full
: Use a full join when you want to return all data fields from both tables but want to create a single row for data that meets the specified join condition(s). When joining disparate sources together where you want to return all the data—for example, when you are joining two customer data sets during a merger of two organizations—a full join is likely to be the right solution.

Summary

Joins can be a fantastic way to add further context to your analysis, as you can add data that didn't reside in the original source. There are challenges with non-unique join conditions—a topic we will cover in future chapters—but hopefully this chapter has helped you better understand how and when to use different types of joins.

CHAPTER 17

Unioning

Most software that works with data demands that you form a single table of data to work from. However, the world is often not that simple, so you'll often have to pull together many tables of data to build that single table. Unioning is a data preparation technique that will help you with this task.

What Is a Union?

You can think of unioning as stacking one data set on top of another. Columns that contain the same content should be unioned as part of the data preparation process. As you'll soon see, this requires the data structures to be very similar.

Let's look at two separate example data sets: York Store Sales (Figure 17-1) and Leeds Store Sales (Figure 17-2).

Store	Date	Product	Sales
York	Jan-2019	Bar Soap	1000
York	Jan-2019	Liquid Soap	400

Figure 17-1. York Store Sales data set

Store	Date	Product	Sales
Leeds	Jan-2019	Bar Soap	800
Leeds	Jan-2019	Liquid Soap	500

Figure 17-2. Leeds Store Sales data set

Unioning these two tables removes the extra set of column headers and stacks the rows of data on top of each other (Figure 17-3).

Store	Date	Product	Sales
York	Jan-2019	Bar Soap	1000
York	Jan-2019	Liquid Soap	400
Leeds	Jan-2019	Bar Soap	800
Leeds	Jan-2019	Liquid Soap	500

Figure 17-3. Unioned result of York and Leeds data sets

The union is usually determined by whether the data fields are:

- Named the same in each data set. The contents of columns with the same name will be stacked in rows under that column name in the resulting data set. Currently, Prep Builder allows only for this type of union.
- Positioned in the same order. The contents of the first column will be stacked on top of the contents of the first column in the other data set, regardless of what it contains. This type of union isn't supported within Prep Builder.
- Manually matched by the user. Some software allows the user to match columns of their own choice before unioning the data sets together. This functionality also isn't supported within Prep Builder.

In the previous example, the tables could have been unioned by either their position or name. In Prep Builder, you can union columns using the other techniques by giving the relevant columns the same name to satisfy the requirements of the first technique.

What If the Data Structure Isn't Identical?

Because unioning in Prep Builder is based on matching column names, when those names are not the same it creates a mismatch in the resulting data set. For example, the same data is captured under the Scent column header in the Leeds Store Sales table and the Type column header in the York Store Sales table (Figure 17-4 and Figure 17-5).

Store	Date	Product	Scent	Sales
Leeds	Jan-2019	Bar Soap	Honey	800
Leeds	Jan-2019	Liquid Soap	Mint	500

Figure 17-4. Leeds Sales Stores with Scent column header

Store	Date	Product	Type	Sales
York	Jan-2019	Bar Soap	Lavender	1000
York	Jan-2019	Liquid Soap	Mint	400

Figure 17-5. York Store Sales with Type column header

Therefore, in a normal union, even though you might assume the column contents should be stacked on top of each other, they will be mismatched in the resulting table (i.e., there will be two separate columns for the same data). When this happens, rows from each table will have nulls where no matching column is found in the other table (Figure 17-6).

Store	Date	Product	Type	Scent	Sales
York	Jan-2019	Bar Soap	Lavender		1000
York	Jan-2019	Liquid Soap	Mint		400
Leeds	Jan-2019	Bar Soap		Honey	800
Leeds	Jan-2019	Liquid Soap		Mint	500

Figure 17-6. Unioned data with mismatched fields

In Prep Builder, you can clearly see these mismatched fields within the Union step (Figure 17-7).

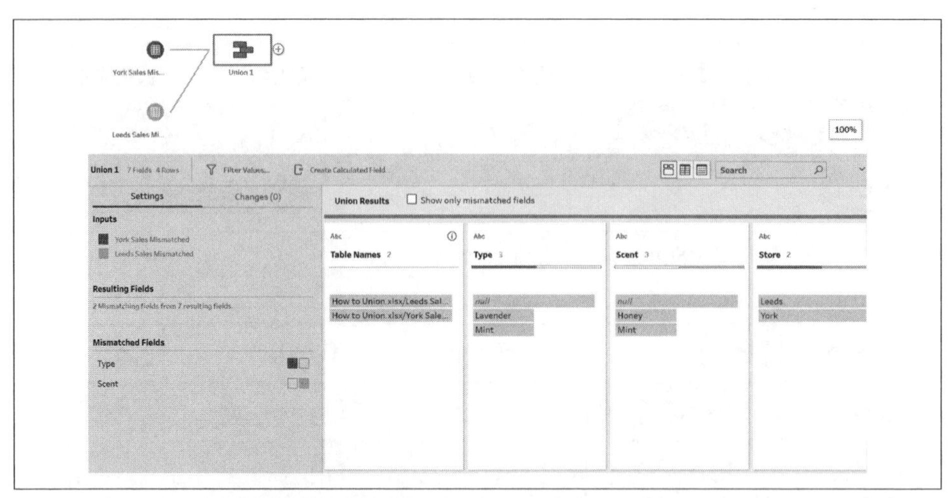

Figure 17-7. Mismatched fields from a Union step in the Profile pane

You can merge the mismatched fields by selecting them and clicking Merge Fields (Figure 17-8).

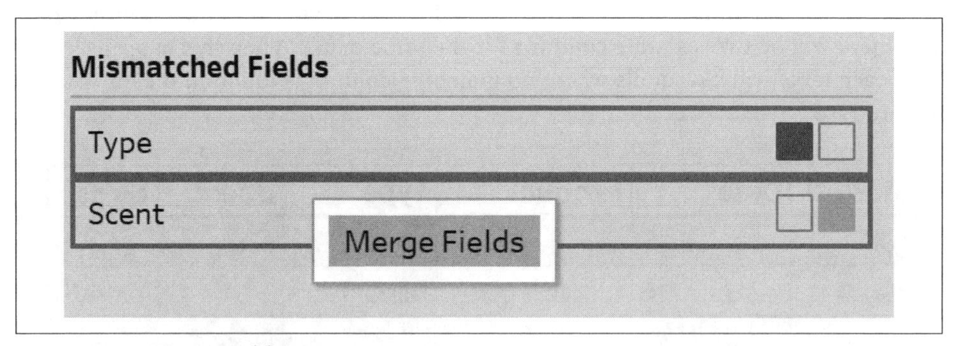

Figure 17-8. Merge Fields option

When to Union Data

So now you have seen *how* to union, but *when* do you deploy the technique? Let's look at some common scenarios where unioning data is a good strategy.

Monthly Data Sets

When working with other teams in your organization or external third parties, you'll often receive files to analyze on a monthly basis. Automated data preparation flows can save you a lot of time shaping the data into a consistent, useful state for analysis. However, if the data needs to build up over time, then you'll need to append the files to each other. Unions are the perfect technique for this. Figure 17-9 shows how you'd use a union to change the York Store Sales data from monthly sales files to a single file for analysis.

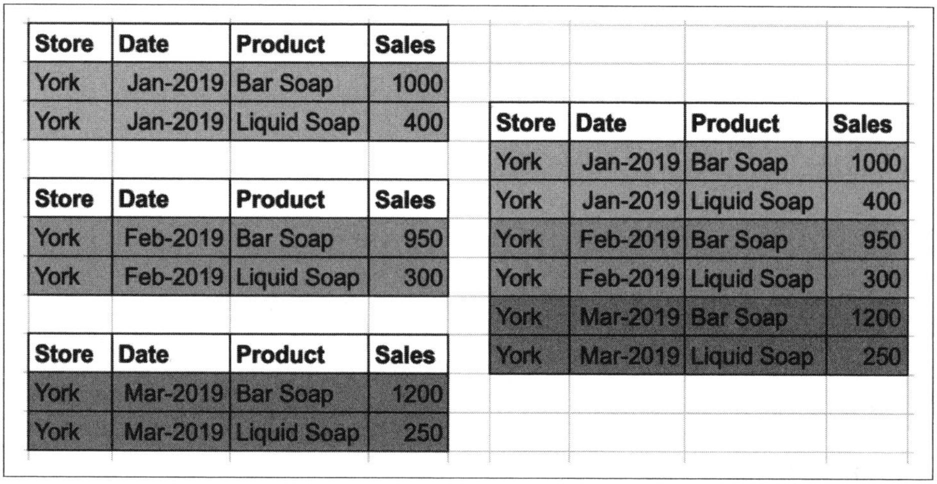

Store	Date	Product	Sales
York	Jan-2019	Bar Soap	1000
York	Jan-2019	Liquid Soap	400

Store	Date	Product	Sales
York	Feb-2019	Bar Soap	950
York	Feb-2019	Liquid Soap	300

Store	Date	Product	Sales
York	Mar-2019	Bar Soap	1200
York	Mar-2019	Liquid Soap	250

Store	Date	Product	Sales
York	Jan-2019	Bar Soap	1000
York	Jan-2019	Liquid Soap	400
York	Feb-2019	Bar Soap	950
York	Feb-2019	Liquid Soap	300
York	Mar-2019	Bar Soap	1200
York	Mar-2019	Liquid Soap	250

Figure 17-9. Monthly data sets unioned into one table

Data Sets from Web Sources

Pulling together similar data sets from web-based sources is another case where unioning files can be useful. For example, you might need to take team rosters from their individual web pages and union them together. You might also consider adding the URL, or sheet name, as a reference in the data set.

In the example shown in Figure 17-10, which uses ESPN's team rosters and Google Sheets' `IMPORTHTML()` function to pull together a full list of players, you can union the two team roster tables together, but you would lose the team each player plays for. When you use the Union step in Prep, Tableau adds the name of the table where the data originated.

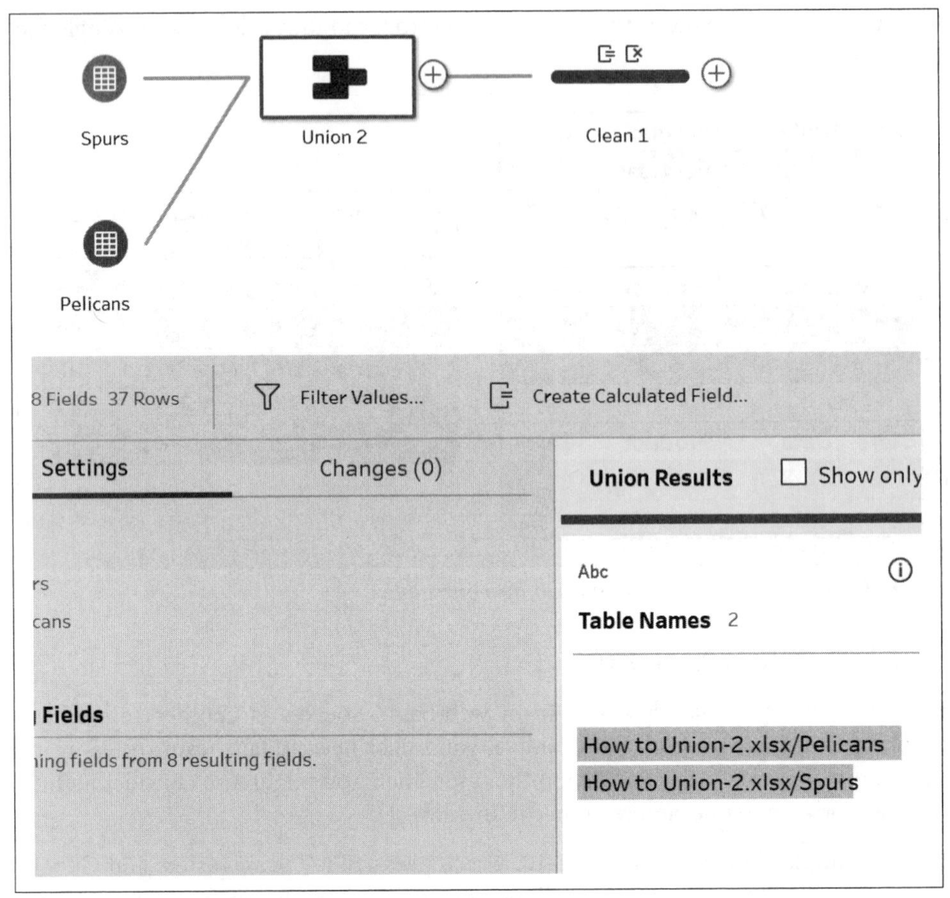

Figure 17-10. Creating a Table Names column as part of the Union step

This allows you to split the team name off from the rest of the table source (Figure 17-11).

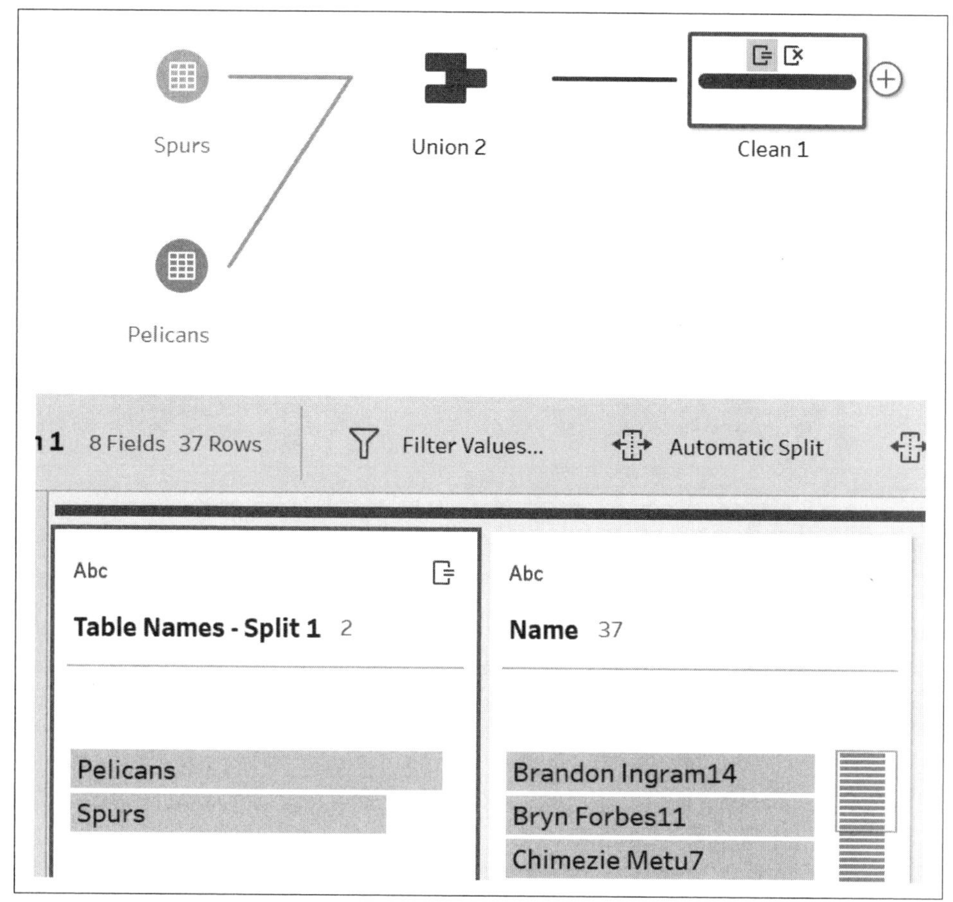

Figure 17-11. Splitting the team name from the table name

Company Mergers

With the rise of cloud-based software, more companies are using similar tools and data structures. This means that when departments and organizations merge, they can union together their data exports. For example, in a merger of two companies that both use Salesforce, using a union to pull together a combined sales pipeline list would be very simple.

Multiple Tables and Wildcard Unions

When unioning parts of your flow, you can add more than two inputs into a single union. However, if you have several inputs to union, it can quickly become tedious to add a single input for each data source. Instead, during the Input step in Prep Builder, you can select "Wildcard union" rather than "Single table." The Wildcard has three sections that allow you to select any folder (including subfolders), file, or sheet that matches the pattern you specify (Figure 17-12).

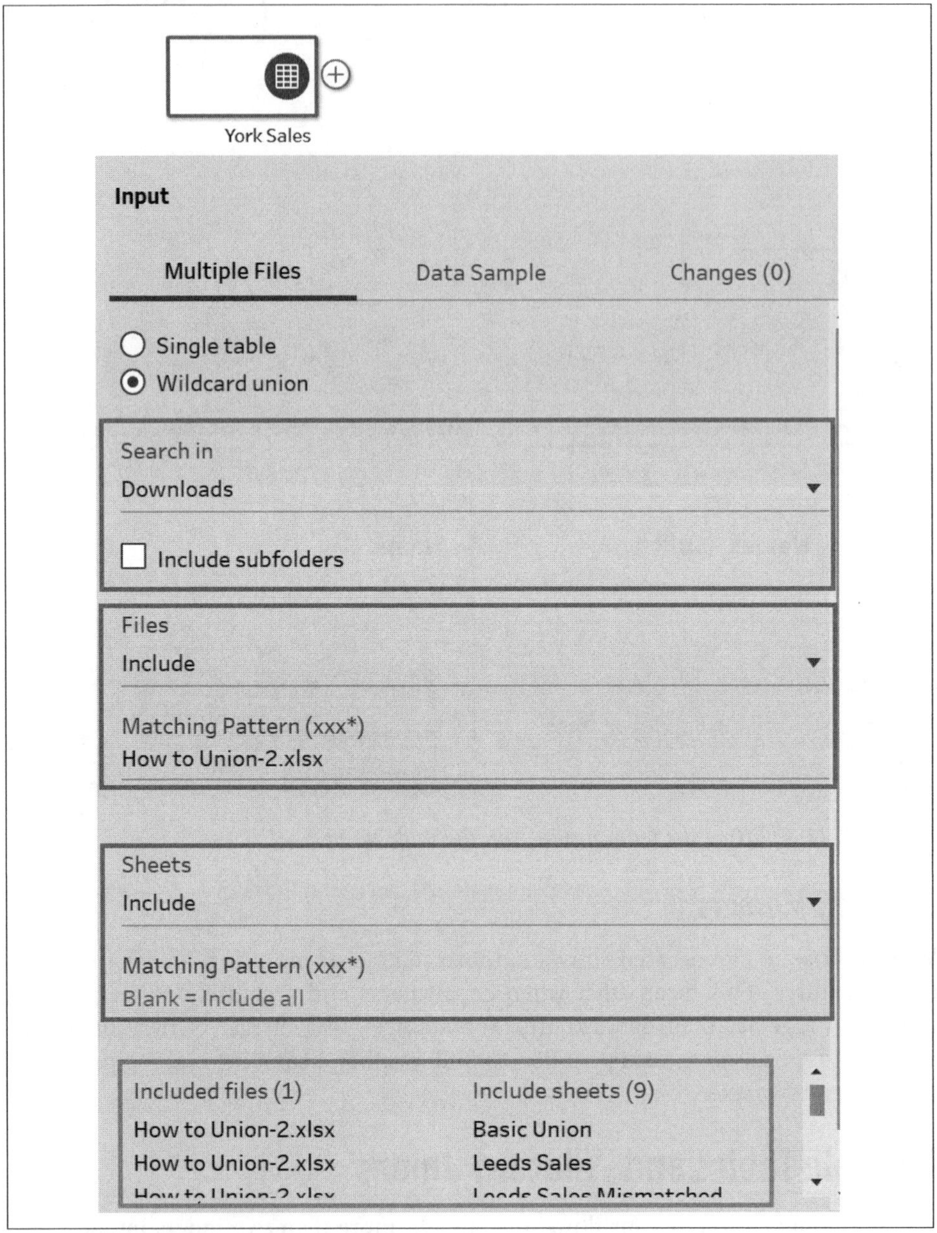

Figure 17-12. Setting up wildcard unions on the Input step

In the sections called out in Figure 17-12, you can see an asterisk (*) beside the Matching Pattern. The asterisk symbol is used as the wildcard character in data circles and acts as a catchall, enabling you to better control what should and shouldn't be captured in the union.

For example, say you have one Excel workbook with three sheets: York Sales, Leeds Sales, and Reference Table. If you wanted to pull the York and Leeds Sales tables together, then in the Sheets section, you would enter **`*Sales`** for the Matching Pattern.

That way, any characters in the sheet name before the word "Sales" would be ignored, but the York Sales and Leeds Sales sheets would be included. Because the Reference Table doesn't meet the criteria, it wouldn't be included in the union or available for analysis. This functionality also future-proofs the flow, as any new sheets that match the specified pattern set would be added to the flow and processed.

Summary

Like joins, unions are a fundamental data preparation technique. Arguably a lot easier to use than joins, as there are no conditions to set up, unioning data simply involves stacking data sets to create a single table for analysis. Finally, using the wildcard union technique on the Input step ensures that as matching files are added to a folder or other location, they will be added to the data set the next time the flow is run, saving you an additional Input step.

CHAPTER 18

Calculations

Your data will hardly ever be perfect for your analysis straight from the source. Calculations are one of the key instruments you have to make the data suitable for answering the questions you and your peers have. Many people fear calculations, but this chapter offers some simple tips that will make them less intimidating and help you unlock their power and flexibility.

What Do Calculations Do in Data Preparation?

At the simplest level, calculations create a new data field or overwrite an existing one. Most calculations use an existing data field as the base to form the new data field. The new or altered field is referred to as a *Calculated Field*, indicating that it did not originate in the initial input data for your preparation flow. New data fields are created as a result of:

- Cleaning (to remove unwanted characters, split names, etc.)
- Arithmetic (sum, average, or multiplication results)
- Merging multiple fields (from dates, email addresses, etc.)

The calculated data fields resulting from these operations become available for analysis or further preparation. Calculations create new fields in Prep unless the calculation is named identically to an existing data field, in which case the original data field is overwritten.

Creating a Calculated Field

There are two main ways to create Calculated Fields in Tableau Prep. First, underneath the Flow pane where either the Profile or Data pane is shown, you'll see the option Create Calculated Field (Figure 18-1).

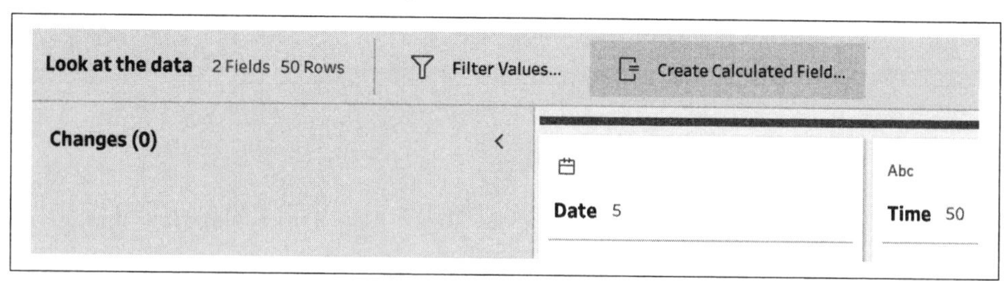

Figure 18-1. Dynamic menu option to create a Calculated Field

Clicking this opens up a separate window where you can create the calculation; the dialog is blank until you specify a calculation to include (Figure 18-2).

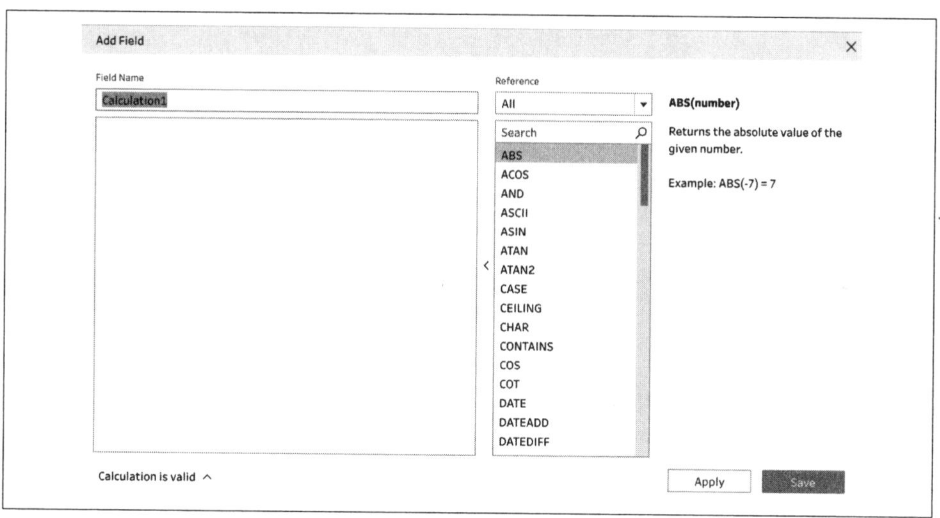

Figure 18-2. Blank Add Field dialog

The second way to create a Calculated Field is by selecting a specific data field. Open the data field's menu (designated by an ellipsis), and you should see the same Add Field dialog as in Figure 18-2, but now with the selected data field listed in the white box (Figure 18-3).

Figure 18-3. Creating a Calculated Field from a specific data field

This is equivalent to clicking Create Calculated Field and then Custom Calculation from the Profile pane menu (Figure 18-4).

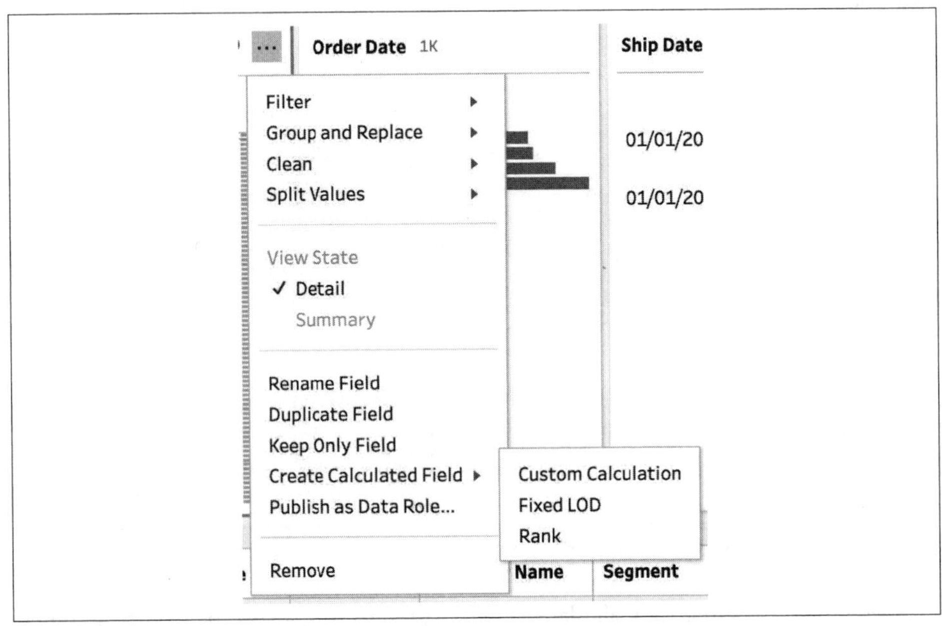

Figure 18-4. Creating a custom calculation from the Profile pane menu

Fundamentals of Calculations

The calculation window shown in Figure 18-2 and Figure 18-3 contains a couple of elements that can really help you learn and eventually master calculations.

The Reference List

The first element is the drop-down Reference list on the right. If you are new to calculations in data tools, this list is a massive help in finding the best function to get the answers you require (Figure 18-5). By entering a function, you are telling Prep Builder what operation to carry out on the data field.

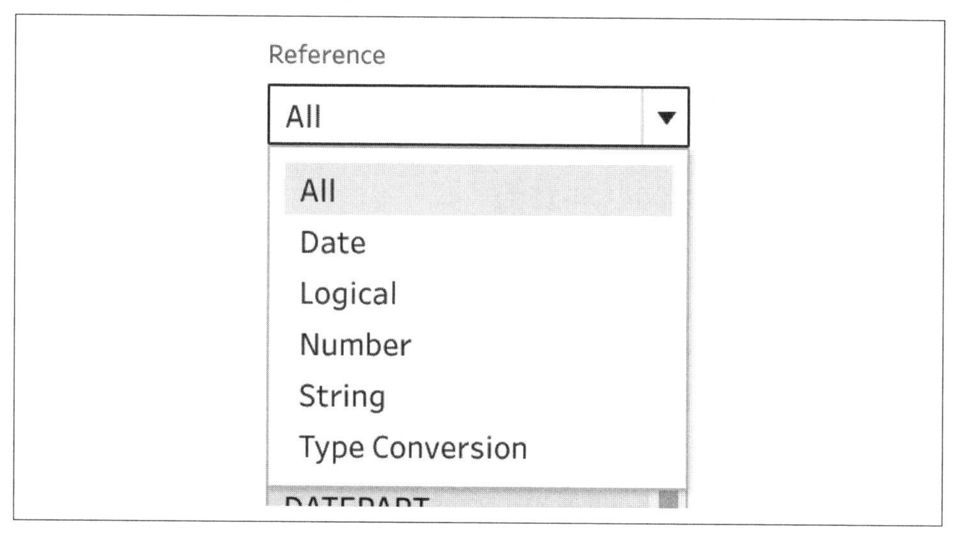

Figure 18-5. Grouping calculation functions to make searching easier

To the right of the Reference list is the second element of support in the calculation window: the syntax, a description, and an example of the function you have selected in the Reference list. Figure 18-6 shows this information for the `DATEPART()` function, which is used to pick out parts of a given date.

DATEPART(date_part, date, [start_of_week])

Returns a part of the given date as an integer where the part is defined by date_part. If start_of_week is omitted, the week start day is determined by the start day configured for the data source.

Example: DATEPART('month', #2004-04-15#) = 4

Figure 18-6. Calculation function details

Let's look at each piece of information in more detail.

Syntax

Even as an experienced Tableau or data software user, you will frequently need to check the syntax of the function as well as the data types needed for the function to work as intended. Here's how to read the syntax:

```
FUNCTION(required input, [optional input])
```

When building functions, keep in mind that you must match all the aspects of the syntax that are not in square brackets—they are not optional. These requirements might include the necessary data types or whether or not the data field is aggregated. Square brackets in the function indicate optional content. Note that when you make your calculation, your data fields will appear within square brackets as well, but this isn't the same thing. Also note where commas appear between the different elements of the function, as you'll need to include them in the calculation you build.

To make writing calculations easier in Prep Builder, you can use spaces, tabs, and even newlines between the different elements. This can make it easier for others viewing your work to understand the calculations.

Description

The description is where most new users will look to understand what the selected function actually does. The names of the functions are often a good clue to their purpose, but the description provides more detail on exactly what task that function is intended for.

Example

The example can be helpful for users at all levels in demonstrating what `date_part` means or how to encode the `start_of_week`. Some of the examples are not the easiest to follow, so searching online is a great way to boost your understanding. Just enter "Tableau" and the function name into a search engine, and you'll find more examples and often a blog post or two.

Building the Calculation

With all this support from the Prep tool, building calculations should be easy, right? Well, not quite. Calculations can be really complex, and a simple typo can break them. Let's look at some examples of both well-formed calculations and potential pitfalls.

When Calculations Go Well

Calculations go well when you use a function, or functions, that Prep considers to be correctly formatted. At the bottom-left corner of the calculation in Figure 18-7, Prep Builder, like Tableau Desktop, indicates that it is valid.

Validation does *not* mean that the calculation is correct in terms of your intended goal. Validating that the data is correct is up to you, as only you know what you were trying to do.

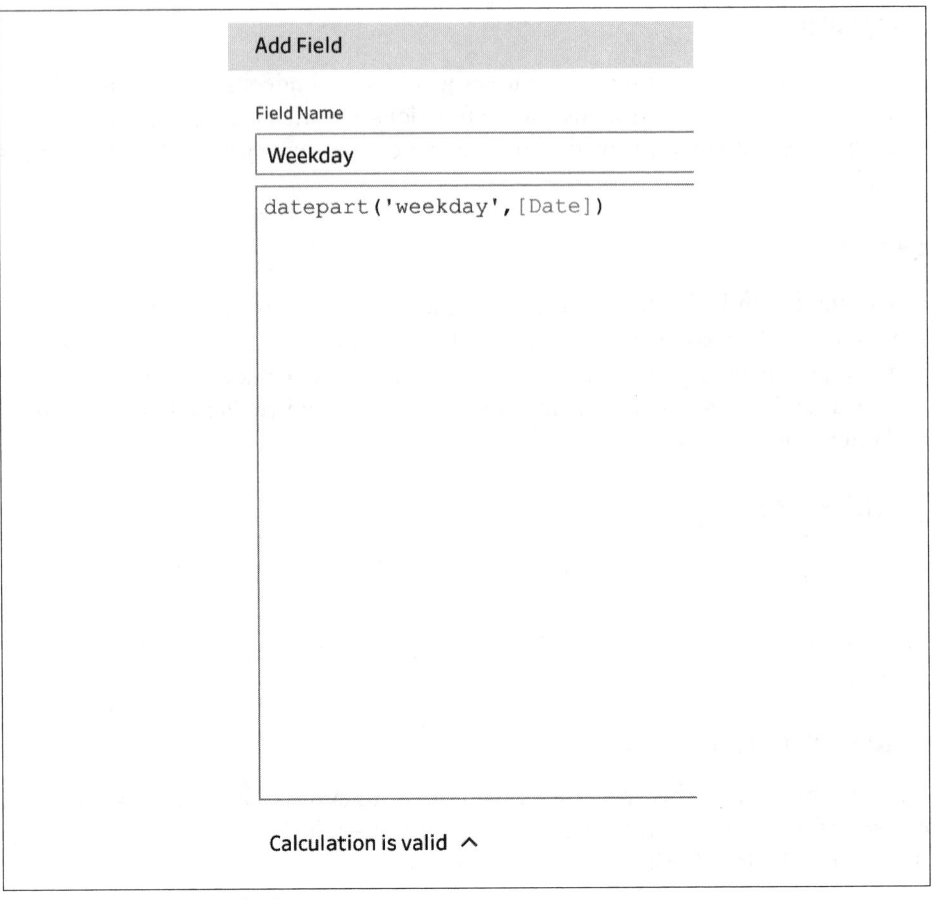

Figure 18-7. A valid calculation

One of the key elements of this screen is the color coding used in the Calculated Field:

- Blue represents valid functions.
- Orange represents the data fields in your data set. If they appear in black you are either missing the square brackets or have misspelled the data field name. Remember, to call the field, you have to match the capitalization and punctuation (e.g., spaces) in its name.
- Gray represents comments, which are a great way to share your logic for others using your data and as a reminder for yourself when you return to the data in the future. You can set comments by adding two forward slashes (//) to the start of a line in the Calculation Editor.

When Calculations Go Poorly

When calculations contain syntax errors, everything appears in red (Figure 18-8).

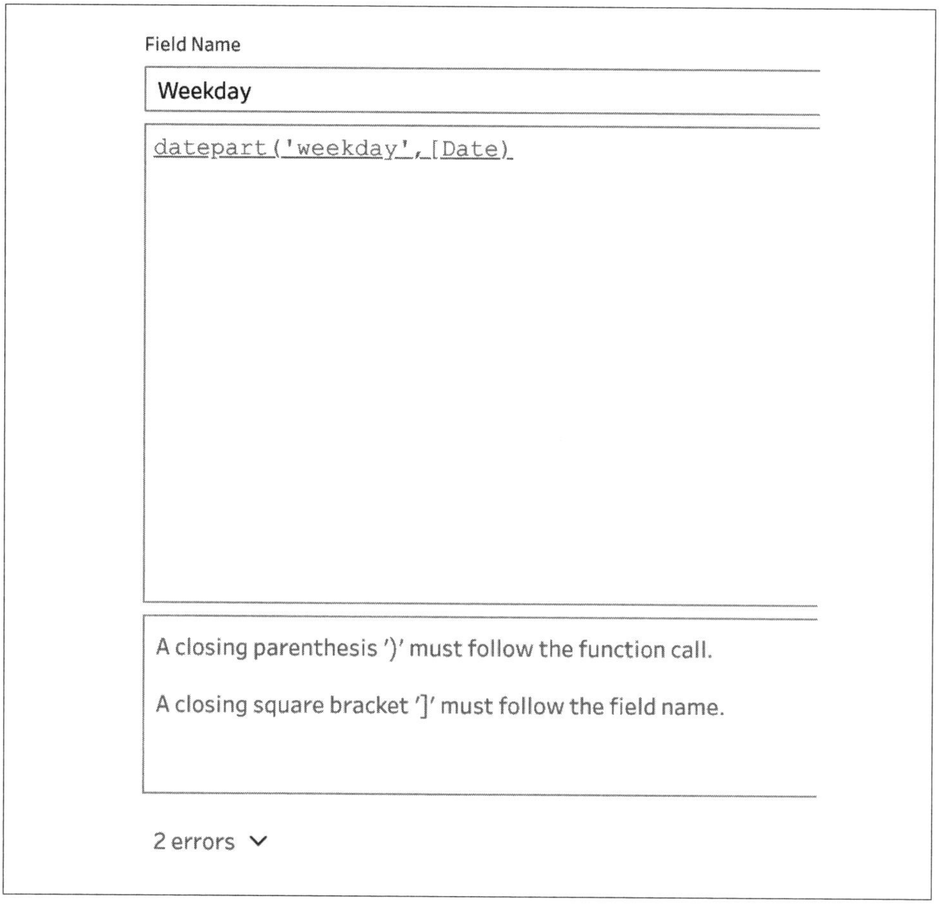

Field Name

Weekday

```
datepart('weekday',[Date)
```

A closing parenthesis ')' must follow the function call.

A closing square bracket ']' must follow the field name.

2 errors

Figure 18-8. Error showing in the Calculation Editor

This can be frustrating because you'll have to find and fix the error to see the helpful color coding just discussed. Therefore, when writing long, logical calculations like `IF` or `CASE` statements, it's a good idea to add an `END` to the calculation periodically in order to spot any errors sooner. If you miss an error in a calculation and click Save or you need to edit the calculation (as discussed next), Prep Builder will clearly indicate when there is an error in a Calculated Field (Figure 18-9).

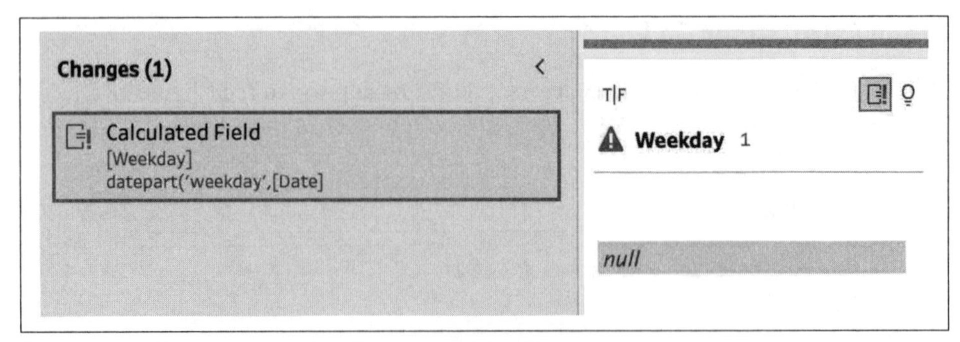

Figure 18-9. Error icons in the Changes and Profile panes

Editing Calculated Fields

If you need to edit a calculation, you can do so from the Changes pane. Clicking the pencil icon (Figure 18-10) opens the Calculation Editor, where you can make the required changes.

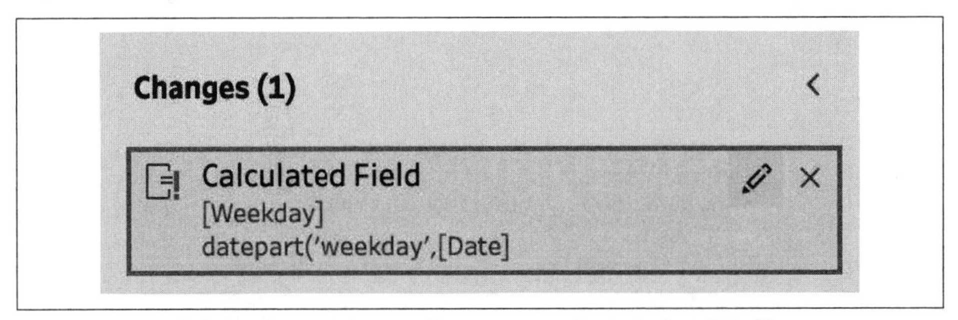

Figure 18-10. Clicking the pencil icon allows you to edit Calculated Fields

Recommendations

One other Prep Builder icon you might have noticed is the light bulb, which is Prep Builder's way of showing recommendations for cleaning processes. The icon appears at the top of the Profile pane for recommendations for the entire data set, or at the top of the data field for recommendations just for that specific item, as shown in Figure 18-11.

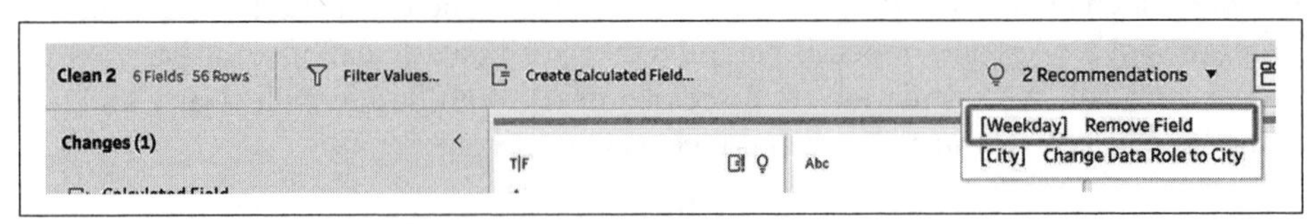

Figure 18-11. The light bulb icon in the Profile pane

Depending on the recommendation, Prep Builder will either open screens for you to implement the change or apply the change automatically when you click Apply. If you click the light bulb icon, you can either select a change if there are multiple recommendations as in Figure 18-11, or simply click Apply to follow the instruction given as in Figure 18-12.

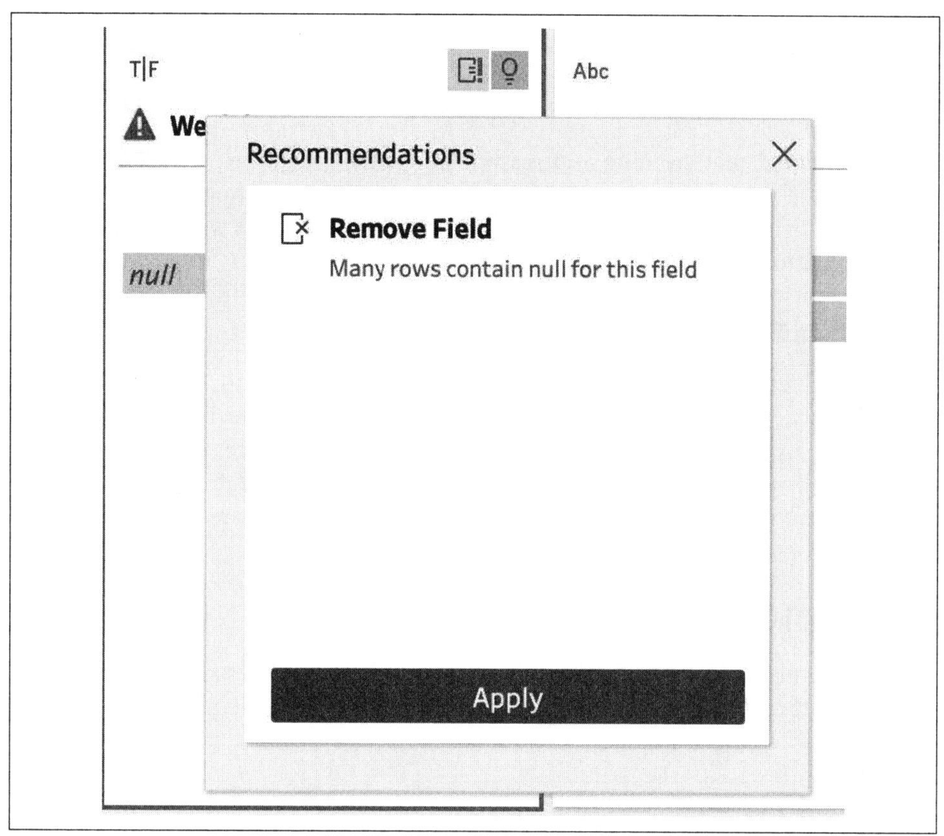

Figure 18-12. A recommendation by Prep Builder

Types of Calculations

So you've seen how to navigate all the aspects of creating a Calculated Field in Tableau, but what are you likely to do with these calculations? In this section we'll look at some common types of calculations to help frame various use cases you may come across.

Numerical Calculations

As Chapter 7 covered, sums, averages, and all other basic arithmetic operations are covered by numerical functions. The main thing to note is whether you are using integers or floats (numbers with decimal points), as Prep Builder will occasionally throw an error if it is expecting the other data type.

String Calculations

Chapter 9 noted that cleaning and manipulating string data fields will be a fundamental part of your data preparation process. Whether it is splitting strings, changing case, or removing unnecessary characters, many string functions will quickly become very familiar to you.

Date Calculations

The way Tableau handles dates is fantastic. From using its own internal calendar to assess what weekday a certain date is to being able to pick apart dates with ease, Tableau is a joy to use compared to SQL or other coding languages where you have to code each element. You'll turn to these functions often to add date intervals or work out the difference between dates. Refer back to Chapter 8 for more functions that are useful for working with dates.

Conditional Calculations with a Boolean Output

Boolean calculations perform very well, as they simply return `True` or `False`. Revisit Chapter 10 for more on using certain functions from the other types of data calculations to return Boolean results.

Logical Calculations

`IF` and `CASE` statements can take a lot of getting used to if you haven't used them before in other tools. Adding logical calculations is important when you're attempting to solve more complex problems. Rest assured that the time you spend practicing will pay you back tenfold.

Type Conversions

Prep Builder expects certain data types for certain calculations. Therefore, you can save a lot of time and reduce calculation complexity by using functions like `Int()` and `Str()` to convert data fields to the correct type (integer and string, respectively, in this example).

Level of Detail and Ranking Calculations

With Prep Builder version 2020.1.3 came the ability to form Level of Detail calculations and the introduction of ranking functionality to the Clean step. Previously, you had to use a number of workarounds to calculate aggregations at a specific level and rank categories based on a measure in the data set.

Level of Detail (LOD) calculations are used to derive aggregations for a metric, for each value in a categorical field, or across the whole data set. The concept for these calculations originated in Tableau Desktop. Normally within Tableau Desktop, the measure is aggregated to the most granular level of dimension(s) in the view, but frequently users wanted the freedom to specify that level of aggregation where it differed from the autogenerated one. As in Tableau Desktop, in Prep Builder LOD calculations are used to set the level of aggregation of a metric. This differs from an Aggregate step in Prep Builder, as it doesn't change the granularity of the entire data set but just one measure. LOD calculations are covered in more depth in Chapter 33.

Ranking calculations allow the user to rank data field(s) based on a measure in the data set. This functionality is also available in Tableau Desktop (as *Table Calculations*) and is again driven by dimensions within the view. As there is no view within Prep Builder, you may need to aggregate the data set to the granularity you want to perform the ranking on. Ranking is described in more detail in Chapter 34.

Both LOD and ranking calculations can be created through either the Custom Calculation Editor or the Visual Calculation Editor. Figure 18-13 shows how to create an LOD calculation using the Visual Calculation Editor.

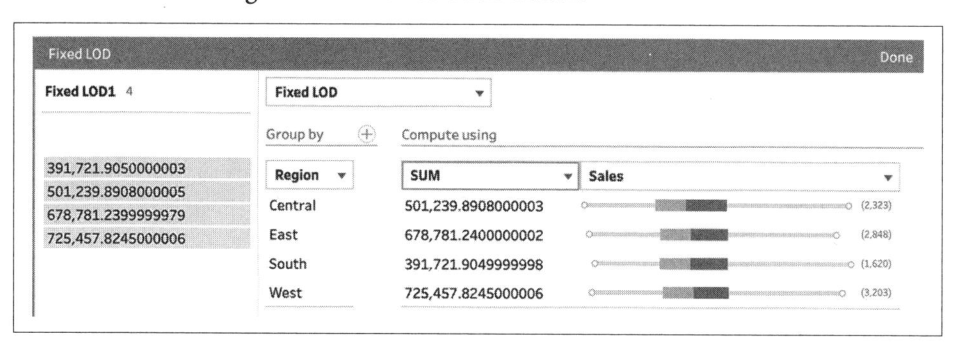

Figure 18-13. Creating an LOD calculation in the Visual Calculation Editor

Within the Visual Calculation Editor, first you'll need to choose the type of calculation you want to perform. You set this in the drop-down menu at the top of the editor.

When you select Fixed LOD, you will be presented with these options:

Compute using
: Select the measure to aggregate and how you want to aggregate it

Group by
: Set the measure to group by, if any

The values you will generate are shown on the left-hand side of the Visual Calculation Editor. The box plot on the right-hand side demonstrates the range of values found in each field in the "Group by" list. Once you have the values you need, click Done, and you'll see a new data field in your data set containing these values.

Summary

For some, calculations are the bane of data preparation, and for others, they are fun logic puzzles. Whichever side of that argument you sit on, calculations are something you will have to use and understand in order to be successful at preparing data. Calculations are vital for cleaning data, adding columns that are required for analysis, and converting fields' data types. The Custom Calculation Editor in Prep Builder, which should be very familiar to Tableau Desktop users, is a tool you will be using a lot. For the more conceptually complex calculations like Level of Detail and Ranking calculations, the Visual Editor can help you better understand the results they will return.

PART IV
Output

CHAPTER 19
Choosing an Output

Prep Builder is built primarily for preparing data for visual analysis in Tableau Desktop. This obviously means Tableau has designed the tool so it's very easy to output the data to Desktop when it is ready. However, the level of simplicity might mean that you miss the optimal output type for the purpose for which you are using Prep Builder (Figure 19-1).

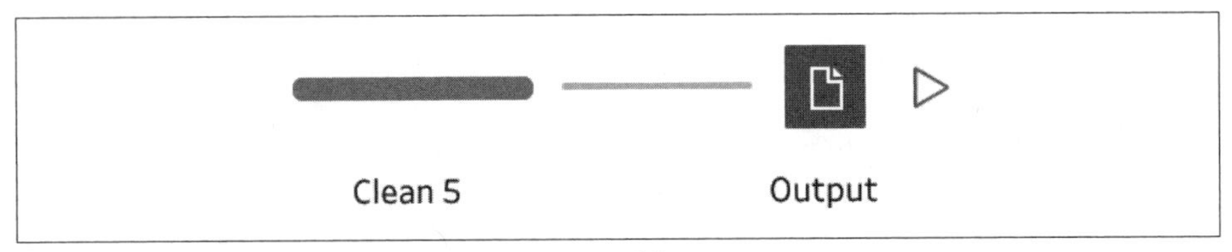

Figure 19-1. The Output step icon in Prep Builder

In this chapter, we will cover your output options in Prep Builder, when you can output data, and other considerations around structuring your outputs.

Types of Output

Within Prep Builder there are four main output types to consider: three files types (shown in Figure 19-2) and one Tableau Server–based option. Each has its own strengths, so let's explore them in turn.

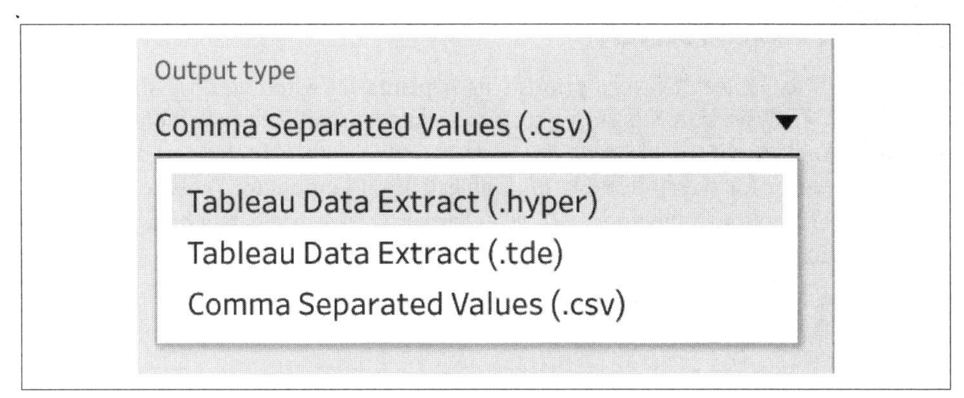

Figure 19-2. File-based output options in Prep

Publish to Files

The following are the three main file output types.

Hyper files

Hyper, Tableau's new form of extract, has made lots of data work faster—in some cases, a lot faster! Hyper files were added to Tableau Desktop and Server in version 10.5. Opening any data extract in Tableau automatically updates the extract to the Hyper format.

If you use the Tableau tools in a version newer than 10.5 (the last version before the naming convention changed to <*Year*>.<*Version*> (e.g., 2018.1), then it's a safe bet that a Hyper file is the best file format for data you'll use in Desktop. Hyper files, with their fast ingestion and analytical query speeds, are optimized for use in Desktop.

TDE files

Before Hyper files, Tableau Data Extracts (or TDEs) were the best file type to use in Desktop. If you are using Tableau version 10.4 or earlier, you will need to output to a TDE instead of a Hyper file.

CSV files

Although Prep Builder was primarily developed for preparing data for use in Desktop, the option to output to a comma-separated value (CSV) file enables you to share your data with users of other data software products. CSV files are also useful for when someone wants a table of data to answer a question they have.

CSV is a basic file format; these files are not optimized for use in data tools, so their performance will be slower than the Tableau-optimized file formats, Hyper and TDE.

Publish to Tableau Server

The Publish to Tableau Server option is used primarily when you are sharing your data and/or analysis with others. Making your data set available for others to use is a key part of self-service data preparation, as it enables other analysts and experts to explore the data for themselves. Using Tableau Prep Conductor (covered in Chapter 21), you can publish the flow to Tableau Server to refresh the data on a schedule of your choosing.

When to Output Data in Prep Builder

Although the end of the flow is an obvious point to output data, there are other options to consider. This section will discuss two: outputting data in the Output step and as a way to preview how your final data set will look in Desktop.

Outputting Data in the Output Step

The Output step offers two main options in terms of outputting the data for further analysis: saving to a file or publishing as a data source. The configuration process differs for each option. Let's look first at saving to a file.

Save to file

As Figure 19-3 shows, there are multiple elements to set up before you can save your output to a file.

Name
: Within large teams and organizations, version control and established file naming conventions are crucial to ensuring that the right data is used for the right purpose. The Name field is set when you click Browse to navigate to your data file in either Windows File Explorer or macOS Finder.

Location
: Like the filename, the location is set as you browse through the file structure of your computer. The default location is your *My Tableau Prep Repository* folder. Remember, Prep creates this file structure when you install it on your computer. If the output is purely for your own use, this location will probably be fine. If you are producing the output for others to use, however, then you will need to change the location to somewhere where they will be able to access it.

Output type
: Though it appears last in the Output step dialog, output type is actually the first thing you should set after you select "Save to file." This is because when you browse the file structure for the Name and Location fields, Prep will search only for that file type you've designated here. In other words, picking whether you

want to output to a CSV, TDE, or Hyper file first will save you from "re-browsing" to reset the file type after you've set the Name and Location fields.

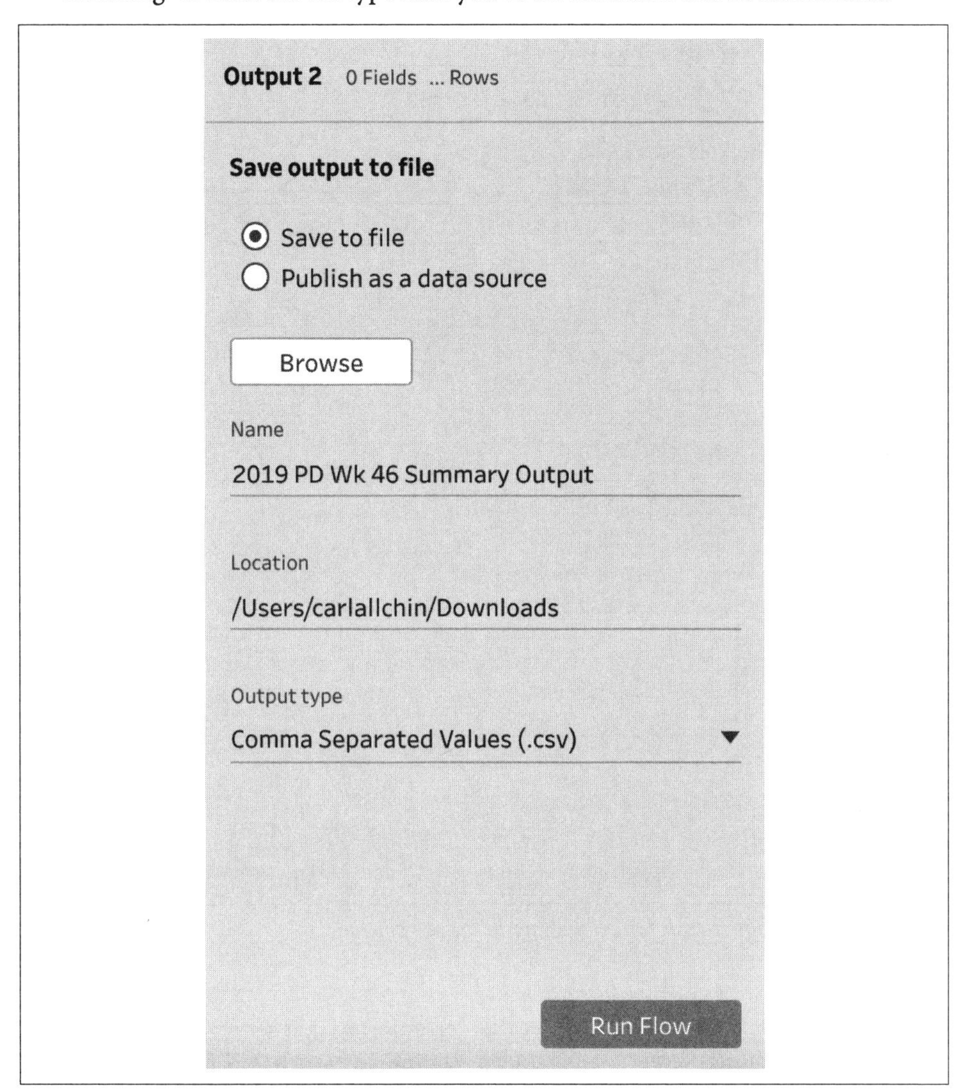

Figure 19-3. Configuration for saving to a file

Publish as a data source

If you aren't saving to a file, then publishing to Tableau Server is the other main option in the Output step (Figure 19-4).

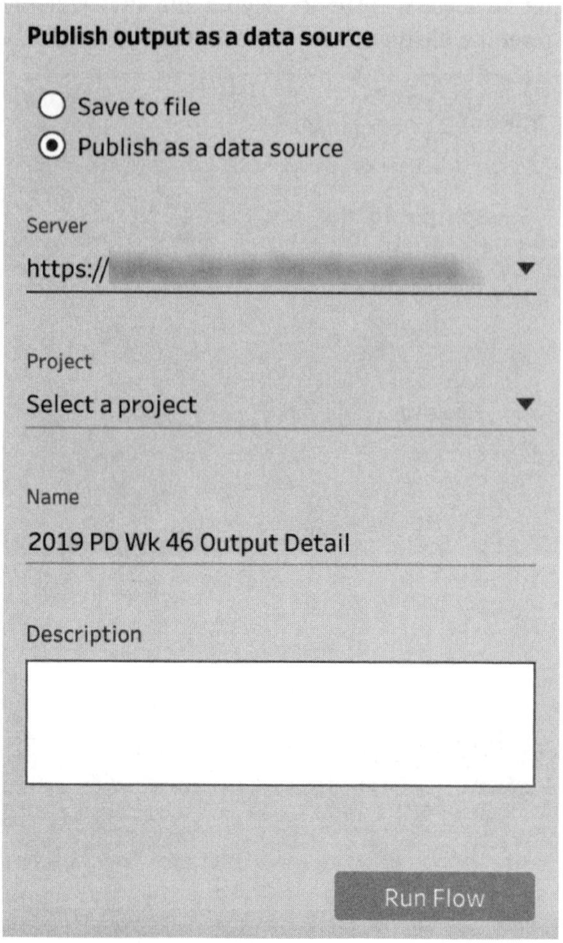

Figure 19-4. Configuration for publishing to a data source

Server
: As most users have just one Tableau server, entering that server's URL is straightforward. Tableau Server and Tableau Online instances are divided into sites that collect and fence off sets of workbooks and data sources. Currently, a Prep workflow can publish only to one site (even if multiple outputs are used within a flow), so you will need to pick where that data will reside.

Project
: Sites can be subdivided into projects. You can think of projects as an individual team's or department's spaces, where a project owner can control the content and access. Choosing the project sets the permissions for the data source.

Name
: As with the Name field for files, version control and naming conventions are key to ensuring the correct data is used for analysis.

Description
: A description of your data source provides more detail than your data source name alone.

Previewing Output Data in Desktop

Instead of outputting data during the Output step, you can output a temporary Hyper file, during any step other than Input or Output, that will open automatically in Desktop. This functionality allows you to preview your data set to determine whether you need to make changes or can finalize that output. Data preparation is often an iterative learning process, and this option allows for fast prototyping. The Preview in Tableau Desktop option is shown in Figure 19-5.

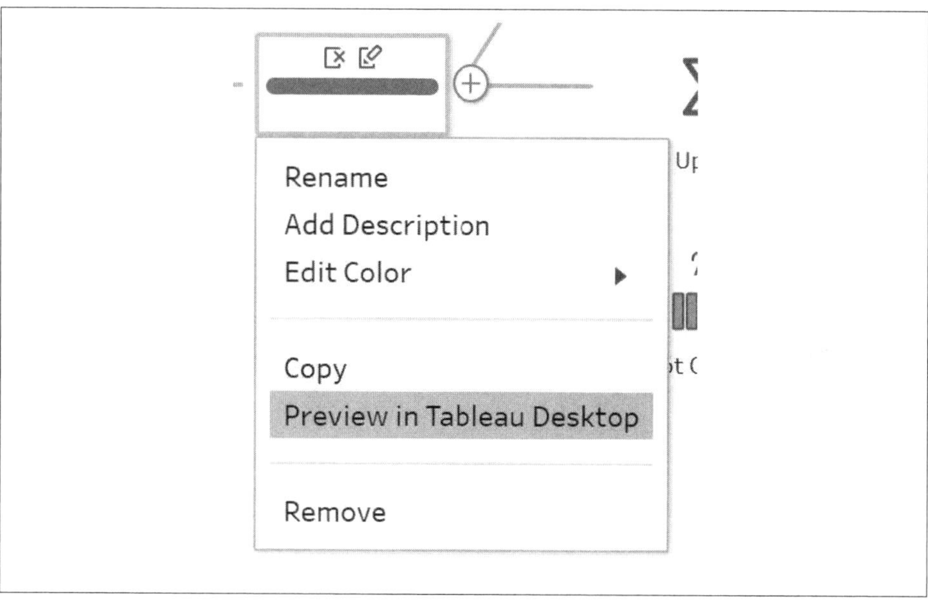

Figure 19-5. Selecting Preview in Tableau Desktop

When you select this option, Prep Builder will build a temporary Hyper file (Figure 19-6).

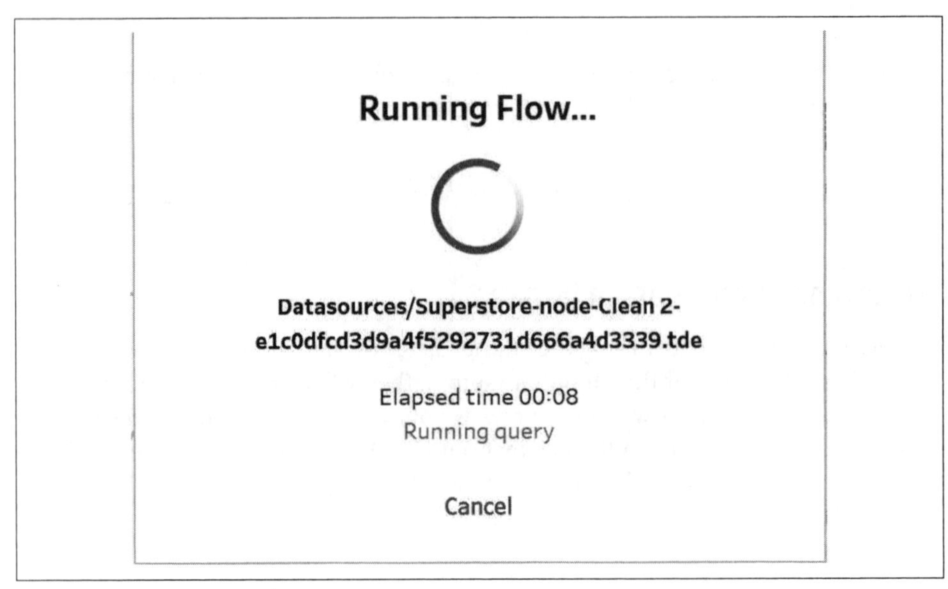

Figure 19-6. Processing the preview

The preview Hyper filename is machine-generated and will appear under the Data tab when opened in Desktop. If you have Tableau Desktop installed on the same computer as Prep Builder, the Hyper file should open automatically (Figure 19-7).

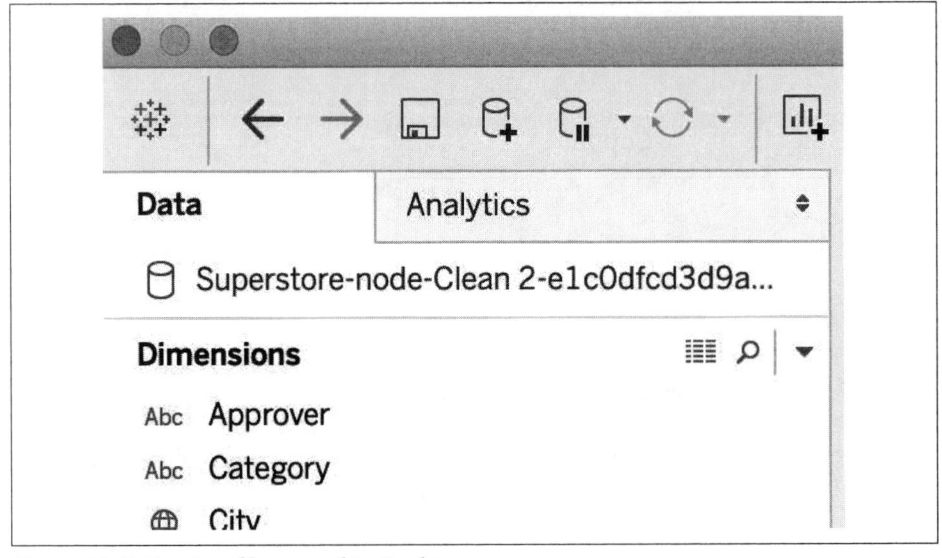

Figure 19-7. Preview file opened in Desktop

One other benefit of the preview in Desktop option is that if you're preparing the data solely for a fast, one-off solution, then this temporary file will more than suffice for your purposes.

Other Considerations for Output Data

Here are some other scenarios to think about when you are publishing output data:

Publishing the same data source to multiple sites/locations
: If you want to publish your data source to multiple sites, you will probably want to add multiple Output steps. As noted earlier in the chapter, it's not possible to publish to multiple server sites in a single Output step (at the time of writing), but you can write to multiple projects.

Updating the data set will clear all previous data
: In the current version of Prep Builder (2020.1), there is no way to partially update a data set. Therefore, take care to ensure that your data sources are available so you can rerun the flow in the future without losing historical data.

Summary

Outputting files from a data preparation tool is the culmination of a lot of hard work. It's important to be sure you are forming those files correctly, that the data is accurate, and that you are not skipping any steps, whether you are saving the output to a file or publishing it straight to Tableau Server.

CHAPTER 20

Outputting to a Database

When Prep Builder was originally released, it was designed to prepare data for Tableau Desktop. But because Prep Builder cleaned and manipulated data so easily and effectively, demand soon grew for the ability to output data not just to TDE or CSV files but also back to the original source of much of the information: databases. This was a significant departure for Tableau, as Desktop had always been a read-only tool, so allowing Prep Builder users to change data permanently opened up both opportunities and risks.

This chapter will cover when and how to write data back to a database and what to watch out for when doing so.

When to Write to a Database

As we've seen in other chapters, messy and multiple data sets can take a lot of time to prepare. If you have spent the time and effort to make the data suitable for analysis, then it is likely you will make it available to others or publish it back to a source where no one else will need to battle it again. This section covers some of the common situations where you might consider publishing to a database.

Clean Data

Dirty data is the reason we have to prepare data in the first place. If you have cleaned the data once, why not load it back to the source? Obviously, you must take care to not remove or filter data that would be useful to others. But if the data set is clean and ready for analysis, there are no potential downsides, and you have write/overwrite permissions on the database, then writing the clean data back is a good way to prevent future rework.

If the data is sourced from a system load, you or Prep Conductor may need to refresh the flow on a regular schedule to prevent dirty data from being pushed back into the clean table as new records get added. You can also work with your data teams to show them the manipulations you have made, and why, as they can probably build these periodic refreshes into the normal load process.

Simplified Joins

Joining tables together can create very useful data sets for analysis. However, setting up the join conditions and join types to do this can be complex. So, if you have already used Prep Builder to join complex tables together, outputting the resulting data set can both share the benefits with others and save them from having to join those data sets themselves. This will also avoid mistakes and rework, making the analysis much easier for users of those data sets.

Staging and Reference Tables

You can use Prep Builder to add a data set to the database to help make analysis easier and more detailed. You can load messy, unstructured data into a database as a *staging table* and work through various stages of improvement until it is ready for use as a production table. Prep Builder can help you process these sources to determine the manipulation and cleaning that they will require.

Another time-saving option is a *reference table*, which you can use to replace long string values with numeric values that are easier to store and faster to process. To create a reference table in Prep Builder, you use the `row_number()` function and an Aggregate step.

Setup for Writing to a Database

No matter what type of data you are outputting to a database, the early setup stages are similar. First, select the option "Write to a database table" (Figure 20-1).

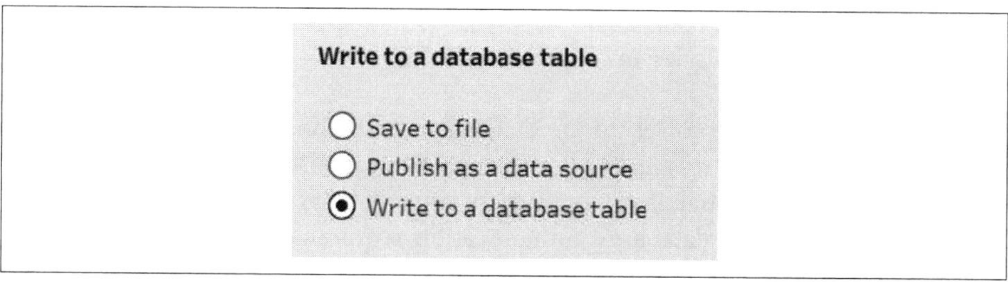

Figure 20-1. Write to Database options in the Output step

You'll see an error symbol, which indicates that the connection has not been completely set up (Figure 20-2). Next, you'll need to select the server to connect to.

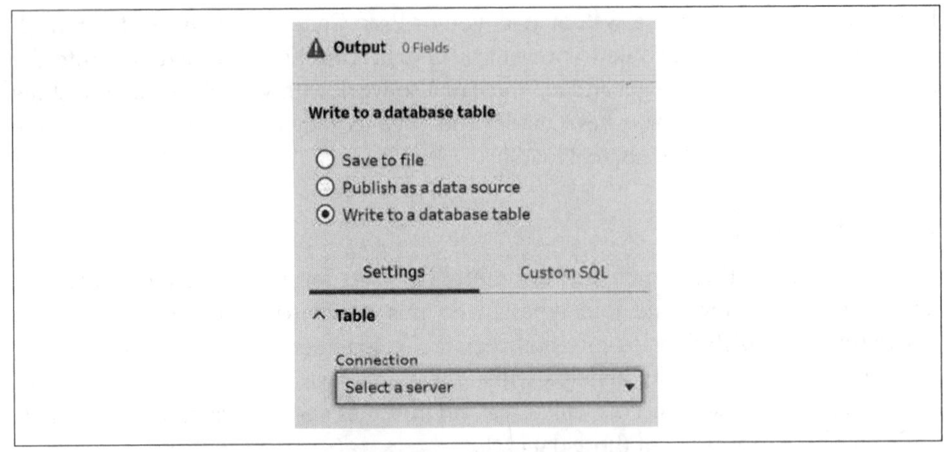

Figure 20-2. Error indicating incomplete connection setup

You will be prompted to select the software the database is running on from the Connection drop-down list. For the example in Figure 20-3, I am using Microsoft SQL Server.

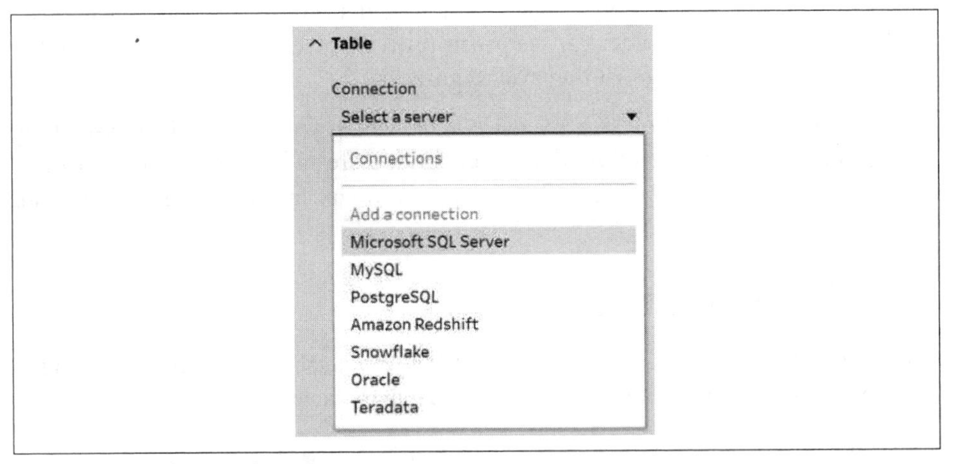

Figure 20-3. Selecting the server to connect to

Next, you'll be prompted to set up the database connection. Here you will enter the server address, or IP address, in the Server box. You can also enter a database in this window, but it isn't essential; if you opt not to, you will be prompted again in the next step. Depending on the database's authentication requirements, you can use either Windows Authentication or a standard username and password that has been set up on the database (Figure 20-4).

Microsoft SQL Server

Microsoft SQL Server

Server:

Database: Optional

Enter information to sign in to the database:

Windows Authentication

Username and Password

Username:

Password:

Require SSL (recommended)

Read uncommitted data

Show Initial SQL

Sign In

Figure 20-4. Setting up the connection to the database with your credentials

In the next step of the setup, you enter the name of the table you're creating and the database where it will reside (Figure 20-5).

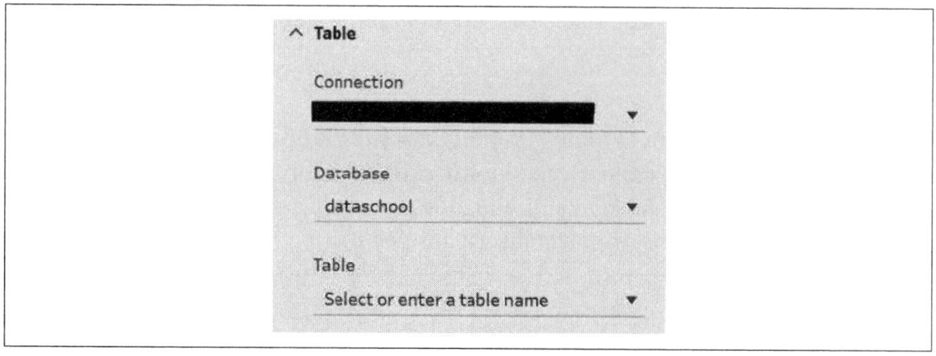

Figure 20-5. Choosing which database and table to write to

If the table does not already exist, you can add its name in the Table section of the configuration (Figure 20-6).

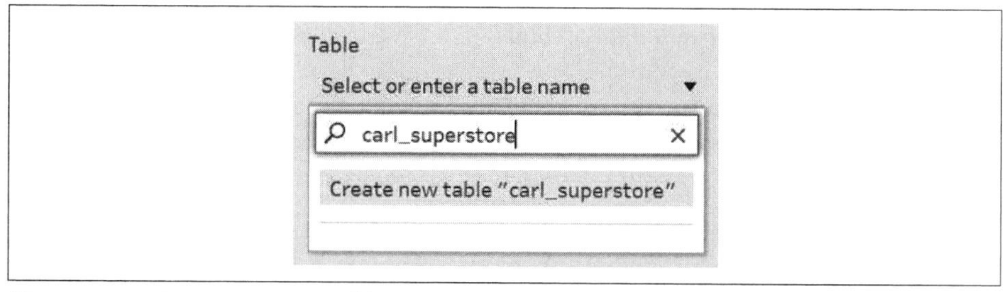

Figure 20-6. Creating a new table to write to

The final step of the setup is to pick how you want to output the table data (Figure 20-7).

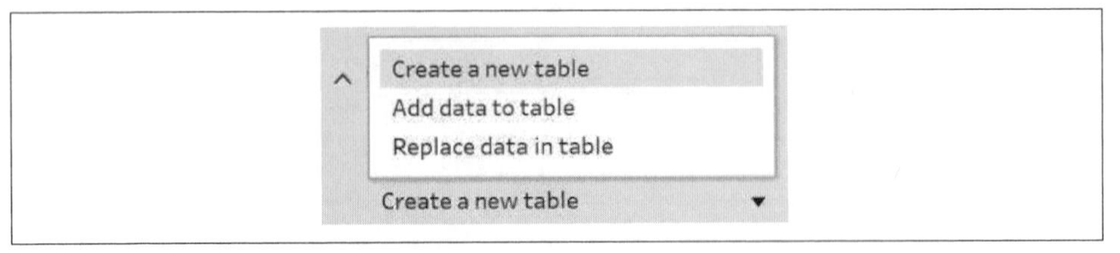

Figure 20-7. Options for table output

You have three choices here:

Create a new table
: Creating a new table means writing a new table that has not existed previously in the database. Databases often have naming conventions based on how they are used within the organization. New tables should follow those naming conventions so they are clear to potential users and thus more likely to be used within the organization. Note that any spaces in the database name will be replaced with an underscore, as most databases do not permit column names containing spaces.

Add data to table
: If you select "Add data to table," Prep Builder will add data to the end of an existing table or create a table if one doesn't exist. This option behaves as an incremental update to the table by unioning the rows being processed within the flow, appending them to the original table (Figure 20-8).

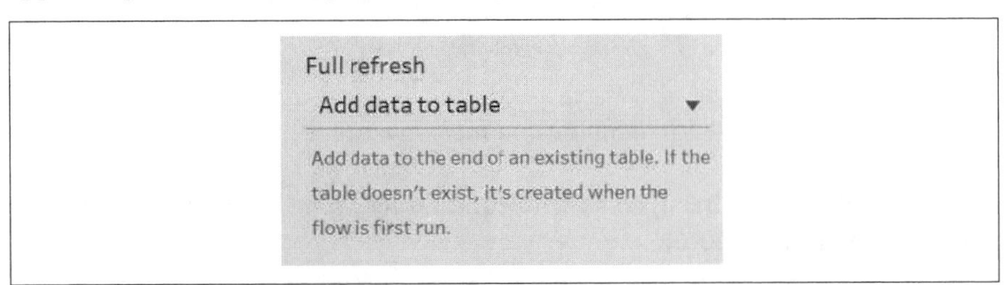

Figure 20-8. Option to add data to a table

If you add a new column as part of the flow, it will be added to the data set for the unioned rows.

Replace data in table
: Selecting "Replace data in table" will overwrite the current data set in the table with the data set processed by the flow (Figure 20-9).

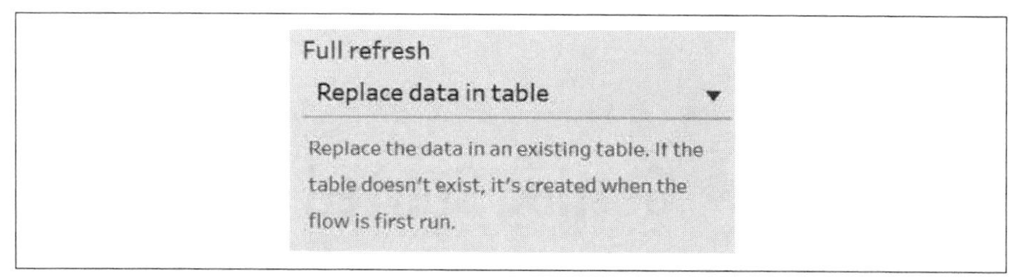

Figure 20-9. Option to replace data in table

What to Watch Out For

The most significant risk of being able to write to a database is overwriting tables accidentally. As with any of your work with data, you are responsible for your actions and the impact they could have. Your permissions on the database will likely be restricted until you have demonstrated the skills and understanding to be trusted with access to the tables and data that would have bigger, potentially catastrophic impacts if incorrectly used or altered.

That said, most organizations control their tools and permissions to ensure that production data sources are not affected until the changes have been tested in a number of environments. Therefore, this "most significant risk" is actually probably the least risky, as it is more tightly controlled.

Prep Builder users need "sandpit" (or "sandbox") space to play and explore in order to learn and develop the skills to write database tables. As covered earlier in the chapter, there are many benefits to being able to write to a database, but you need the proper skill set before you can take advantage of those benefits. The sandpit allows you to write database tables in a space where they won't be seen as production tables and used accidentally by others, and where you can alter or remove the tables without having to worry about those changes affecting the organization.

Summary

The ability to write to a database is a strong skill to add to your data preparation toolbox. There is a risk of overwriting or deleting unrecoverable data, but generally the database permissions will prevent you from doing so. By using Prep Builder to write to a database, and regularly scheduling refreshes through Prep Conductor, you can create and productionalize solutions in Prep to deliver clean, refreshed data to end users.

CHAPTER 21

Getting Started with Tableau Prep Conductor

When Tableau Prep Builder was first released, many people finally had the chance to build data preparation flows to remove the tedious and repetitive task of cleaning and merging data sets to enable valuable data analysis. Once a flow is created, it reruns each time a user clicks the run icon in Prep Builder. For a single flow this is simple, but for vast and differing data sources, it often involves multiple preparation jobs. This is where Prep Conductor comes in: you can build a flow in Prep Builder but then schedule it to run on Prep Conductor when needed.

This chapter looks at when you might need to use Prep Conductor, the capabilities it offers, and, finally, how to actually use Prep Conductor.

When to Use Prep Conductor

Prep Conductor is primarily used to run an uploaded flow on a set schedule. This has many benefits:

- The flow doesn't need to be manually opened and run each day.
- Errors are logged on the server rather than just on an individual's computer, so you can troubleshoot more easily if a flow fails.
- Prep Conductor likely runs on a computer that has more processing resources than the data author's computer, so preparation flows will run faster.
- Prep Conductor acts as a central repository for all flows, promoting reuse.
- The flow can be downloaded and maintained by any person with proper permissions.

These benefits help build more robust processes around your organization's data preparation work. The time savings they bring enables you to collate more data, spend more time on the analysis, and work more closely with the organization to develop further and deeper data solutions.

How to Get Prep Conductor

Prep Conductor is part of the Tableau Server product, whether Server is hosted by your organization or on Tableau Online. At the time of writing, Tableau Server/Online is part of the Creator subscription package, which also includes Prep Builder. To use Prep Conductor, you will need to pay for the Data Management add-on, apply the new license key, and restart the server. There is a minimum purchase of 100 users for Tableau Server or 25 users for Tableau Online. From there, a Prep Conductor process will be present on any "node" of the server that has a Backgrounder process running on it. The Backgrounder is the workhorse of Tableau Server, updating data extracts and Prep flows, so ensuring there are enough of them for the volume of tasks is a key part of Tableau Server administration.

Prep Conductor can be turned off in the server by the server administrator. You'll find this option on the General tab in the Settings menu (Figure 21-1).

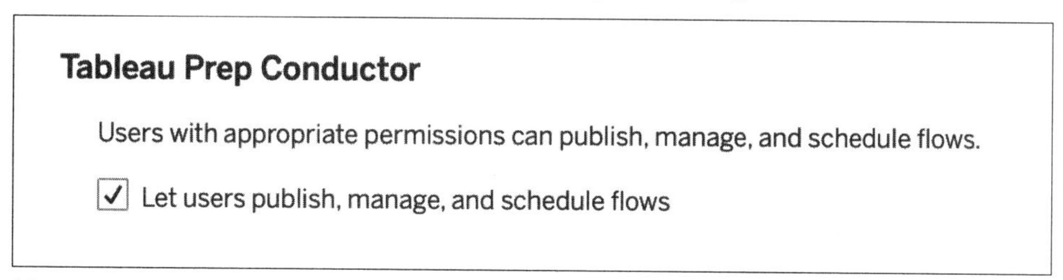

Figure 21-1. Server controls to enable or disable Prep Conductor

Loading a Flow to Prep Conductor

Here is the process to load a flow onto Tableau Server:

1. Connect to Tableau Server in Prep Builder (Figure 21-2).

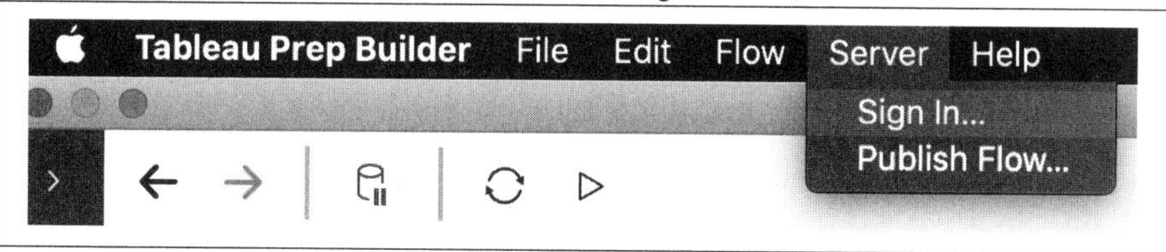

Figure 21-2. Signing in to Tableau Server in Prep Builder

2. Enter the server connection details. This will be the web address of Tableau Server/Online. Then click Connect (Figure 21-3).

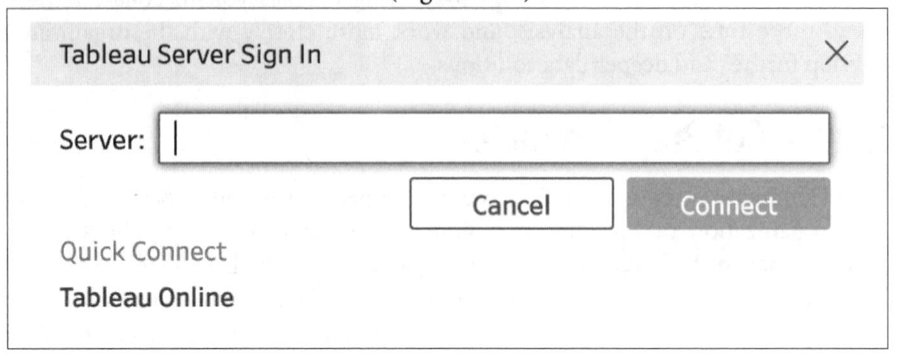

Figure 21-3. Entering the server address

3. Enter your server credentials and click Sign In (Figure 21-4).

Figure 21-4. Entering Tableau Server credentials

4. Select the item in Tableau Server/Online you want to publish the flow to (Figure 21-5).

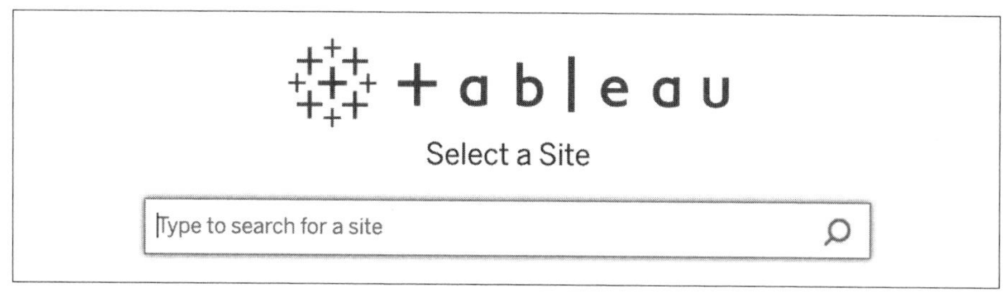

Figure 21-5. Selecting a Tableau Server/Online site

5. Once the flow is ready for publishing, go back to the Server/Online menu at the top of the Prep Builder screen and select Publish Flow (Figure 21-6).

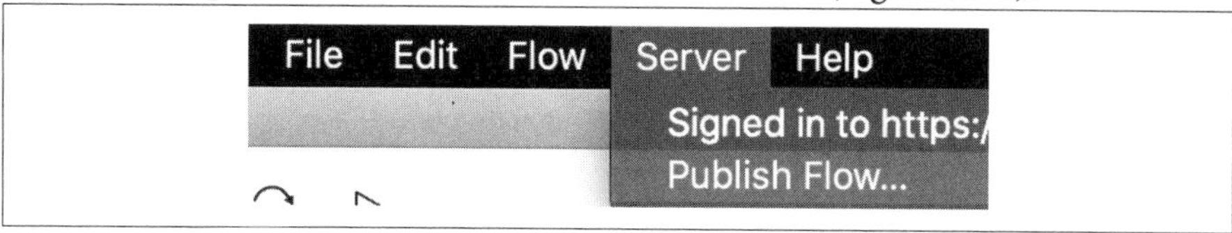

Figure 21-6. Publishing a flow in Prep Builder

6. Pick the project on the server to publish the flow to, name the flow, and add any other description that may assist the users of the server (Figure 21-7).

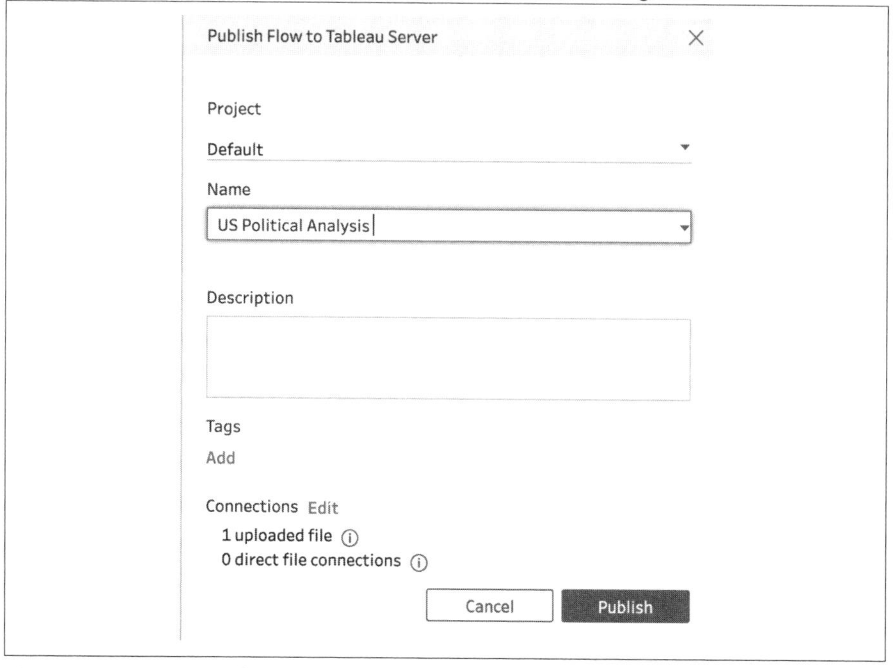

Figure 21-7. Server publishing setup in Prep Builder

Once the flow is published to Tableau Server/Online, the next stage of the process is to set up the scheduling of the extract. After clicking New Task in the Scheduled Tasks tab in Prep Conductor, you are given the option to set up a schedule for each output of the flow, or all of them (Figure 21-8).

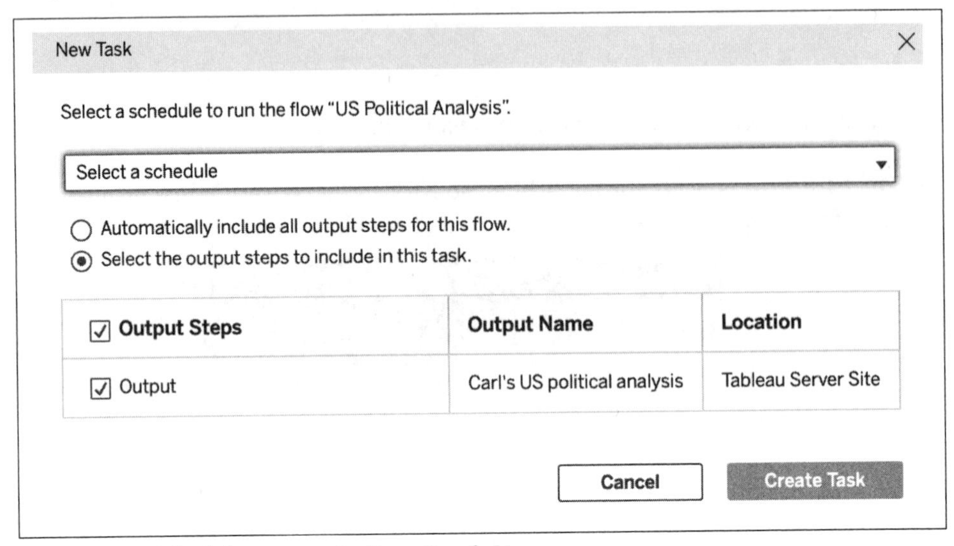

Figure 21-8. Scheduling menu in Prep Conductor

The schedules available are managed by the server administrator. Additional schedules can be added, but these are controlled to ensure that flows are run primarily when fewer other processes and users are accessing the server (Figure 21-9).

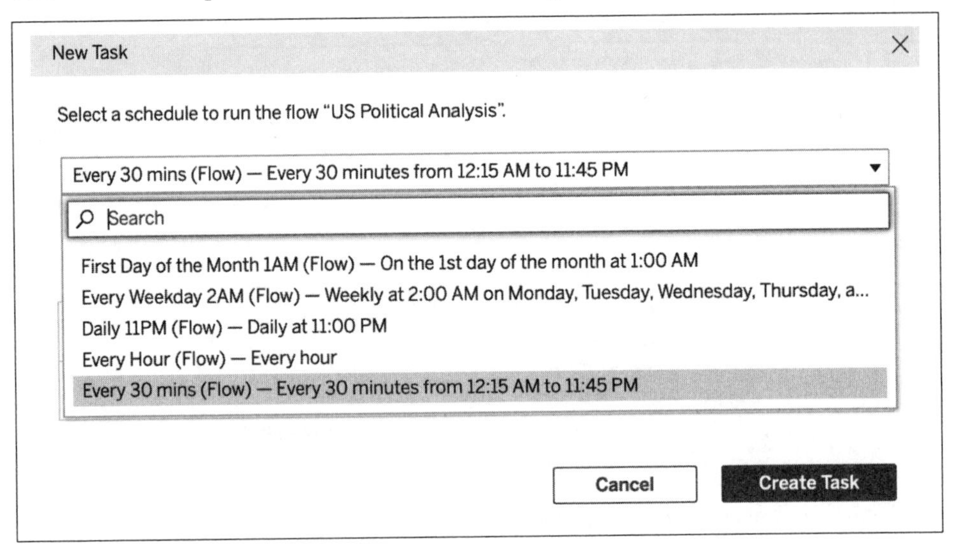

Figure 21-9. Schedule choices in Prep Conductor

Once you've set up the schedule, you can edit it in the Scheduled Tasks tab of the flow in Prep Conductor by clicking the ellipsis icon next to the schedule name (Figure 21-10).

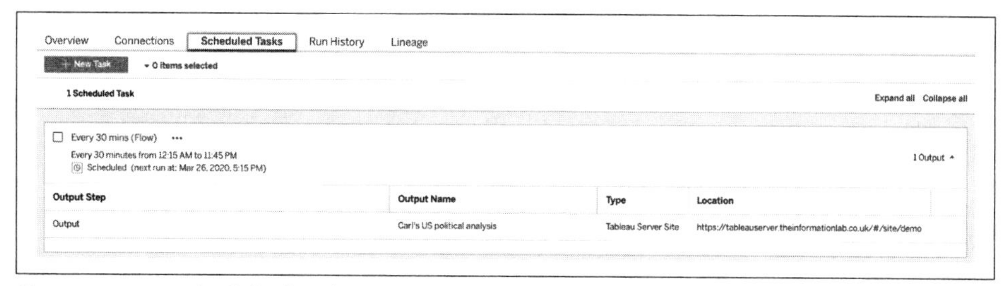

Figure 21-10. Schedule detail in Prep Conductor

If you no longer need the scheduled task, you can also delete it by clicking the ellipsis icon in the Scheduled Tasks tab (Figure 21-11).

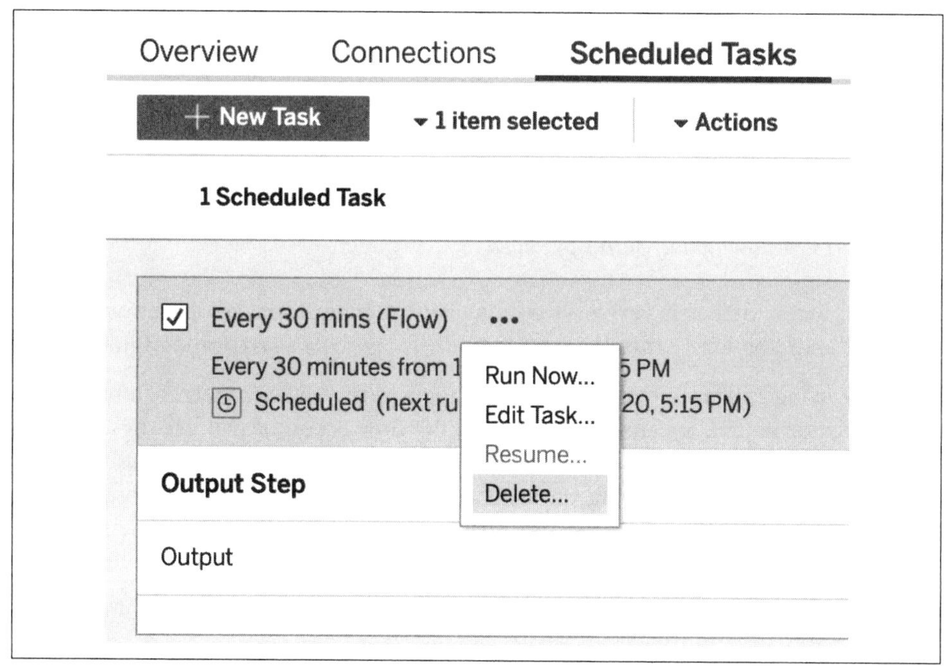

Figure 21-11. Deleting a schedule in Prep Conductor

Once the process runs, the Status column on the Overview tab shows whether or not the scheduled flow has run successfully. In the instance shown in Figure 21-12, it has.

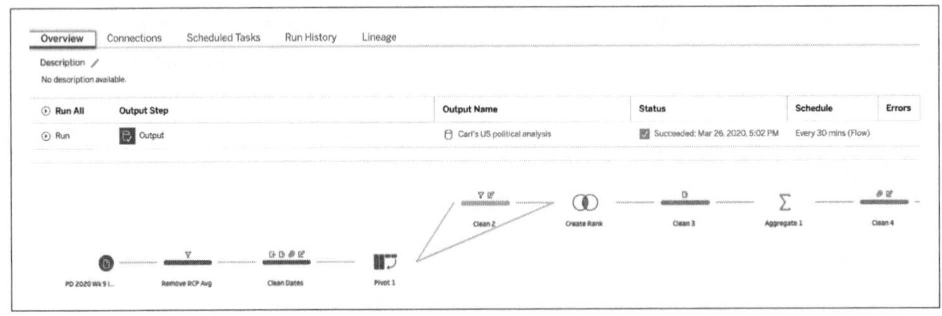

Figure 21-12. *A flow with an update status of Succeeded*

If you have multiple runs to review, then you can check whether or not the runs were successful in the Status column of the Run History tab (Figure 21-13).

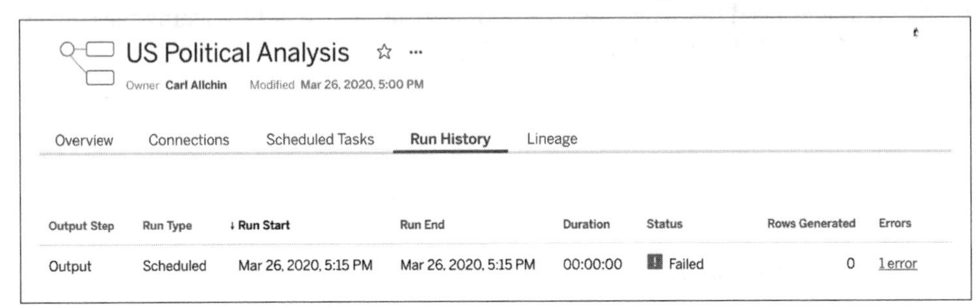

Figure 21-13. *A flow update status of Failed*

Prep Conductor makes it very easy to manage the main tasks associated with your flow by identifying where there are issues, and why, with clear error messaging.

If you are using a data source that requires a login, you may wish to embed the access credentials for it. This means that each time the flow is run, it will use these details. Some organizations will provide what is commonly known as a *service account* for these details rather than relying on an individual to update their password or link access to that specific person. To embed the credentials, click the ellipsis menu on the Connections tab of Prep Conductor for the flow you want to embed the credentials in and select Edit Connection (Figure 21-14).

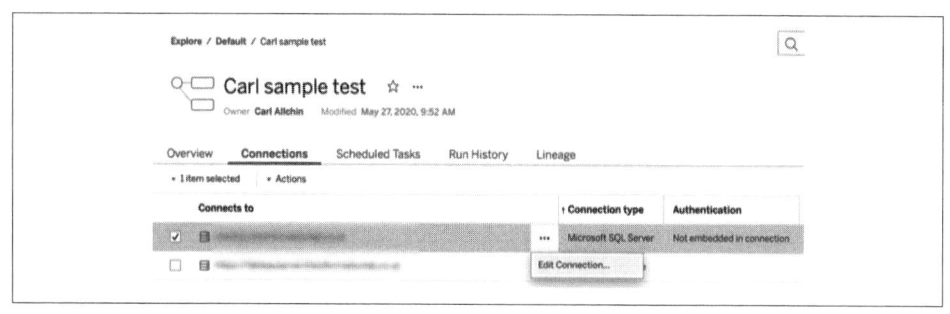

Figure 21-14. Editing connection details, like login details, in Prep Conductor

You'll be presented with a screen to change the server for the data connection as well as the login details for that connection (Figure 21-15). This is where you can embed the credentials so they are saved for use in each future run.

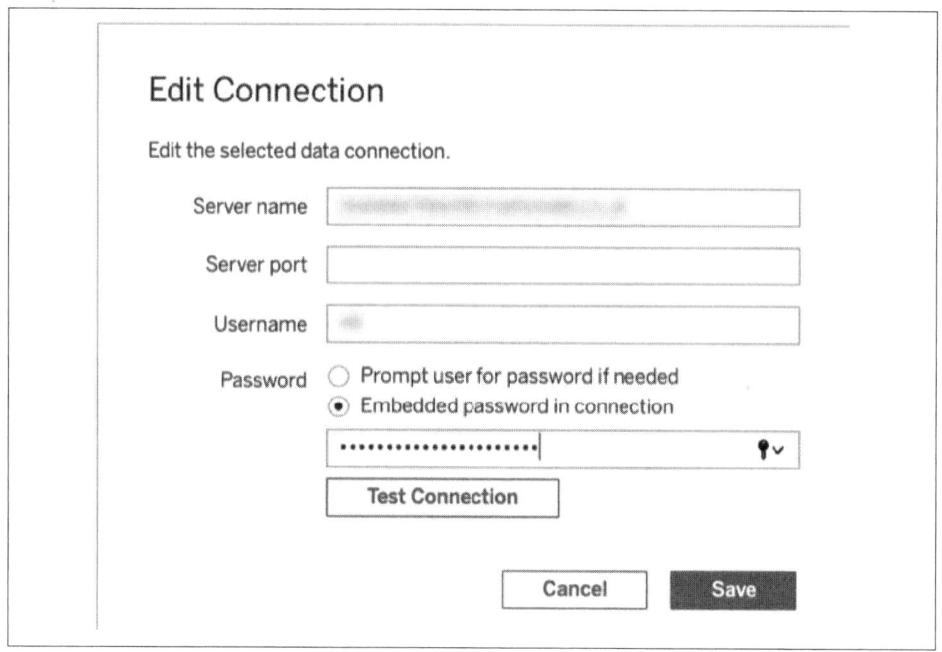

Figure 21-15. Embedding credentials for a data source in Prep Conductor

Click Save to store the details for the next run of the Flow. If you want to check that the details are properly configured, you can click Test Connection.

Other Benefits of Using Prep Conductor

Prep Conductor can benefit you in ways other than scheduling flows and recording whether or not they ran successfully. For example, the Connections tab demonstrates both the inputs and outputs of the flow (Figure 21-16).

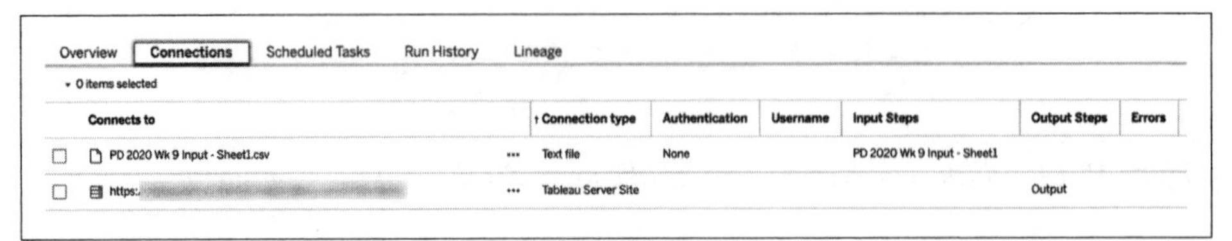

Figure 21-16. Connections tab in Prep Conductor

Here you can see the type of file as well as whether authentication is required to use the input.

Finally, the Lineage tab indicates where the output is used within Tableau Server. It shows you which workbooks rely on the output as well as the sources that feed it (Figure 21-17).

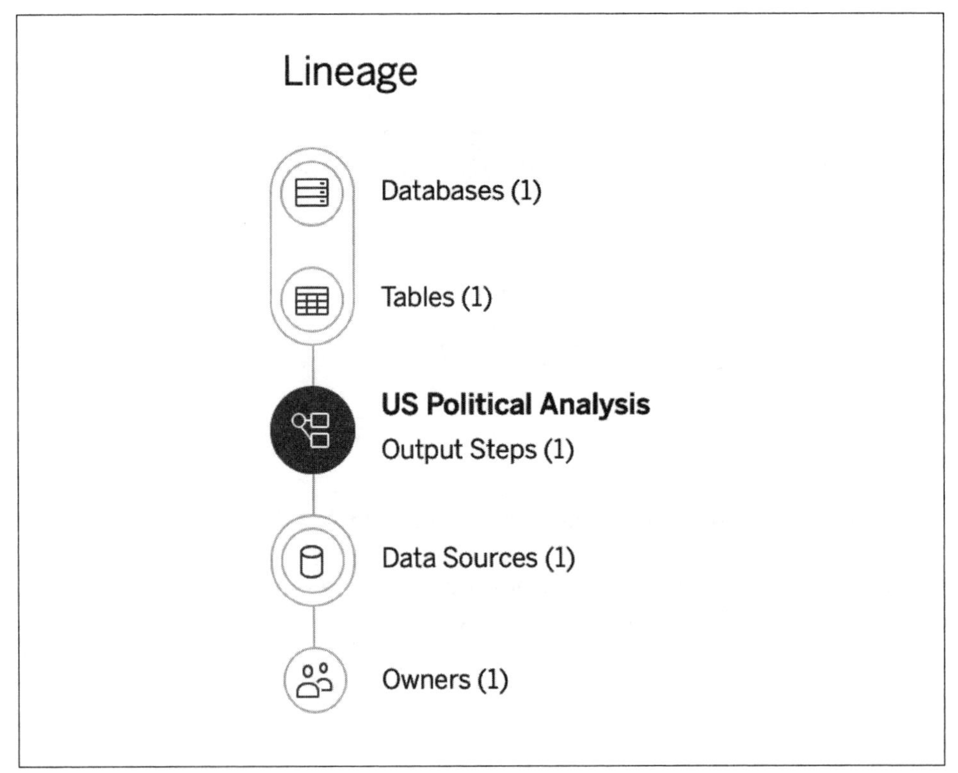

Figure 21-17. Lineage diagram in Prep Conductor

In other words, the Lineage tab gives you an overview not only of what relies on the flow but also what it relies on. Tying these aspects together can make data administration much easier. If edits need to be made, you can see exactly what is and isn't going to be affected.

Summary

Prep Conductor is a data management tool that utilizes the power of Prep Builder's flows and automates their output. Using Prep Conductor will save you even more time than using Prep Builder alone. And efficiency isn't the only benefit—it also gives you clarity on inputs, outputs, and errors that will help you manage data complexity.

PART V
Cleaning Data

CHAPTER 22

Creating Additional Data

The challenge with data preparation is often in cleaning and removing columns or rows of data, but doing so isn't always necessary. When preparing data sets for analysis, you'll often need to create additional data through certain techniques. This chapter will go through situations when you might or might not need to create data and then will cover the various data creation techniques in Prep.

When Not to Create Data

The aim of the data preparation process is to reduce the level of manipulation required during the data analysis. Therefore, anything you can do during data prep to simplify the "data work" of analysis is worthwhile. For example, any column or row that is not in the data set, but should be for your analysis, needs to be created. There are multiple techniques to do this, but first let's explore where you *shouldn't* be creating extra columns or rows.

Dynamic Calculations in Desktop

Tableau Desktop is designed to allow you to conduct analysis at the speed of thought —that is, to deliver answers as quickly as you can think of the questions. Part of this design are three types of calculations that factor in the data fields within your analysis as you build it onscreen:

Calculated Fields

Tableau Desktop calculates the measures used within a view based on the dimensions that are also present in the view. If you create a calculation for total sales, `sum(Sales)`, Desktop will return the total sales of the entire data set if there are no dimensions in the view. If there are dimensions in the view—for example, one called Product—Desktop will return the total sales for each product in the data set. Changing the dimension(s) in the view ultimately changes how total sales is being calculated. Therefore, unless you want this value to be set, you shouldn't build this type of calculation in Prep Builder.

Table Calculations
: These calculations use the dimensions within the view to create additional analysis. As the dimensions (or their positions) change on the view, the Table Calculations update accordingly (Figure 22-1).

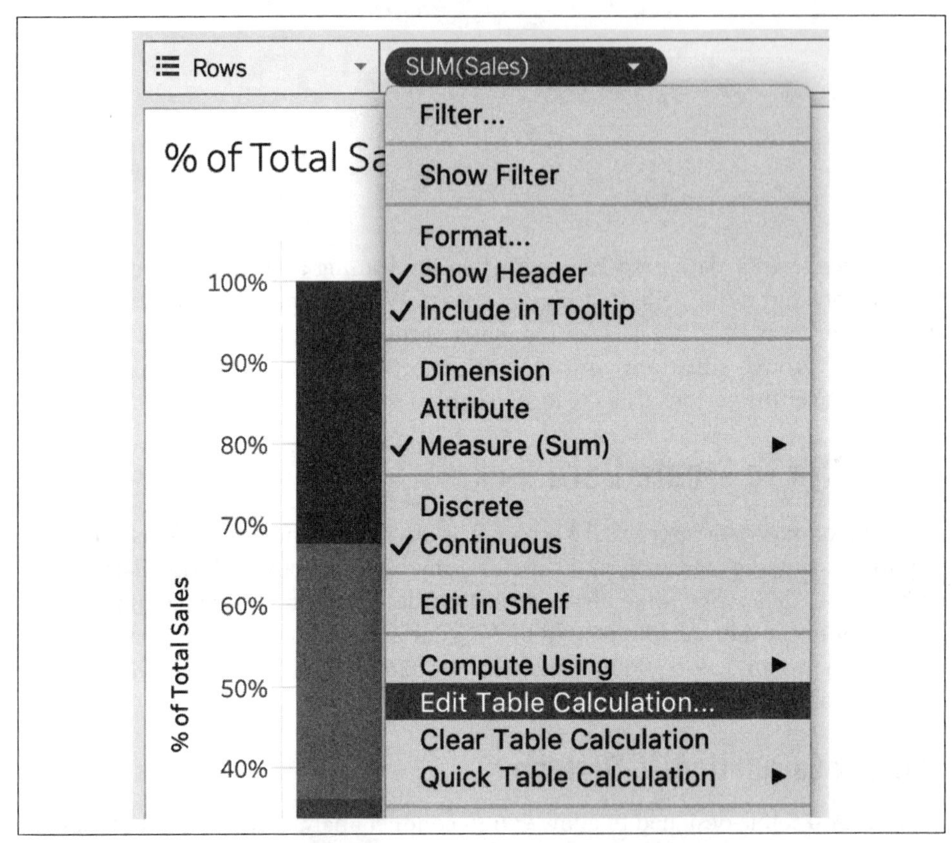

Figure 22-1. Setting up a Table Calculation in Tableau Desktop

Level of Detail calculations
: There are two types of Level of Detail (LOD) calculations that update as different dimensions are brought into the view on Desktop: exclude and include. LOD calculations work by excluding or including the specified dimensions in the calculation, thereby changing the level of detail at which Tableau processes the aggregation of measures.

If you want calculations that adapt with the dimensions on the view, then these needn't be set in the data preparation stage. However, if you want the contents of your columns to be consistent and cemented in place, then they should be. Working with data fields that hold measures is likely to involve the Aggregate step in Prep, where you set how those measures are being calculated (for example, total sales at the region level).

Duplicate Records from Joins

One way to create additional rows is to join data sets together (Figure 22-2). If you join a data set to another where the granularity is identical, or a value is appended, then you are unlikely to create duplicate rows of data if there are matching data fields. However, if you have different levels of granularity, or the join condition fields do not match, then you risk creating duplicate rows of data.

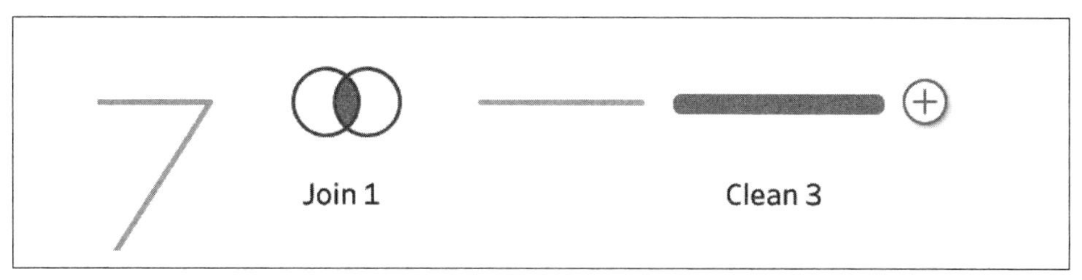

Figure 22-2. A Join step in Prep Builder

In other words, if the granularity of the data set being added is higher, the join is likely to add rows by duplicating records from the original data set. Any form of unintentional duplication will hinder your analysis.

Joins are covered in more detail in Chapters 16 and 32.

So what approach should you take to create additional columns or rows?

Creating Additional Columns

As this section will discuss, there are multiple ways to create additional columns in a data set.

Using Calculations

Creating calculations adds a new column to the data set (unless that calculation has the same name as an existing data field, in which case the original field will be overwritten). Calculations are either cleaning or data creation functions:

Cleaning
: Cleaning calculations split longer strings or transform existing data fields into new columns that are more useful for analysis. In Figure 22-3, we would need to create a separate column for the British Pounds (GBP) value once it is converted from US Dollars (USD). As the column currently is formed, there is no way to use the conversion rates. If you are confident that your calculations are correct, then it's fine to overwrite the unclean data set using a calculation with the same name. If you have any doubt, though, I'd recommend giving the calculation a different field name so it's easier to validate the changes taking place. You can always change the field name or filter out the unclean field after validation.

Figure 22-3. A data field that should be split into multiple columns

Creation
: These are new functions you create to calculate ratios, counts, and totals involving a combination of different fields in order to make analysis easier or more thorough.

Any calculations you apply during data preparation can save end users from having to figure out how to achieve the same result in Desktop, lowering the technical barrier to entry for working with the data set.

Pivoting Rows to Columns

You can create additional columns from existing rows using the Rows to Columns pivot option (Figure 22-4).

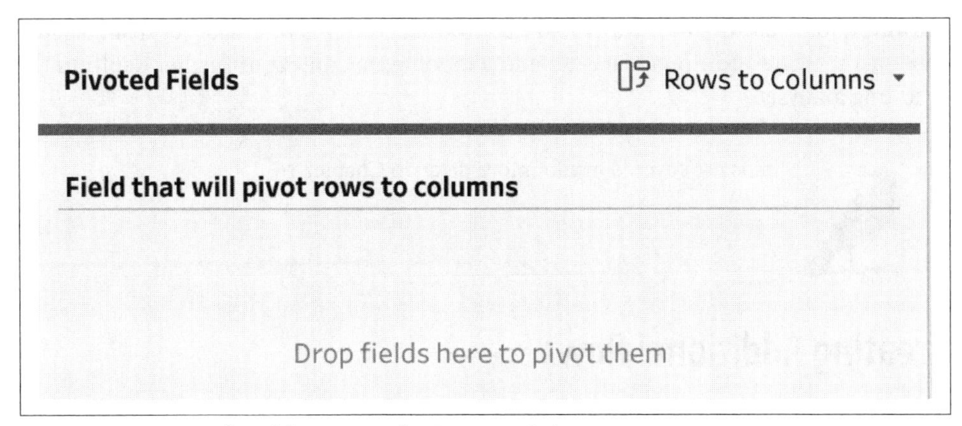

Figure 22-4. Pivoted Fields pane in the Rows to Columns setup

Before the pivot, your data will often have different rows holding different measures in the same column. To make analysis easier in tools like Tableau Desktop, you can set up the Pivot step to transform rows into columns, which will transform each measure within a row into a column of its own.

This type of pivot is covered in more detail in Chapter 14.

Joining Data Sets

As mentioned earlier, a Join step (Figure 22-5) can create complexity if granularity is mixed or join conditions are not clear.

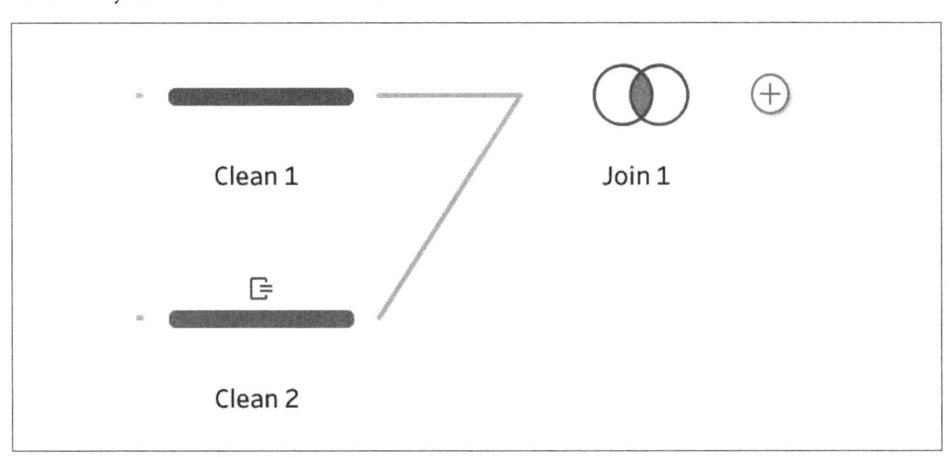

Figure 22-5. A Join step within Prep

However, joins are the best way to add additional columns to a data set from a separate data set. The additional data can add a lot of extra context or further depth in the resulting analysis.

Joins are covered in a lot more detail in Chapter 16.

Creating Additional Rows

This section will cover multiple techniques for creating additional rows in your data set.

Pivoting Columns to Rows

Just like the Rows to Columns pivot functionality, the Columns to Rows pivot can create multiple additional rows that will be useful for analysis (Figure 22-6). The Columns to Rows pivot is often used to transform a column of multiple dates into two columns: one for the Date data field and one for the corresponding values. This type of pivot creates a lot more rows, but it makes analysis much easier, as Tableau offers a lot of flexibility and functionality once dates are in a single column.

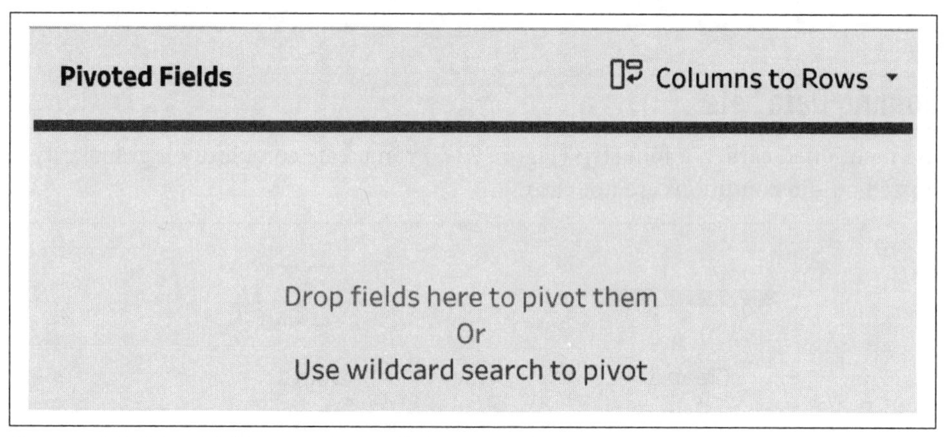

Figure 22-6. Pivot Fields pane in the Columns to Rows setup

Unioning Data Sets

Unions (Figure 22-7) are frequently used to stack data sets with very similar structures on top of each other. This adds new rows for each additional data set used. Unions are often used to add different weekly or monthly data sets together to create a single file, which enables easier analysis.

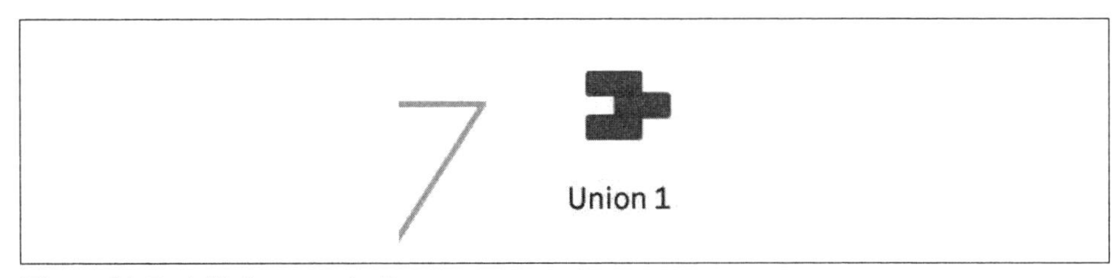

Figure 22-7. A Union step in Prep

Scaffolding Data Sets

The final technique to add rows, scaffolding, is often used to fill in missing dates or records in cases where you require a record value, even if it's a null or zero. This involves adding your core data set to a *scaffold* that includes all the records you require. Scaffolding is a more complex technique, so there is an entire chapter devoted to it: Chapter 39.

Joining Data Sets

Although scaffolding involves using joins, as just described, you can also use joins directly to create additional rows. Though this carries the risk of creating duplicates, it can be beneficial for certain use cases. For example, if you have one data set with sales targets at a regional level and another data set for the list of stores in that region, you could split the target across those stores by joining those data sets together. The difference in the granularity of the data sets means that the less granular data set (i.e., the targets) will be duplicated across the stores. A calculation can then correctly allocate the target across the stores.

Summary

Many of the techniques covered in this chapter are described in more detail throughout the rest of the book. There are other ways to create additional columns or rows in your data set, but these will probably be the main techniques you use. Although you might lose a bit of flexibility of the data set by precalculating some measures, doing so in the data preparation stage can save hours of research and work for end users who lack the technical knowledge or skills to handle this task themselves.

CHAPTER 23

Filtering

One of the most important factors when cleaning data is deciding whether the data:

- Can be cleaned up
- Should be ignored
- Should be removed

If you decide on the latter option, then you need to filter your data set. This sounds like a very easy decision to make but it shouldn't be, especially if you are preparing data for others to use. Being certain that you or the end users won't need this data going forward is difficult. If you are positive the data isn't needed, don't remove it except as the last step in the data prep process before publishing, after you've considered the following:

- Does that data give the user context on other data points?
- Is the data messy but manageable? Just because the data might be hard to tidy up doesn't mean it couldn't have value to end users.
- If the business logic changes—that is, the user has a different business experience —will the data suddenly have meaning?

With those caveats in mind, let's explore what a filter is and where to use one.

What Is a Filter?

A filter allows you to keep (a *Keep Only* filter) or remove (an *Exclude* filter) data from a data set. Once you have decided what you would like to keep or get rid of, you have several types of filters to choose from:

- Selection
- Calculation
- Wildcard
- Null values

There are also two different forms of filters that can be applied within each type:

- Data field filters remove columns (fields) of data.
- Data value filters remove rows of data.

Different Types of Filters

Let's dig into each type of filter to understand how and when to use it.

Selection

Selection is the most basic form of filtering. The experience is different depending on whether you want to filter out data fields or data values.

You can filter out data fields (columns) in multiple places within Prep Builder, but the metadata grid is ideal. You can easily select and deselect the checkbox options as required. The metadata grid is available on the Input step, as well as Clean steps throughout the flow (Figure 23-1).

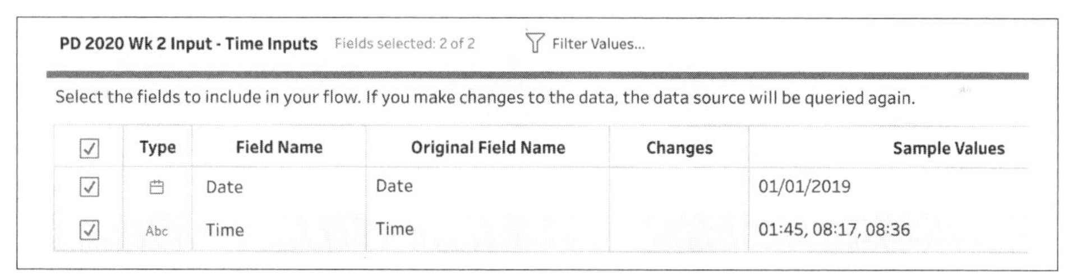

Figure 23-1. The metadata grid in Prep

Columns can also be removed through the ellipsis menu at the top of a data field (Figure 23-2).

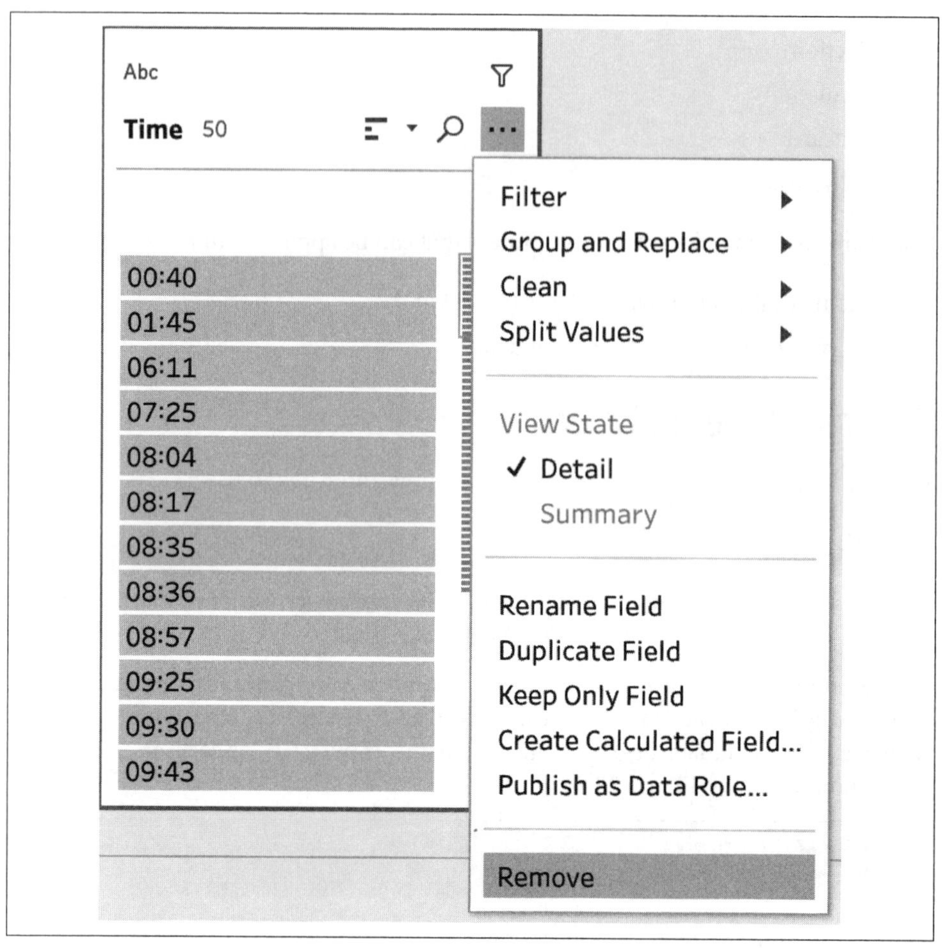

Figure 23-2. Removing an individual column through the data field menu

For data value filters, you select, through a range of actions, what to apply the filter to. You can do this in multiple places within Prep Builder, as outlined next.

In the Profile pane

Within the Profile pane, the distribution of instances of a value is indicated by the gray bar, as discussed in Chapter 11. By right-clicking on the bar, you can keep or exclude its value (Figure 23-3). To select multiple values, drag your mouse over the items or hold down the Ctrl or Command key while clicking each item.

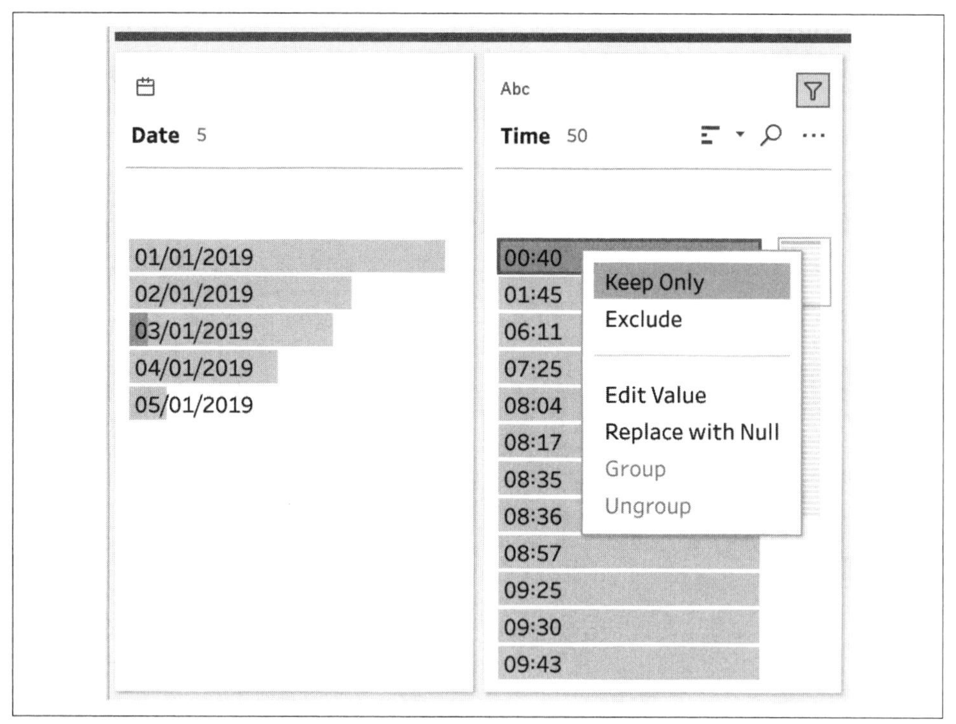

Figure 23-3. Filtering by selecting records in the Profile pane

In the Data pane

Within the Data pane, you can select any value at the bottom of the screen to filter out all other values (Keep Only) or remove the selected value (Exclude), as shown in Figure 23-4.

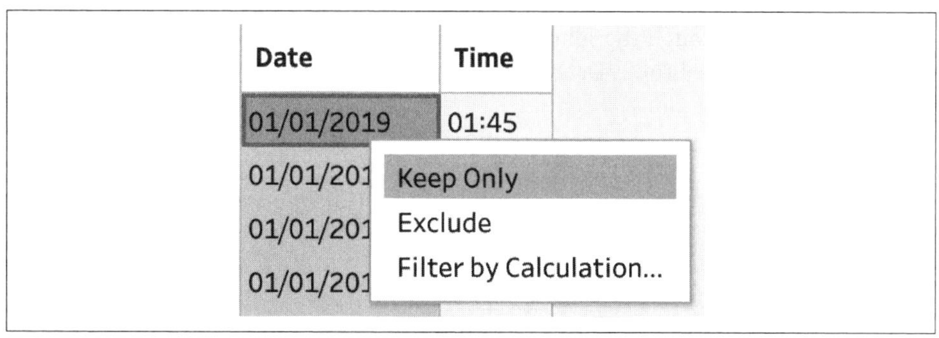

Figure 23-4. Using the Data pane to filter

From a data field

By clicking on the ellipsis menu when mousing over the data field in the Profile pane, you can choose the filter option Selected Values; this allows you to select the values you want to Keep Only or Exclude (Figure 23-5).

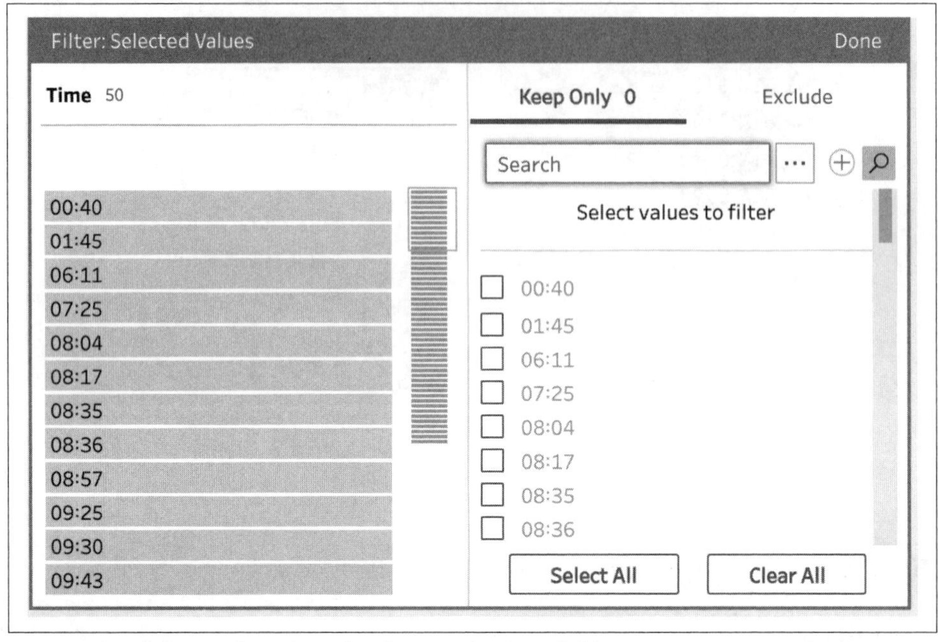

Figure 23-5. Filtering by Selected Values in the data field menu

Calculation

After selections, calculations are the most common form of filter. The easiest way to set these filters is through the Filter Values icons located on the gray bar separating the Flow pane and the Profile/Data pane (Figure 23-6).

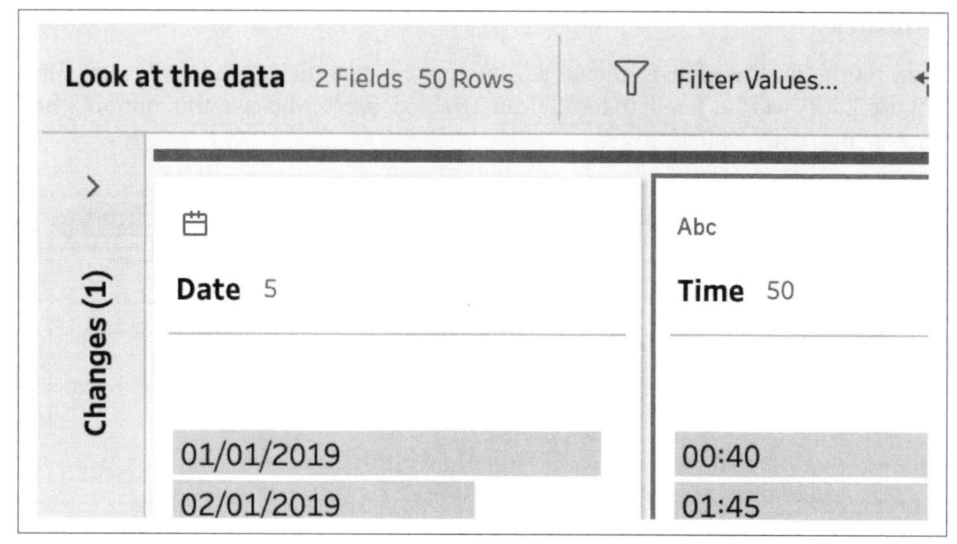

Figure 23-6. Filter Values icon in the dynamic options bar

Selecting Filter Values opens the Add Filter dialog (Figure 23-7). Notice this differs slightly from Add Field dialog that appears when you create calculated fields from scratch.

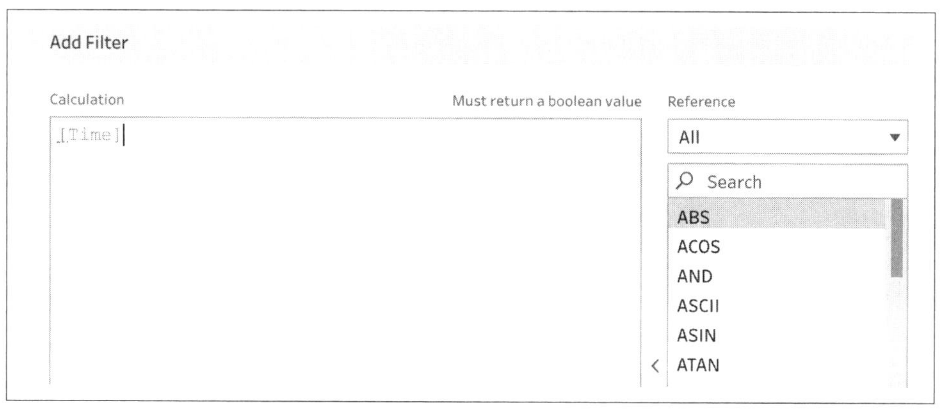

Figure 23-7. Add Filter dialog

Creating a Boolean calculation will leave the `True` values in the data set to use in future steps. The `False` values will be removed from any potential output. You can also trigger this type of filter through the ellipsis menu at the top of a data field in the Profile pane. The only difference with this method is that the selected data field will be automatically added to the Calculation Editor.

Wildcard

Also found in the ellipsis menu at the top of data fields, the Wildcard filter (Figure 23-8) will be familiar to Tableau Desktop users who use this option when filtering by a discrete field.

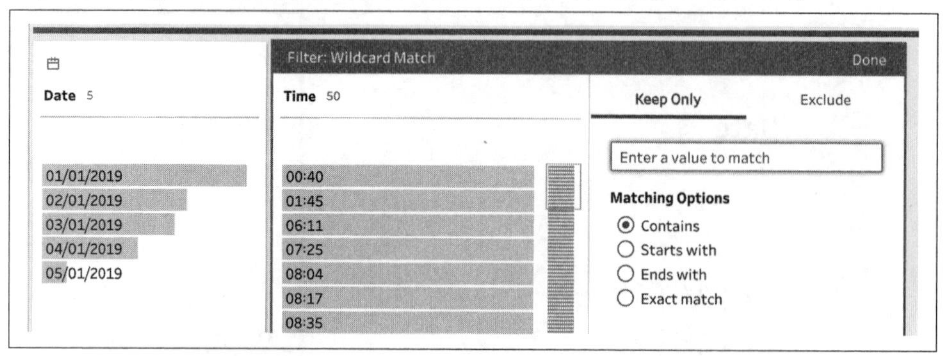

Figure 23-8. The Wildcard filter

You'll be prompted to enter a value within the selected field. The wildcard means the value you enter doesn't have to exactly match the value found in the data field; that is, it can be only a substring of the full value. You have four options to specify where the substring should appear:

Contains
: The substring can appear anywhere within the string.

Starts with
: The substring has to match the first characters of the value being assessed.

Ends with
: The substring has to match the last characters of the value being assessed.

Exact match
: The substring must be a perfect match for the entire value in the data field.

If the match is found, the row of data for that value will continue to be used in the data set if the Keep Only filter option is selected; otherwise, that row is removed from the data set. Exact matches will be filtered out if you choose the Exclude filter option.

Null Values

In the Null Values option (again triggered from the ellipsis menu at the top of a data field in the Profile pane), you select whether you want to keep only rows that contain null values or keep those with non-null values (Figure 23-9).

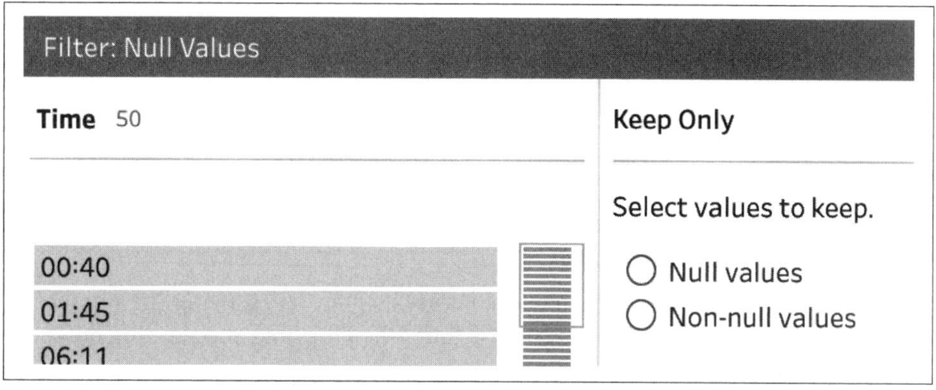

Figure 23-9. The Null Values filter

When to Filter Out Columns

Removing columns of data is a data preparation step that you shouldn't take lightly. Once you've made this decision, there is very little (apart from a confusing join from an earlier step) that will bring this data back from oblivion.

That said, being able to remove unnecessary data fields is valuable in the following cases:

- Removing blank columns. When using Excel or text files, you will often find apparently blank columns, which often have a space lurking in just one of the rows. Remove these and don't look back!
- Removing columns that are mostly nulls. Just because certain rows are null doesn't instantly mean that you should remove the full column. Is there a way to populate those rows of data using the merge functionality? Is there a reason why that column *should* contain nulls? If so, leave the column in place; otherwise, filter it out.

When to Filter Out Rows

As with columns, there are many reasons for filtering out rows of data. Taking care not to remove data that others might find useful in their analysis is important. Remember, by filtering values out of a data set, you remove all the rest of the information within that row.

Some of the reasons for filtering rows of data out of your data set include:

- Removing data errors. It's impossible to correct erroneous data if you don't know what the value should be. In that case, removing the data record is the only option.
- Removing rows outside of scope. Your data set might be on the last complete year of transactions, in which case there is no point to having multiple years of data. Filter out the additional years.
- Improving performance. As mentioned in Chapter 12, reducing the number of rows improves Prep Builder's response time when forming and running flows.
- Securing data. You may be in the situation where certain end users shouldn't have access to certain records. Within Tableau Server, you'd set up User Filters to manage data security, but Prep Builder doesn't have that feature. Therefore, you would want to use a filter to remove the confidential records and then output a new data set that's safe for users to access.

Summary

The filter may be the most used cleaning technique in data preparation, so this chapter could have been much longer. This chapter has covered the basics, and Chapter 37 goes into more advanced filtering techniques. It's crucial to practice filtering and get comfortable with what you are removing from the data set. Deploying the right filtering technique will not only prepare your data for analysis but also future-proof your flows for updating that data.

CHAPTER 24

Removing Data During Input

You can simplify a lot of your data preparation work by making changes to the initial data connection, which is created and shown in the Input step in Prep Builder. However, you often know that certain elements in your data set need to be changed or removed even *before* you input it into Prep Builder. This chapter will cover some considerations for removing data at that early stage and how you might go about doing so.

Changing Your Data Set Before Loading It

Data sets are proliferating and growing rapidly, so you have to think carefully about what is actually being loaded into the tool. Any input data will be loaded into your computer's memory, so any effort to reduce the amount of data that has to be processed will be useful. Prep Builder will sample the data set on the initial load, processing the full data set only when you run the output.

For Prep Builder, the initial connection is the Input step, but Prep Builder doesn't load in all of the data instantly. Prep will load the *metadata*—the data about the data—in the Input step first. This helps end users in two ways:

- Providing a quick overview of the data
- Preventing slow load times, since Prep Builder isn't having to process all of the data

Deselecting the fields that you and the end users do not require will save that data from being processed by Prep Builder (Figure 24-1).

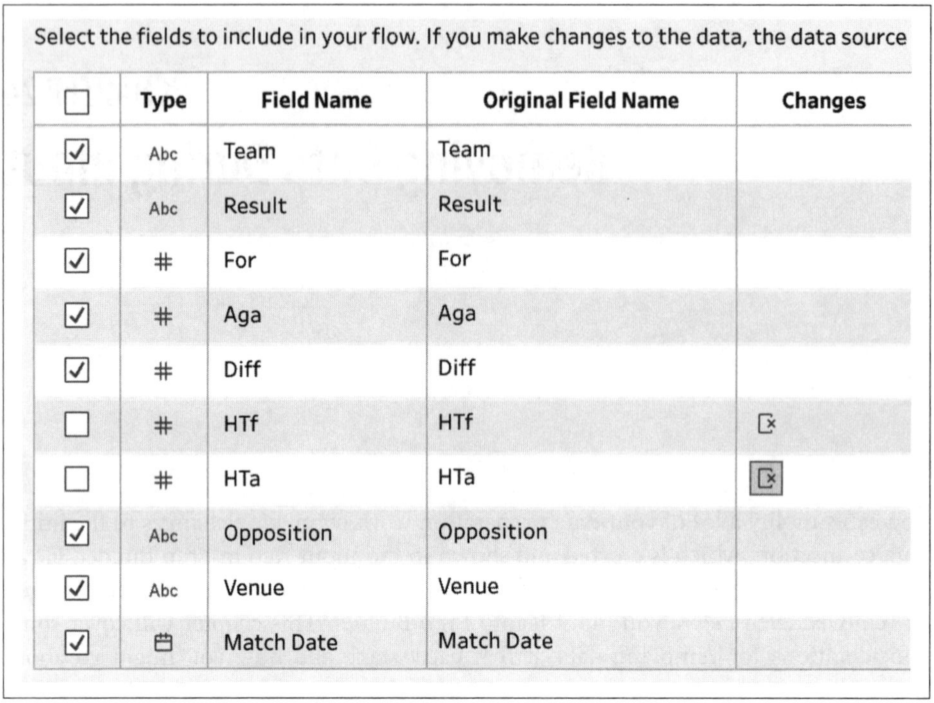

Select the fields to include in your flow. If you make changes to the data, the data source

☐	Type	Field Name	Original Field Name	Changes
☑	Abc	Team	Team	
☑	Abc	Result	Result	
☑	#	For	For	
☑	#	Aga	Aga	
☑	#	Diff	Diff	
☐	#	HTf	HTf	
☐	#	HTa	HTa	
☑	Abc	Opposition	Opposition	
☑	Abc	Venue	Venue	
☑		Match Date	Match Date	

Figure 24-1. Deselecting some fields in the metadata pane of the Input step

Making changes to your data set here can also save you time and reduce complications later in the data preparation process.

Slow Performance, Slow Build, Slow Output

As already mentioned, performance is a big reason to remove unnecessary data whenever you get the opportunity. But it isn't just the faster processing speed that helps, it's that the software can keep up with how you want to work, show you what you might want to do next, and then enable you to make those changes. Nothing is more frustrating than the wasted time of watching a load screen.

Reducing the number of rows and columns can aid performance and decrease the amount of data that Prep Builder has to process. The cleaning, reshaping, and merging processes are more efficient with less data, as it's easier to spot what is required once there are fewer rows to assess, not to mention it's faster to process that data once those changes are made. You can reduce the number of rows by applying filters on the Input step and in many other steps throughout the flow. You can reduce the number of columns by deselecting them within the Input step or excluding them in any of the steps throughout the flow.

Not only are the input processing and build time affected by unnecessary data, but the output processing time is also affected. Flows may be run very frequently through either Tableau Prep Builder or the server-based Prep Conductor tool. Prep Conductor enables you to schedule your flows to run on a regular basis (Figure 24-2).

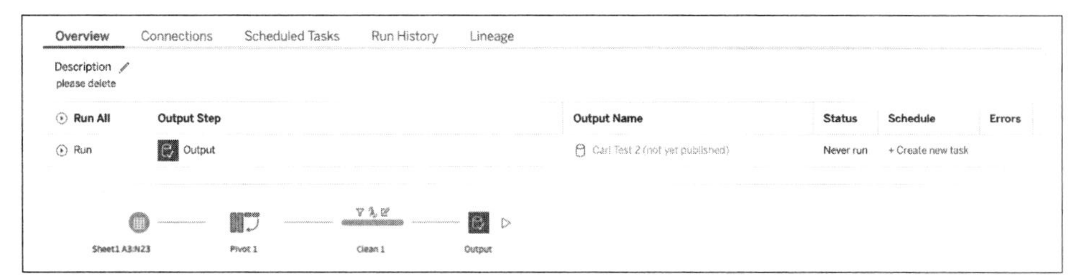

Figure 24-2. View of Prep Conductor on Tableau Server

After publishing a flow to Tableau Server, you can set a schedule through Prep Conductor if that schedule is available (Figure 24-3).

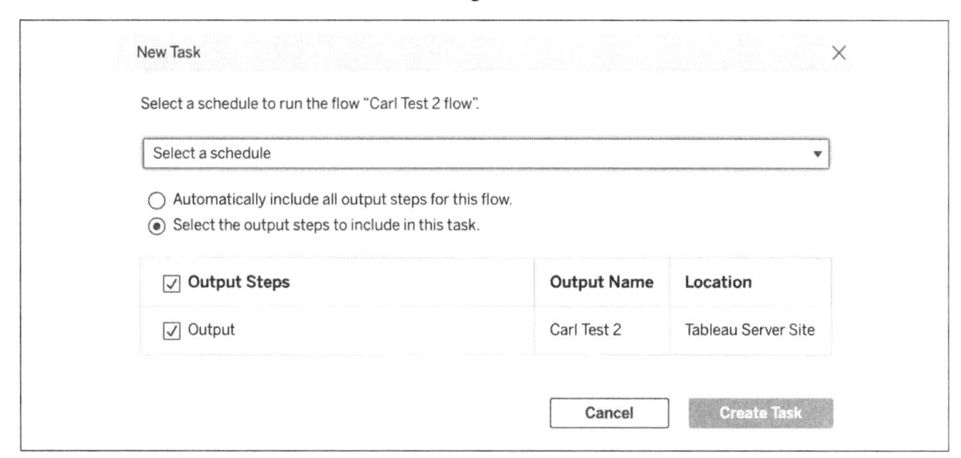

Figure 24-3. Setting up a refresh schedule on Prep Conductor

The schedules available to you are determined by your server administrator (Figure 24-4).

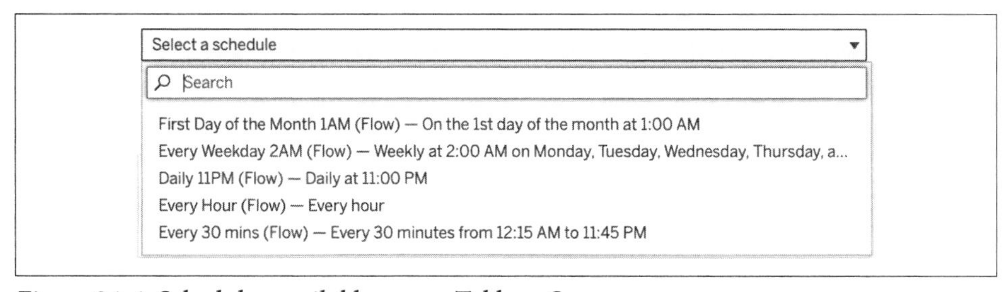

Figure 24-4. Schedules available on my Tableau Server

If you have a very frequent schedule, like every 15 minutes or each hour, any unnecessary data can add up and place an unnecessary load on the server. This can reduce performance for others and rapidly accrues when hundreds of data sources are being processed each day on large data sets.

Removing Columns

The majority of changes made during the Input step are to the data fields, or columns, of your data set. Simply deselecting the column for the data field in the Input step will prevent it from being loaded by Prep Builder. Do this only if you're sure the column contains nothing that you need. Having messy data in a column is not a sufficient reason to remove it. If any of that data could be useful to you or your end users, then you can clean that data in subsequent steps.

Completely null columns are a likely candidate for removal at this stage. These commonly occur in Excel files where a user has spaced out the data for formatting reasons. The data will likely appear with a column header like F4 for the fourth column in the data set. If all of your data field names appear this way, you may need to use the Data Interpreter to find the column headers in the Excel worksheet. You can manually rename the data field names in the Input tool, but this task can be laborious.

The Data Interpreter is an option within Prep Builder and Desktop to pick up the data tables that you cannot easily access by just deleting or renaming fields. Take the file in Figure 24-5 as an example.

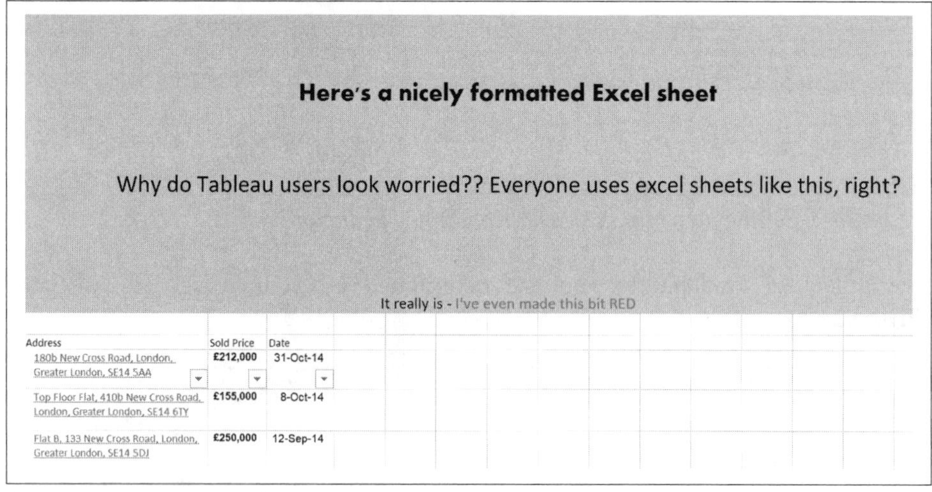

Figure 24-5. A formatted Excel spreadsheet

Loading this data set in Prep Builder would show the Input step in Figure 24-6.

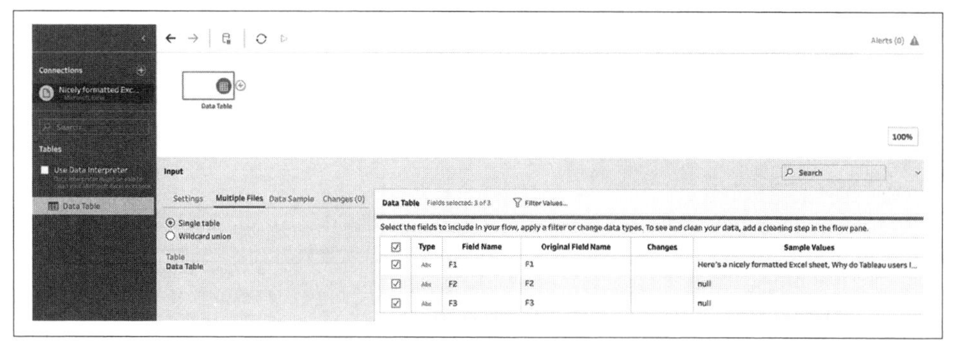

Figure 24-6. Input step showing issues with formatted spreadsheet

Here you can see the data hasn't loaded as expected. The address, price paid, and date are the columns we would need for analysis, but they are unobtainable with the current setup. Using the Data Interpreter allows Prep Builder to find the data table in the spreadsheet. Clicking the Data Interpreter option in the Connections pane displays the data table as another option (Figure 24-7).

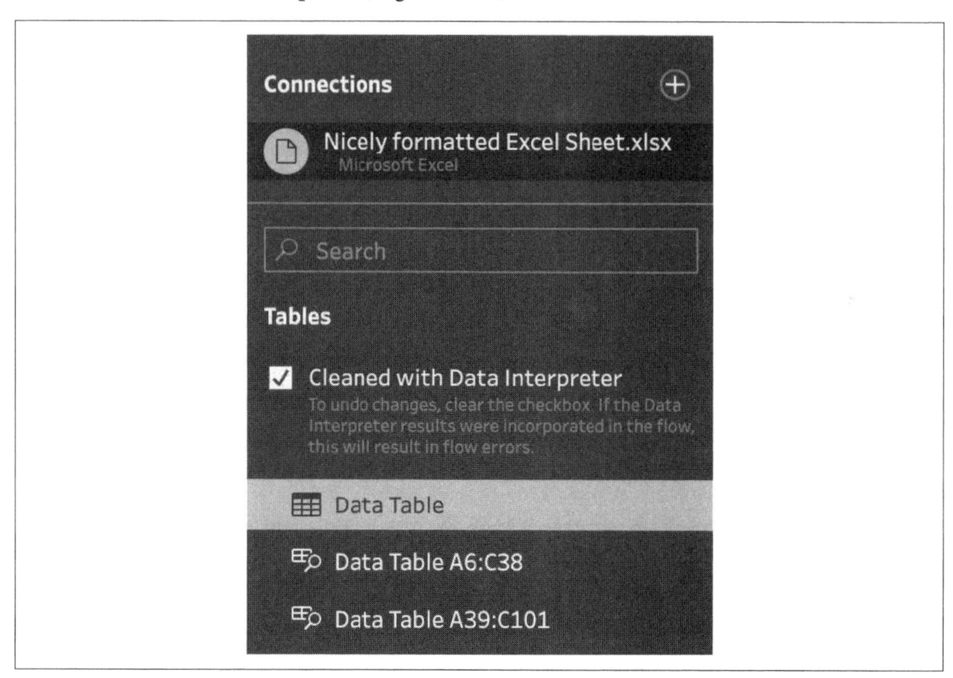

Figure 24-7. Using the Data Interpreter to access additional data inputs

To use the new tables available for input, you can remove the existing Input step and replace it with the table(s) you need. In this example, the data set is much cleaner when using the data from cells A6 to C38 (Figure 24-8).

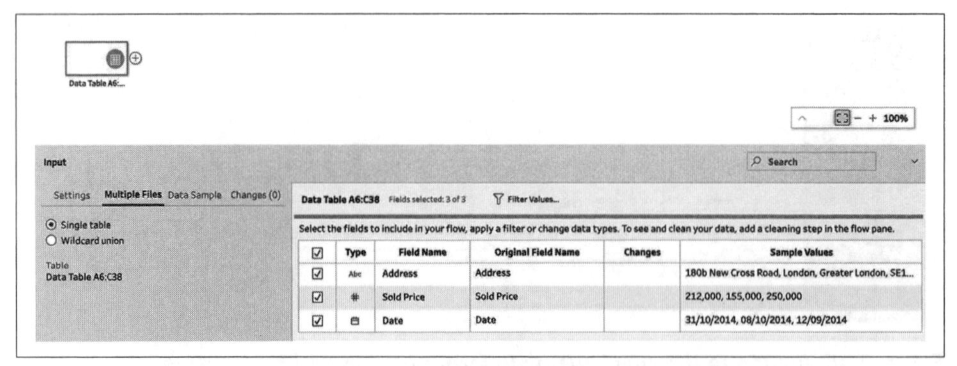

Figure 24-8. The Input step using the data resulting from the Data Interpreter

Using a combination of the Data Interpreter and selecting the data fields in the Input step will enable you to return the data you need.

Removing Records

By changing the data type of a column, you can change the values in the data set. If a column contains any non-numeric characters and you change the data type of that column to a number, Tableau Prep will replace the non-numeric values with nulls. Let's look at an example.

To change the data type, click on the Data Type icon in the metadata grid within the Input step. In Figure 24-9, we're changing from "String - default" to "Number (whole)."

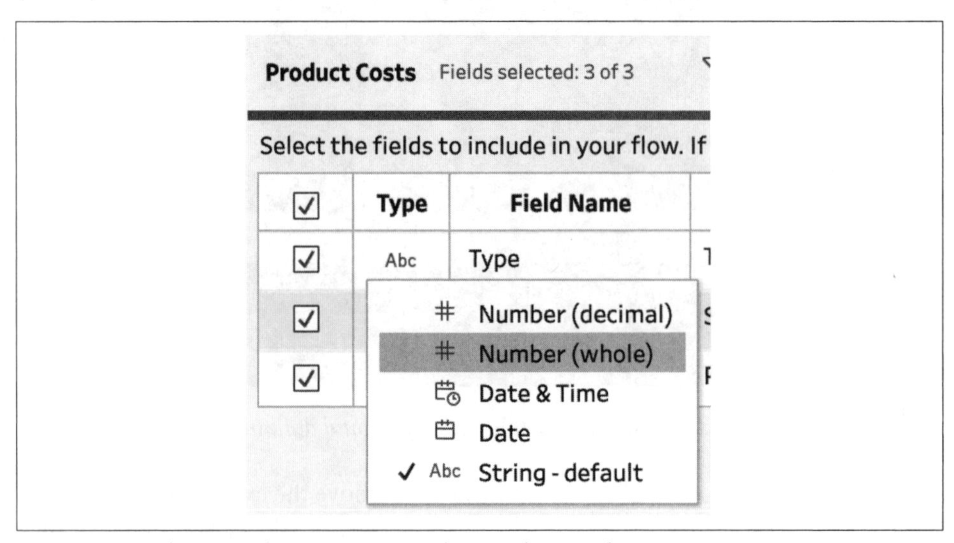

Figure 24-9. Changing data types in Prep's metadata grid

However, the sample values shown in the metadata grid have changed to null from their previous string values (as seen in Figure 24-10) .

☑	Type	Field Name	Original Field Name	Changes	
☑	#	Type	Type		null

Figure 24-10. Result of the change in Figure 24-9

If you do not need those records with a null result within the flow, they can be filtered out too. Simply click the null value in the Profile pane, and you can choose to exclude it. This process can also happen in the Profile pane. Let's look at another example, this time the Value column from Preppin' Data's 2019: Week 2 challenge (*https://oreil.ly/EGUmO*), where the data set has been split into two tables, creating a second set of headers. Figure 24-11 shows the data field before the data type conversion.

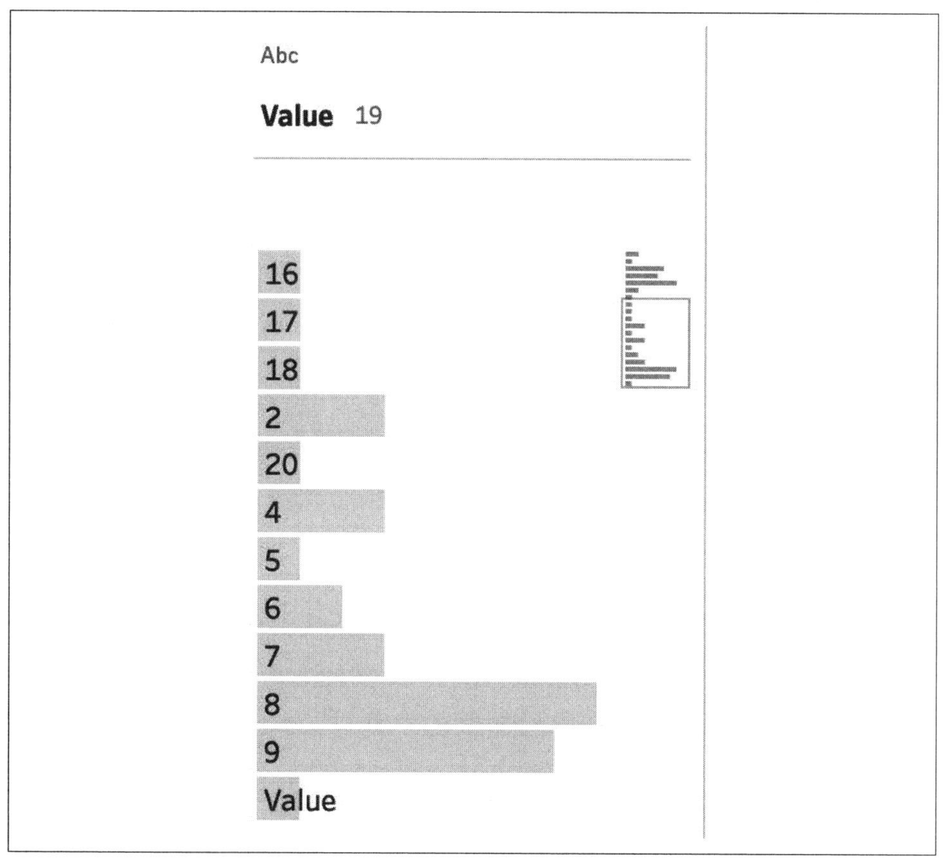

Figure 24-11. Sample of the values in the Value data field

Changing the data type of the Value column from a string to a number changes the string value to a null (Figure 24-12).

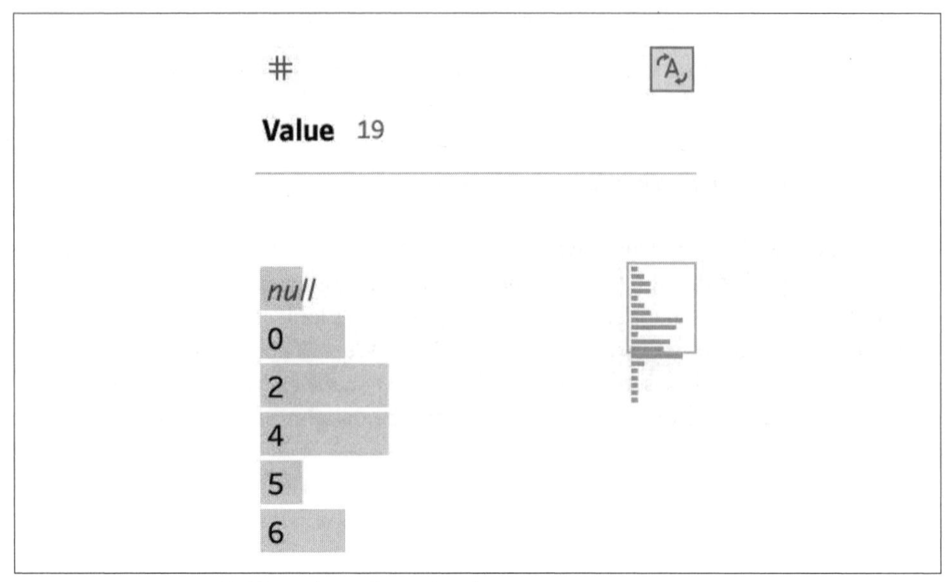

Figure 24-12. Data field resulting from the data type conversion

Summary

With most data sets you work with, you may notice only a slight difference in performance or efficiency after removing data, either by excluding fields or filtering out rows. However, the cumulative effects of these data removal actions over time will speed up your analysis, allow faster access to shared data sources, make the remaining data easier to prepare, and free up vital server resources.

CHAPTER 25
Splitting Data Fields

One of the most common actions you will take when preparing string data for analysis is to split a string field into its subparts. Splitting was briefly mentioned in Chapter 9, which covered the basics of working with strings if you need to refresh your memory about the data type. Splitting is required for many reasons, such as operational systems picking up data and outputting unique IDs for each record or squeezing records together to fit them into a specified database table. The human brain is fantastic at spotting patterns in data (that's why we create visual analytics, after all), so you can often spot the need to split data fields (columns) by just looking at the data set.

For example, in Figure 25-1, we can see that we probably need to split the Product Code field on the left into three separate columns (on the right) in order to help us analyze this data set.

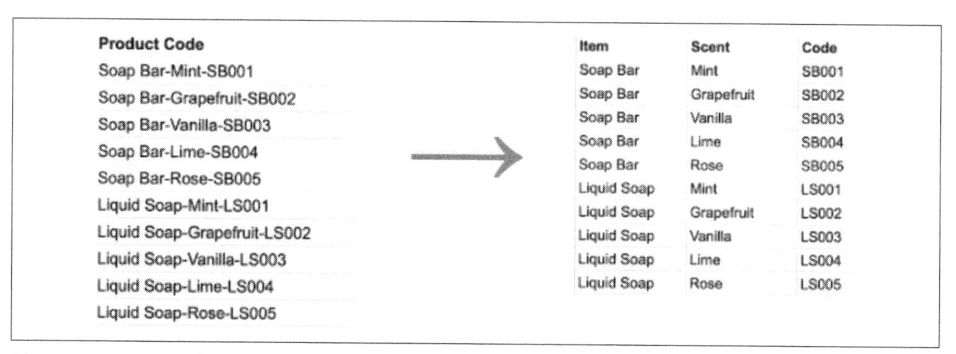

Product Code
Soap Bar-Mint-SB001
Soap Bar-Grapefruit-SB002
Soap Bar-Vanilla-SB003
Soap Bar-Lime-SB004
Soap Bar-Rose-SB005
Liquid Soap-Mint-LS001
Liquid Soap-Grapefruit-LS002
Liquid Soap-Vanilla-LS003
Liquid Soap-Lime-LS004
Liquid Soap-Rose-LS005

Item	Scent	Code
Soap Bar	Mint	SB001
Soap Bar	Grapefruit	SB002
Soap Bar	Vanilla	SB003
Soap Bar	Lime	SB004
Soap Bar	Rose	SB005
Liquid Soap	Mint	LS001
Liquid Soap	Grapefruit	LS002
Liquid Soap	Vanilla	LS003
Liquid Soap	Lime	LS004
Liquid Soap	Rose	LS005

Figure 25-1. Result of splitting Product Code field

Basic Splits

Splitting data in most data tools is very easy; Prep Builder is no different. Simply choose the data field you want to split, click the ellipsis in the top right of the field, select Split Values, and then click Automatic Split. Prep Builder will split up the field using what it believes to be the most appropriate logic (Figure 25-2).

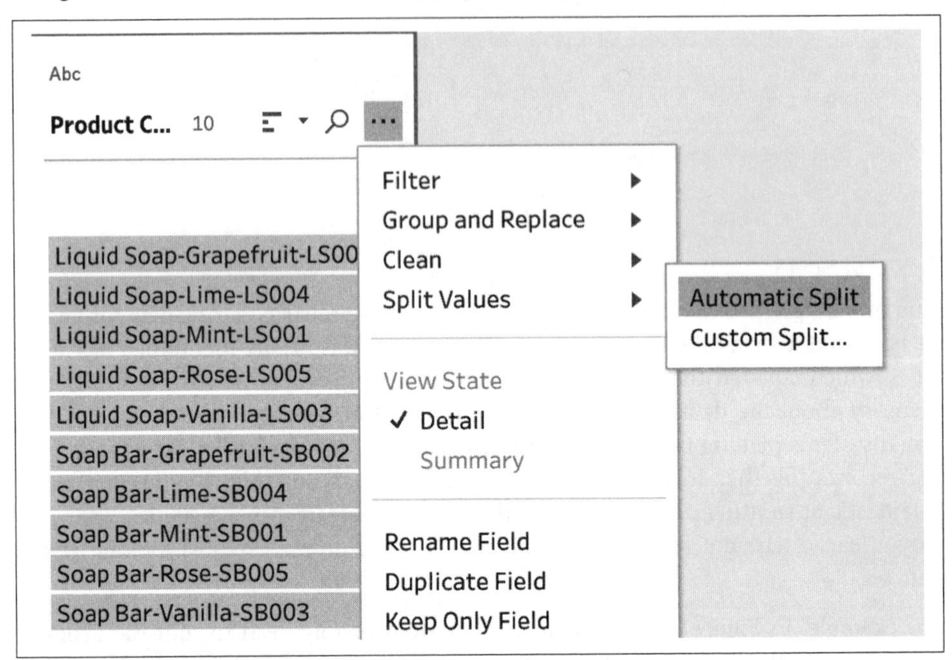

Figure 25-2. Selecting Automatic Split from a data field's menu

In this case, the automatic split has worked as desired (Figure 25-3).

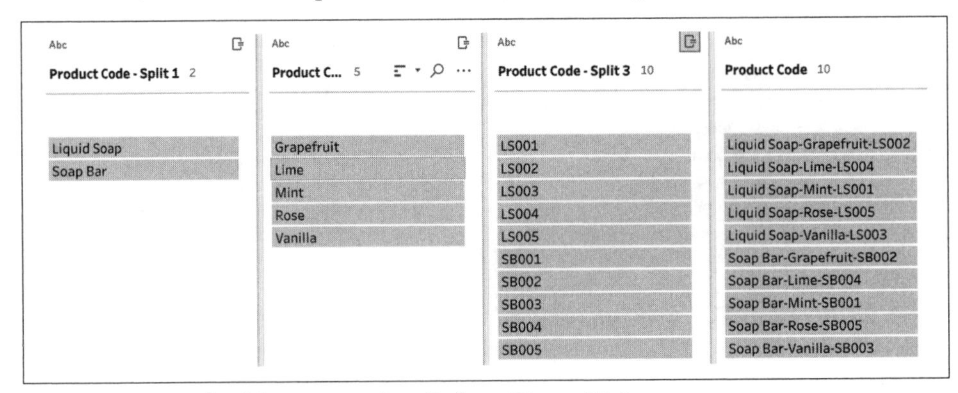

Figure 25-3. Result of the automatic split from Figure 25-2

To perform this task, Prep Builder actually writes three Calculated Fields, which you can always edit if you want a slightly different result (Figure 25-4).

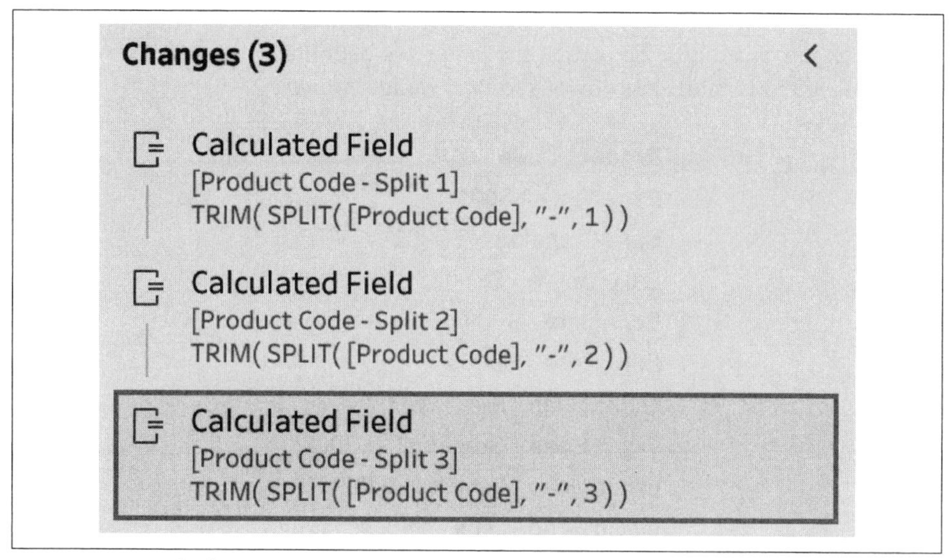

Figure 25-4. Calculations in the Changes pane resulting from an automatic split

You will find these calculations in the Changes pane in Prep Builder. You can also edit them from here, or alternatively learn how Prep completed the task you set it (Figure 25-5).

Field Name

Product Code - Split 3

```
TRIM( SPLIT( [Product Code], "-", 3 ) )
```

Figure 25-5. Opening one of the split calculations in the Calculation Editor

This formula is splitting the Product Code field on the hyphen (-) separator and pulling back the third part (i.e., what comes after the second hyphen but before the fourth, which in this case doesn't exist). The resulting values are then trimmed, which means any leading or trailing spaces are removed, as these can wreak havoc when you are matching text values.

Advanced Splits: When Automatic Splits Don't Work as Intended

Let's tweak the data a little bit so that the pattern of delimiters (the character we split the data field by) is a little less obvious to Prep Builder (Figure 25-6).

Product Code
Bar - Mint - SB001
Bar - Grapefruit - SB002
Bar - Vanilla - SB003
Bar - Lime - SB004
Bar - Rose - SB005
Liquid Soap - Mint - LS001
Liquid Soap - Grapefruit - LS002
Liquid Soap - Vanilla - LS003
Liquid Soap - Lime - LS004
Liquid Soap - Rose - LS005

Figure 25-6. Revised data set with unclear delimiter for Prep Builder

Now the Product Type (first part of the Product Code) for Soap Bar no longer includes the word *Soap* and each field now includes spaces around the hyphens. With this irregular pattern, Prep no longer gives us the results we are looking for (Figure 25-7).

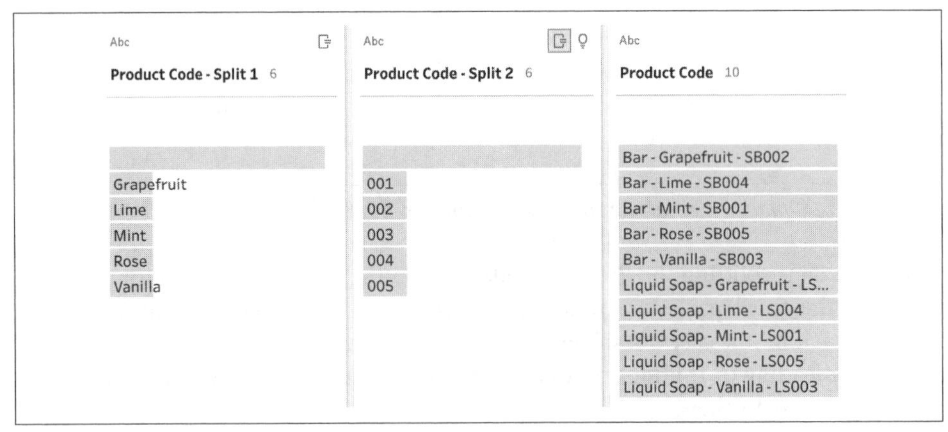

Figure 25-7. Result of an automatic split of the data set in Figure 25-6

The Product Type has disappeared altogether in the data set resulting from this split. Notice all those blank records? They are the result of the word *Soap* being removed from the Soap Bar rows; note there are no blanks resulting from the Liquid Soap rows. Let's look at the calculation Prep is writing for this split (Figure 25-8).

Field Name

Product Code - Split 2

```
TRIM( SPLIT( SPLIT( SPLIT( SPLIT( [Product Code],
"Liquid", 2 ), "Soap", 2 ), "-", 3 ), "LS", 2 ) )
```

Figure 25-8. Using the Calculation Editor to understand the split issue

Well, that's not clear, is it?

These situations are where your human eye can take over, and you should choose the Custom Split menu option instead of Automatic Split. By setting up the Custom Split to work on the hyphen, we're back to getting the results we want from this data set (Figure 25-9).

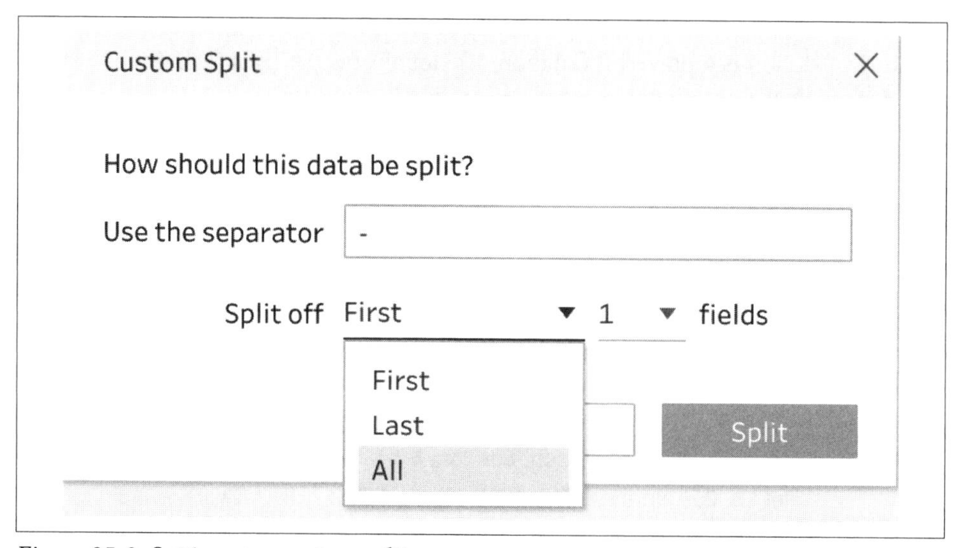

Figure 25-9. Setting up a custom split

In the Custom Split dialog, you set both the delimiter and what you want returned from the split. In Figure 25-9, selecting All ensures that all the data is returned even if you expect a certain number of columns to be created (Figure 25-10).

At the time of writing, the All option is limited to splitting 150 items.

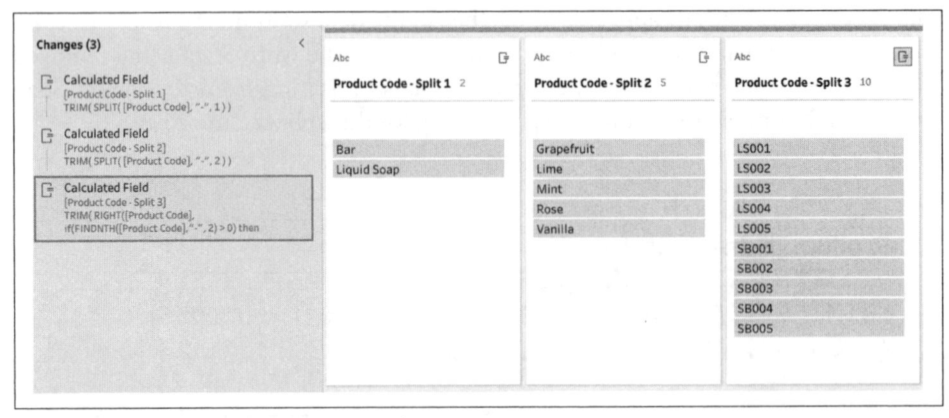

Figure 25-10. Result of the custom split

The Custom Split option can save you from having to write a number of complex calculations. In this example, the third calculation Prep Builder has built is quite complicated, so you could simplify it if you wanted, but there is really no need to unless your flow's performance is lacking.

When Not to Split Data

Although `SPLIT()` is a powerful function, it's not always the right technique for your task. Let's look at a few scenarios where this might be the case.

Address Data

In Figure 25-11, the address details are separated by commas. However, it is difficult to define an appropriate split because the street part of these addresses is inconsistently formatted, with varying numbers and punctuation. Addresses are typically difficult to split and likely will require more complex calculations and logic to align the correct parts to the correct columns.

Name	Address	Reason
Jim	15 Big House, Wimbledon, London, SW1 4PW	Supports Arsenal
Jan	Flat 18, 250 Northern Road, Stockport, SP12 7FE	Swore at a pigeon
Lit	15 Big House, Wimbledon, London, SW1 4PW	Likes the Bulls
Joh	The Willows, Old Street, Brighton, BR31 7ON	Drinks Macro Lager

Figure 25-11. Address data separated by commas

If you want to tackle this problem, have a go at the 2019: Week 46 Challenge (*https://oreil.ly/y4Sx_*), which you can use splitting to solve. The solution is available on the Preppin' Data website for you to check your flow, although there is often more than one right answer; we care mainly about whether you've achieved the data set in the output given, not how you arrived at it.

No Clear Delimiter

If there is no clear and logical separator to split by, then the split technique is definitely not appropriate. In this circumstance, more advanced string functions like `FIND()` or `FINDNTH()` might be a good approach. These functions allow you to find out if a character or substring exists within a string by its position. You can then pair these functions with `LEFT()`, `RIGHT()`, or `MID()` to split out the parts of the field required. These functions were covered in more detail in Chapter 9. Another option in this situation is to use regular expressions (or regexes), which will be covered in Chapter 31.

Summary

Ultimately, `SPLIT()` is a great first function to investigate when breaking up data columns to aid your analysis. Prep has a couple of great options, but might not always deliver the correct solution, in which case you'll need to build the logic yourself. Fortunately, `SPLIT()` generates the calculation logic and makes the calculation available in the Changes pane for editing, so it is easy to make the changes you require.

CHAPTER 26

Cleaning by Grouping Data

Data preparation would not be necessary if we always had someone else curating a perfect data set for us. However, we can (and sadly often have to) clean the data ourselves. As mentioned in Chapter 9, one of the most common challenges you'll face in data prep is cleaning up string data—for example, standardizing the string values enough to be able to count instances of values even when they have typos. One technique can come in especially handy for this scenario: grouping. This chapter will cover what grouping is and how to use the built-in grouping tools in Prep Builder.

What Does Grouping Mean?

Grouping means applying logic to (mostly) string data fields to recognize a common characteristic among them, such as their meaning or intended value. For example, we might expect the following data items to be grouped together:

- Edinburgh
- Edenburgh
- Edinborough
- 3d!nburgh

As humans, we can recognize that all these different names probably all refer to Edinburgh, Scotland (especially if the column were called City Name). But data software does not see this data the same way, so we have to give it some direction for how to handle these different collections of characters.

Why Use Grouping

Grouping is a technique you need to learn for multiple reasons.

Improving Accuracy

When they hear the term "data," most people seem to think about system-generated data being fed into databases. After a year of data analysis, most data workers would consider themselves lucky to work with such data! Most data sources people use are still manually compiled. Even leading equity research agencies manually input corporate results from PDFs to build the data sources for their analysis. And therein lies the problem.

Manually entering data increases the risk of mistyping a letter or number. Add in modern-day deadline-driven culture to this mix, and you are looking at a potentially very messy data set. If you haven't invested in autogenerated data, you probably also don't have a heavily resourced IT team who will clean up those mistakes. Fortunately, grouping allows you to combine together incorrect and correct values to form a fully correct data set.

Navigating the Data Hierarchy

"Drilling into data" is an expression you are likely to encounter as you work in data. This refers to starting at a high-level (less granular) view of the data and gradually focusing on lower-level (more granular) views, often by filtering out what is not of interest to keep the data set manageable. Hierarchies in tools like Tableau allow you to "drill down" like this, but the hierarchies aren't always set within the data. You can mitigate this by using groups to build higher or intermediary levels between existing data fields in your data set. Common hierarchies can include:

- Time (years, months, weeks, days, etc.)
- Geographical (country, state, city, etc.)
- Organizational (office, department, team, etc.)
- Operational processes (call centers, teams, products)

Groups can be used to pull together lower-level entities to create a new higher level. This differs from aggregation, which is about changing the granularity of records in terms of measures. Grouping, by contrast, is about forming consistent data values within a string data field. This can help make aggregations easier within both Prep and Desktop. The flexibility of grouping techniques means that you can update your analysis quickly in response to organizational changes like a merger or a team/management restructure (Figure 26-1).

City
York
Leeds
Newcastle
Sheffield
Leicester
Northampton
London
Southampton

Region	Team
North	York
North	Leeds
North	Newcastle
North	Sheffield
South	Leicester
South	Northampton
South	London
South	Southampton

Figure 26-1. Creating a region hierarchy of teams

Smoothing Reorganizations

Reorganizations used to be a nightmare to handle from a data perspective. Organizations might decide to change management structure or restructure a few teams' sales territories without looping in the data team. Building significant and robust reference tables to reflect and apply this organizational restructuring often takes too long. Grouping allows us to handle these challenges on an ad hoc basis in the short term (Figure 26-2) until more formal reference tables can be built.

Region	Team
North	York
North	Leeds
North	Newcastle
North	Sheffield
South	Leicester
South	Northampton
South	London
South	Southampton

Region	Team
North	York
North	Leeds
North	Newcastle
Midlands	Sheffield
Midlands	Leicester
Midlands	Northampton
South	London
South	Southampton

Figure 26-2. Reorganization of Region field

Grouping Techniques

Grouping and replacing values in Tableau Prep is one of its most useful functions, and it's also very easy to use. As this section will describe, there are three main techniques for grouping values, which allows Prep to handle a large range of use cases.

Manual

Manual grouping is great for a simple, quick fix. However, for a larger, more complex challenge, this technique will soon become frustrating.

Manual grouping allows you to apply your own logic to a data set by selecting data points to group together (Figure 26-3).

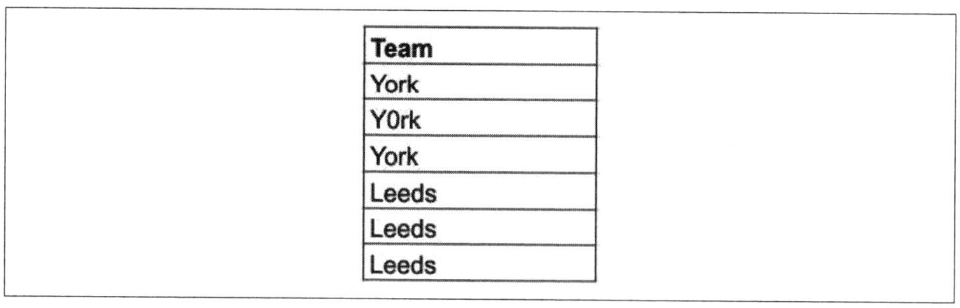

Team
York
YOrk
York
Leeds
Leeds
Leeds

Figure 26-3. Team mismatches that would be resolved by grouping

Clearly, in this example we want to make the second row of data the same as the other instances of York. Each tool has its own method for this, but in Prep Builder the solution is very simple:

1. Use the Profile pane to quickly see odd values in the data set (Figure 26-4).

Abc

Team 3

Leeds

YOrk

York

Figure 26-4. Viewing an outlier in the Profile pane

2. Select the values you would like to clean/amend by holding down the Ctrl key or the Command key on Mac (Figure 26-5).

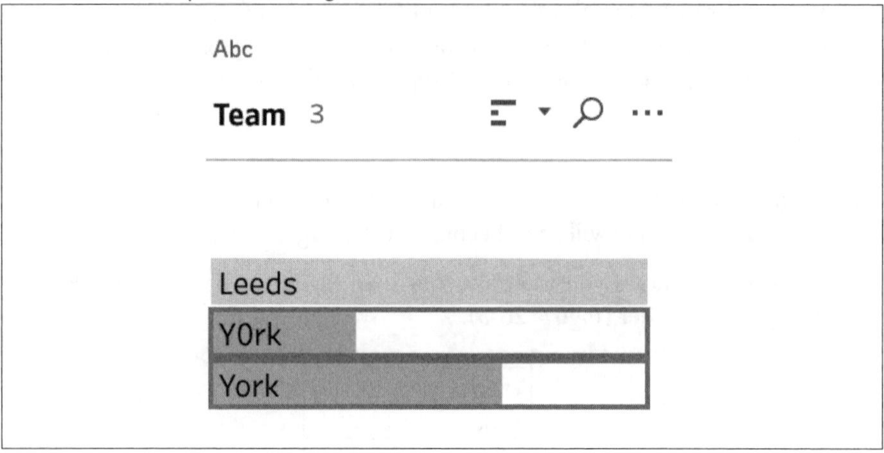

Figure 26-5. Selecting multiple values in Prep Builder

3. Right-click to bring up the menu (Figure 26-6).

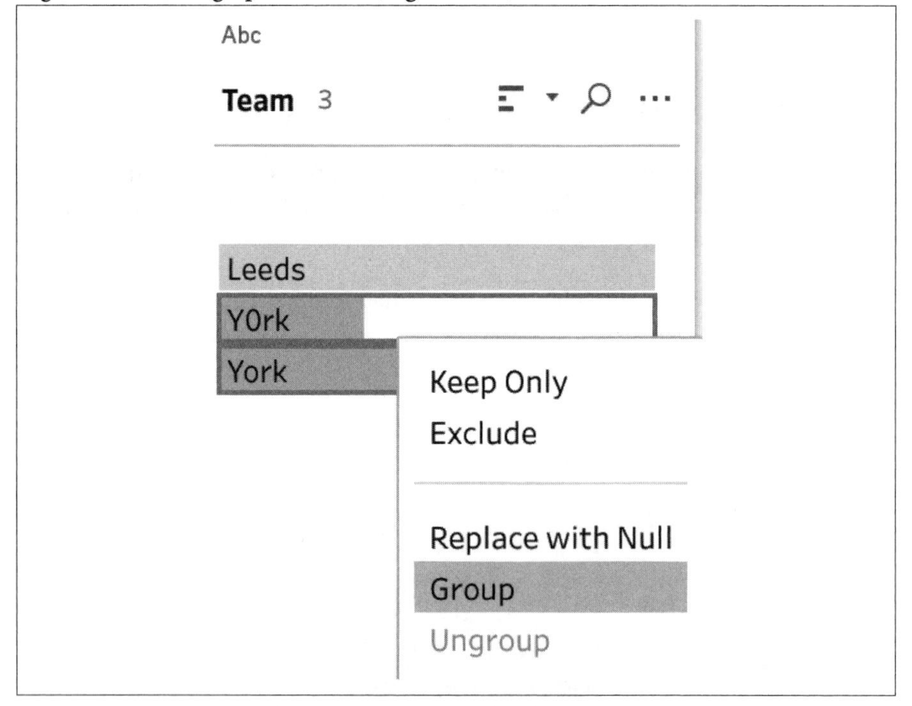

Figure 26-6. Menu of options for grouping multiple values

4. Select Group to form the new grouping. Keep in mind that the first data value you select will become the name of the group. This is easy to change, though: simply double-click the grouped value to rename it (Figure 26-7).

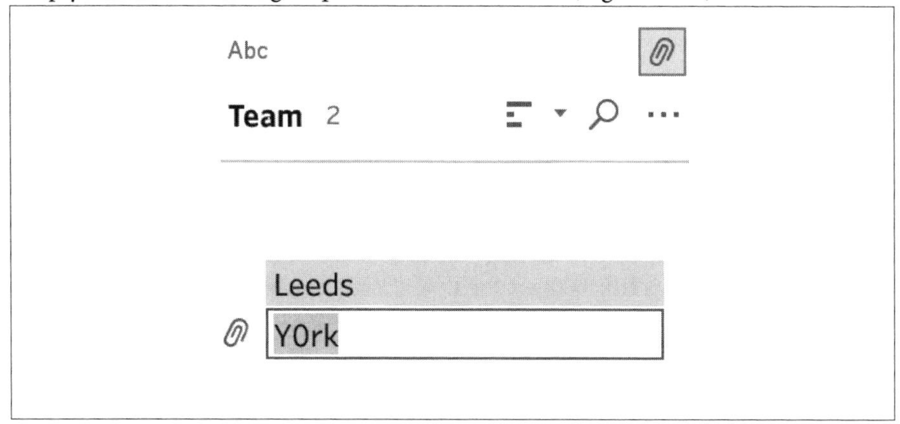

Figure 26-7. Renaming a group in the Profile pane

When you edit the group (which you can do from the Changes pane), you can see which data values have been grouped together. This view shows you how any data values that match the grouped strings will also be grouped (Figure 26-8).

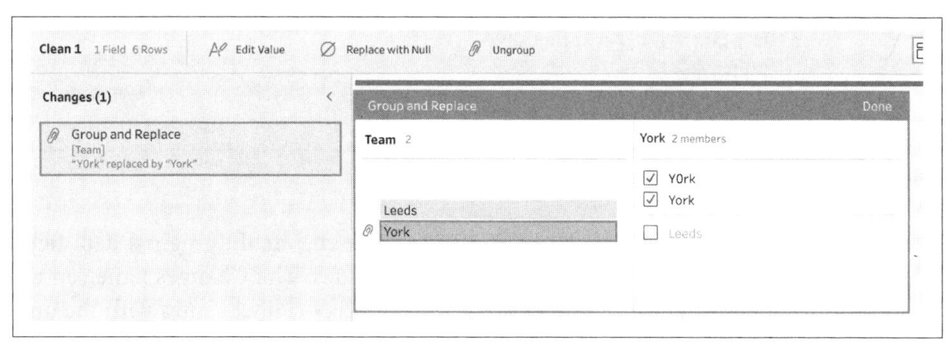

Figure 26-8. Values that are part of the York group

The easiest way to create a group within Prep Builder is to double-click a value in the Profile pane. This allows you to change all instances of that value in the data field (column). Keep in mind that the value must be the same data type as the data field.

Calculations

There are many ways to clean data values with calculations, but grouping cleaned data values is normally accomplished in a couple of ways.

IF statements

`IF` statements are a type of *logic statement*. Logic statements are commands to tell the software to work through a series of steps based on a true/false condition. In the example shown in Figure 26-9, the calculation is searching for the incorrect name `'Y0rk'` and converting any matching values to `'York'`. If the value isn't `'Y0rk'` then the calculation simply returns the value that already existed in the field.

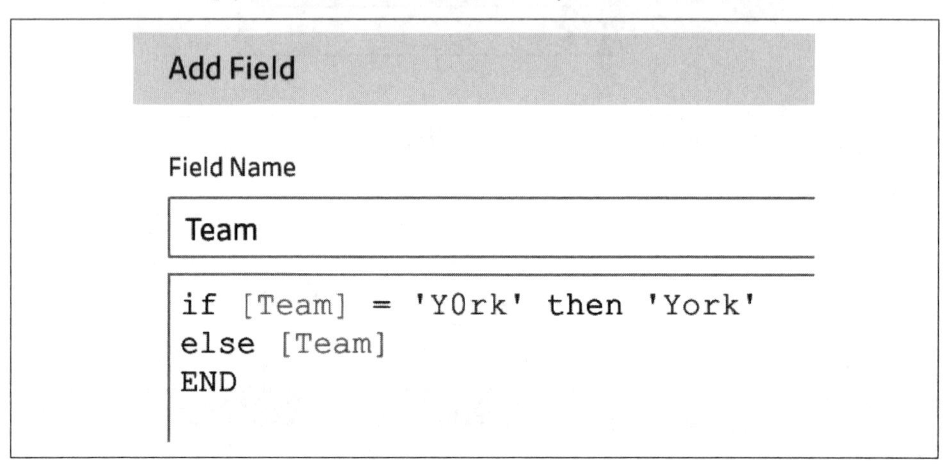

Figure 26-9. Calculation to correct the `'Y0rk'` value

Remember, if you give the new calculation the same name as the original data field, the results of the calculation will overwrite the old values. The Changes pane can act as an audit trail, but duplicating the field to compare the original values with the output values can be a useful sanity check. Completing the calculation in Prep Builder rather than waiting to complete the task in Desktop means this process has to be processed just once, rather than each time the data set is used. You can future-proof the calculation by adding in potential values and their corresponding conversions before they appear in the data set if you know they may pop up.

REPLACE() functions

`REPLACE()` functions will replace a character in a string data value with the character captured within the calculation. In Figure 26-10, the `REPLACE()` function is removing any zeros found in the data values and replacing them with lowercase letter os.

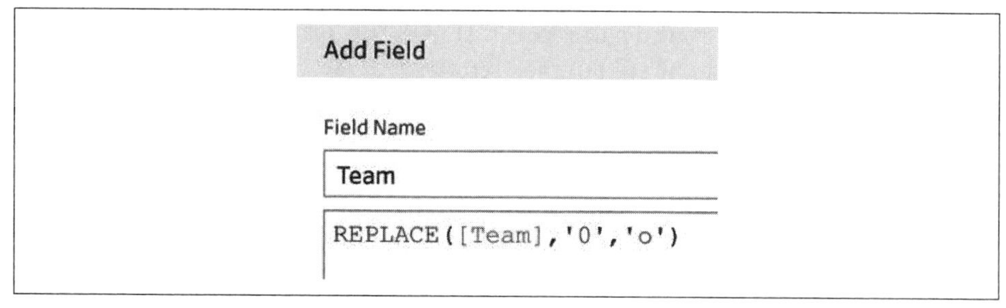

Figure 26-10. REPLACE() function in Prep

While you can create IF statements with many conditions (the longest I have written has 78 conditions), this is quite an arduous task. Each new variation you find that needs to be replaced requires another condition. REPLACE() statements can help here if only certain characters need to be removed. Any alphanumeric character that needs to be replaced can't exist in any other string. This technique is fine in our example, as none of the town names needs to include a zero, but if we had an *e* instead of an *o* in York (i.e., "Yerk"), we wouldn't be able to replace it without also accidentally turning Leeds into "Loods."

Built-in Functionality

Some data preparation tools have their own built-in capabilities for grouping similar strings together. Prep Builder has pulled together many of the best features and made it very easy to use them. In addition to manual and calculation-based grouping, Prep Builder provides these three methods of grouping:

- *Pronunciation* grouping is based on how the string would sound if spoken in English. Similar pronunciations are grouped together.
- *Common character–based* grouping analyzes letters and sorts them into an order. Similar ordered strings are then grouped together. The original position of the characters isn't important.
- *Spelling-based grouping* evaluates the differences in characters used in strings and determines the changes required to spell all the terms similarly. The difference between the common character- and spelling-based grouping techniques is the position of common characters; the spelling option, unlike the common characters one, requires many letters to be in a similar order.

Using the Preppin' Data 2019: Week 2 (*https://oreil.ly/EGUmO*) data set, let's take the City field example (once the Data Interpreter has been used to import the relevant data; see Chapter 24 for more details on the Interpreter). Here the data field contains a number of spellings for each city. In this data set, I was just expecting values of

Edinburgh and London, but Figure 26-11 shows the list of values that actually appeared in the column at the start of the exercise.

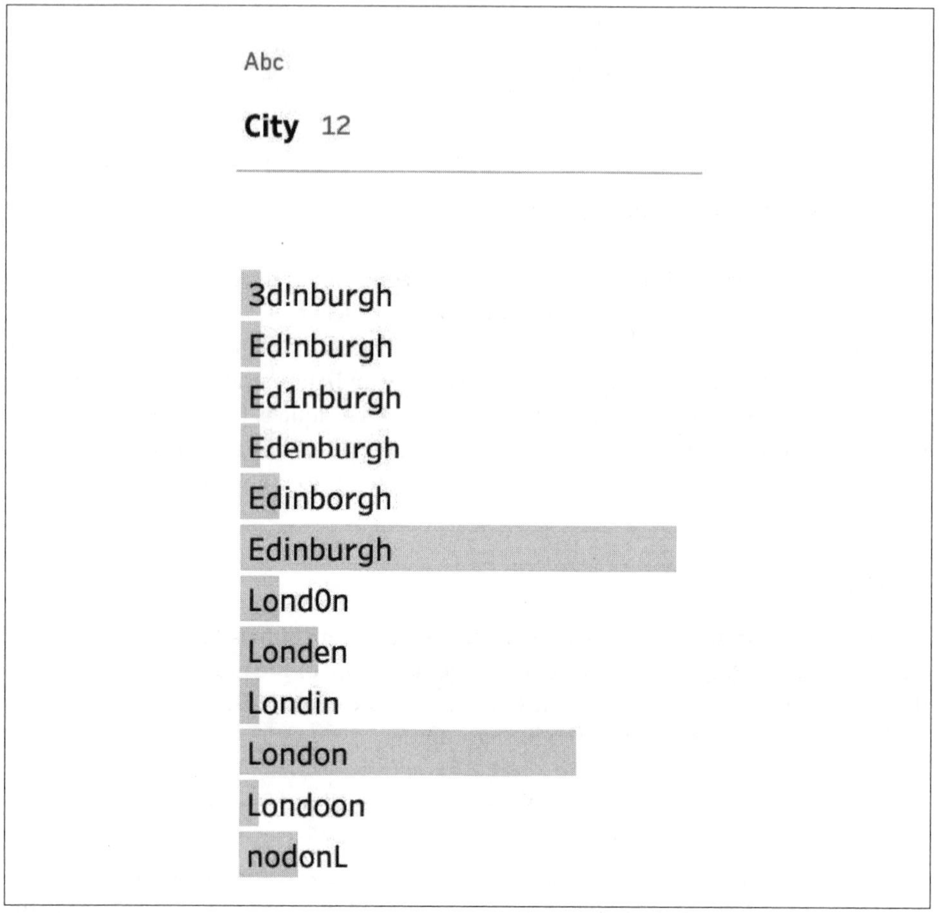

Figure 26-11. List of values in the City field in the 2019: Week 2 challenge

Let's see what happens as we progressively use each of the built-in functions in the Group and Replace menu. First, using Pronunciation cleans up a lot of the data, as you can see in Figure 26-12.

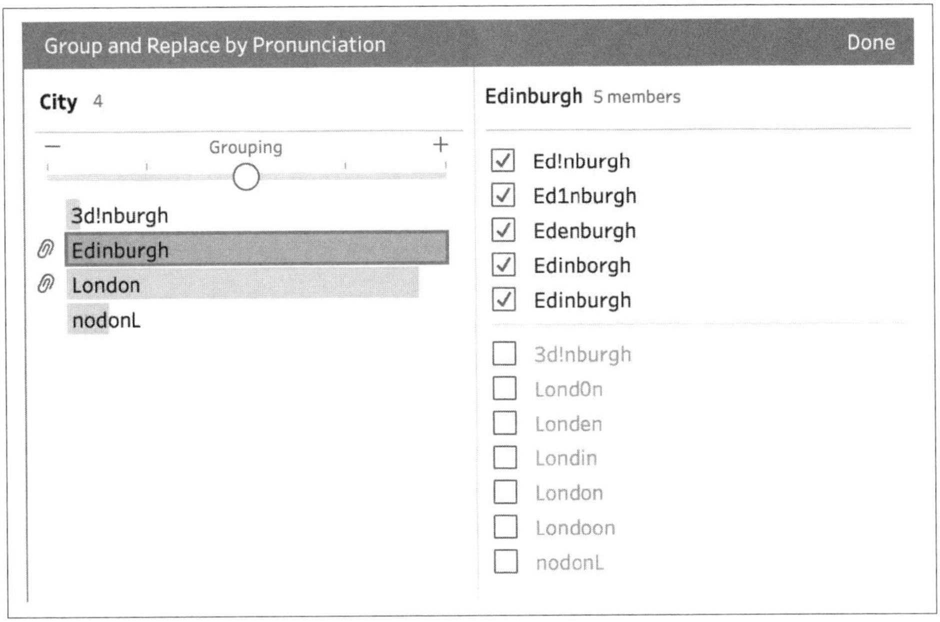

Group and Replace by Pronunciation
Done

City 4

Grouping

3d!nburgh
Edinburgh
London
nodonL

Edinburgh 5 members

- [x] Ed!nburgh
- [x] Ed1nburgh
- [x] Edenburgh
- [x] Edinborgh
- [x] Edinburgh

- [ ] 3d!nburgh
- [ ] Lond0n
- [ ] Londen
- [ ] Londin
- [ ] London
- [ ] Londoon
- [ ] nodonL

Figure 26-12. Applying the Pronunciation Group and Replace option

However, not everything has been correctly grouped together. Like the Profile pane, the Group and Replace functions use histograms to show the dominant values. The correct spellings of Edinburgh and London—and the most common in this set—have formed the two majority groups. By selecting one of those two groupings, in this case Edinburgh, you can see the values that Prep Builder has grouped together. If any of these values is incorrect, you have two options:

- Change the sensitivity of the Group and Replace function by moving the dot on the plus/minus Grouping scale. Move the dot toward the minus sign to reduce the sensitivity of the algorithms making the groupings. This means more data is likely to be added into one group. Conversely, move the dot toward the plus sign to increase the algorithm's sensitivity so that less data is included.
- Manually deselect values. By unchecking a selection in a grouping, you can remove this value, and all the related records, from the grouping.

Because the pronunciation option works on how the letters are spoken in English, the 3 at the start of 3d!nburgh or the rearranged letters of nodonL are not close enough to be considered a match. So, first save the progress made thus far by clicking Done in the Group and Replace controls, and then apply the Group and Replace functions again to the resulting field, this time selecting Common Characters. The resulting data is one step closer to the desired output (Figure 26-13).

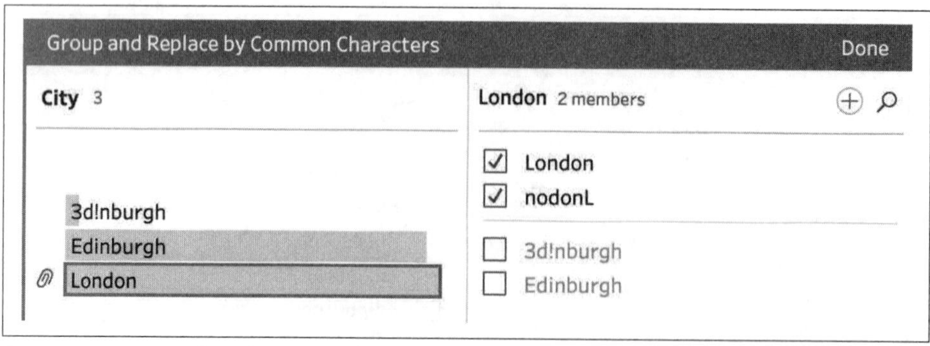

Figure 26-13. Results of grouping by common characters

Because nodonL contains the same letters as London, the two terms are grouped together. Again, as London has the highest number of records, nodonL is added to the London grouping. To save this further progress, click Done. Finally, let's address the last mismatch using the Spelling Group and Replace option (Figure 26-14).

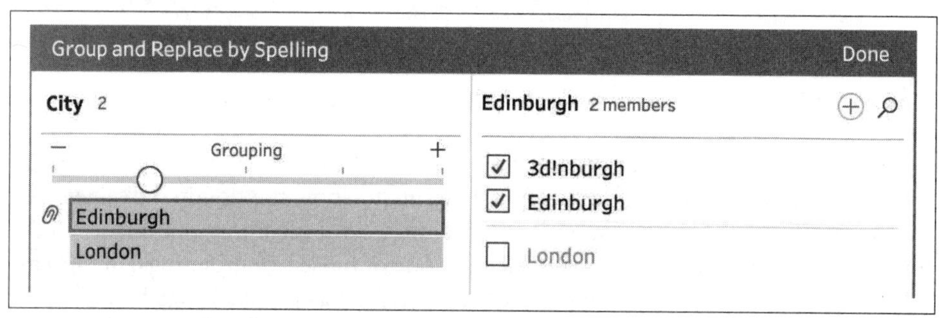

Figure 26-14. Results of grouping by spelling

Here you can see that 3d!nburgh has been grouped with the rest of the Edinburgh values, as there are only two characters that differ from the correct spelling.

Writing a calculation to cover all of these examples would take a lot of time, so this Group and Replace functionality can save you a lot of time and effort. Even though this data field is now ready for analysis, you must be careful if you plan to join this field back to a source data set where the values are still in their original "messy" state. Joining, covered in Chapter 16 and Chapter 32, must be applied to like-for-like values; therefore, this data set would no longer completely join to its earlier form. Understanding what is needed of the data set will determine where in the workflow to perform each data preparation task. Luckily, with Prep Builder, it is quick and easy to adjust this ordering as you iterate through your solution.

Summary

Before using Tableau Prep, grouping like strings together involved either long `IF` calculations or a lot of manual selections. Unfortunately, these methods would rarely update with the introduction of varying values. Prep's use of Group and Replace algorithms to group strings together not only makes the initial preparation easier but also future-proofs the flow.

CHAPTER 27
Dealing with Nulls

Nulls, or the absence of data, are a fickle beast within data preparation. Experienced data preppers will know almost instinctively how to deal with them, or at least how to manage the challenges that come with a data set containing them. Novice data preppers do not have the same set of use cases or experience to draw on, though, so this chapter covers the basic considerations for working with a data set containing nulls.

What Is a Null?

The absence of data is not the same as a zero, a new row, or a space, all of which are actually values. Nulls appear in data sets for many reasons, including:

- They are the result of mismatched fields in a union.
- They are the result of mismatched fields in a left, right, or fullOuter join.
- There's no original data entry for that record, but other data points in the set exist (i.e., the other fields are not null).

Prep Builder shows the number of null records for any data field in its Profile pane (Figure 27-1).

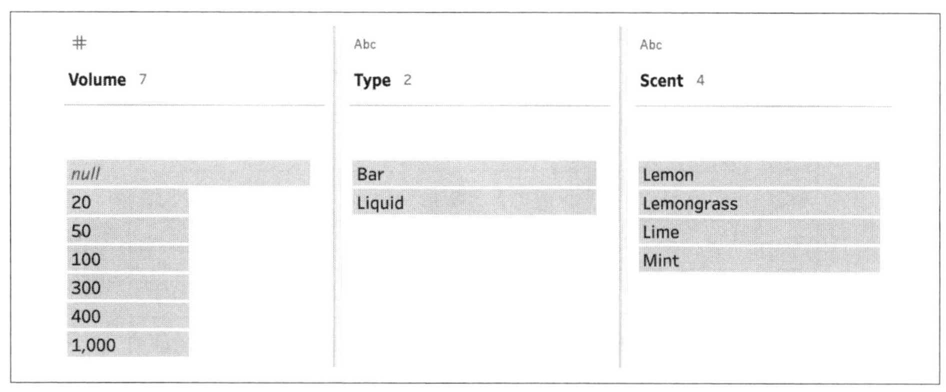

Figure 27-1. Nulls in the Volume data field

Now that you know what they are and where to find them, let's look at when it's acceptable to have null records in your data set and when you should remove or replace them.

When Is a Null OK?

To understand when a null value is acceptable, let's assess the impact of one. The most common situation where nulls can affect your analysis is when you are averaging values. Figure 27-2 shows a simple data set that has nulls, and Figure 27-3 shows the same data set but with zero values instead of the nulls.

Type	Scent	Volume
Bar	Mint	100
Bar	Lemon	50
Bar	Lemongrass	20
Bar	Lime	
Liquid	Mint	1000
Liquid	Lemon	400
Liquid	Lemongrass	
Liquid	Lime	300

Figure 27-2. Soap Scent data set with nulls

Type	Scent	Volume
Bar	Mint	100
Bar	Lemon	50
Bar	Lemongrass	20
Bar	Lime	0
Liquid	Mint	1000
Liquid	Lemon	400
Liquid	Lemongrass	0
Liquid	Lime	300

Figure 27-3. Soap Scent data set with zeros

If you use both these tables in Prep Builder in exactly the same way to calculate averages, due to the null values, they will produce two different outputs. For example, adding an Aggregate step after the Input step and calculating the average volume for each type of product generates very different results. When you apply this technique to the data set with nulls, the values returned for Bars and Liquid are 56.6 and 566.6, respectively (Figure 27-4).

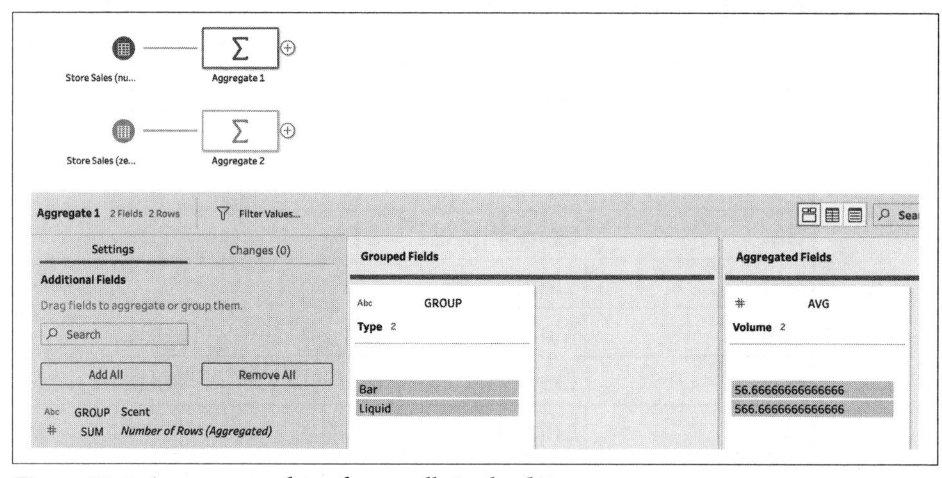

Figure 27-4. Averages resulting from nulls in the data set

However, applying the same technique to the table containing zeros gives results of 42.5 for Bars and 425 for Liquid product types (Figure 27-5).

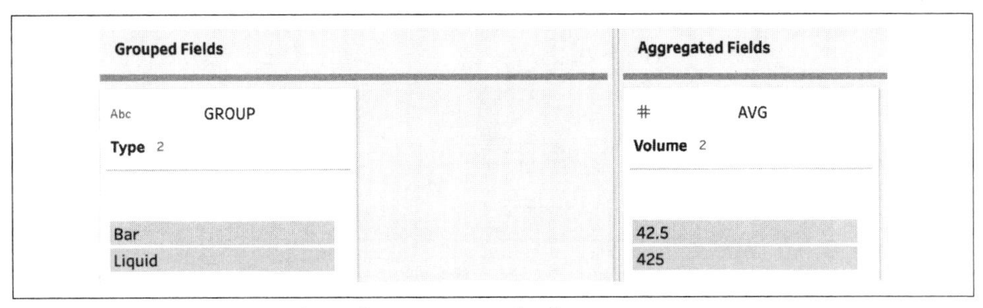

Figure 27-5. Averages resulting from zeros in the data set

So what is the math in this situation and why does the result differ? Well, to calculate the averages, Prep Builder is summing up the volumes and dividing by the number of rows. For both tables, the sum of the volumes is 170 for bars and 1,700 for liquid. However, Prep Builder does not count null values in the row count for the average. This means that volume sum is being divided by 4 for the table with no nulls (as each type of soap has four rows), but being divided only by 3 for the table with nulls (as each type of soap has only three non-null volumes).

This is an important point to consider, because if a null is correct (i.e., a record didn't occur and shouldn't have), it is the perfect entry for the record in the respective data field. However, if a null is present but it's due to a value not being recorded when it should have been, then the null should be replaced with a zero. Situations like this occur when a product is stocked in a shop but not sold. Nulls should be used where a product isn't stocked within a certain store and, therefore, never had the possibility of being sold.

How to Remove or Replace a Null

As a data prepper, you may have to decide that the null cannot remain in your data set. You have a few options in this case.

ISNULL()

If your analysis doesn't allow for a null value to be in place, you may need to filter out the row. `ISNULL()` is a fantastic function that allows you to simply assess whether a record has a null in a given field. `ISNULL()` returns a Boolean result of `True` or `False` so you can filter out just the `True` values, thereby removing the nulls (Figure 27-6).

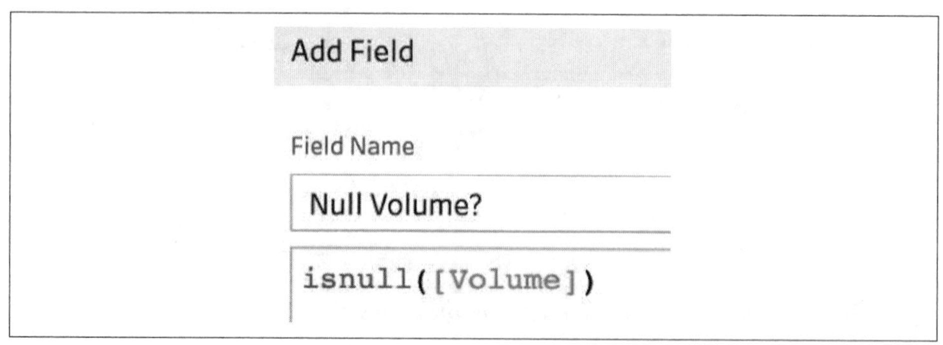

Figure 27-6. Applying the `ISNULL()` function to the Volume data field

In this example, each row with a null in the Volume data field will return `True`. Filtering out this data produces the result shown in Figure 27-7.

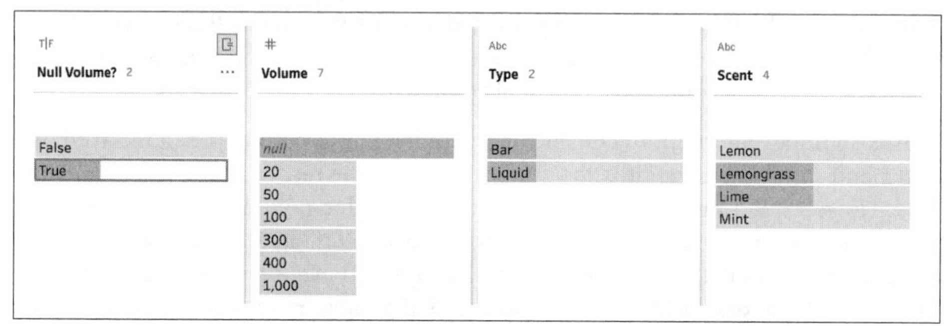

Figure 27-7. Result of `ISNULL()` calculation with null values highlighted

Removing columns full of nulls is a wise choice too, as it is very unlikely that they are useful for any reason, especially if the column doesn't have a name and returns `F`*x* (*x* is a number) when Prep Builder doesn't find a field name in the data set.

ZN()

`ZN()` (short for "zero if null") replaces a null value for the specified data field with a zero. This is useful when you want the row to be considered in aggregations, especially averages as in the previous example (Figure 27-8).

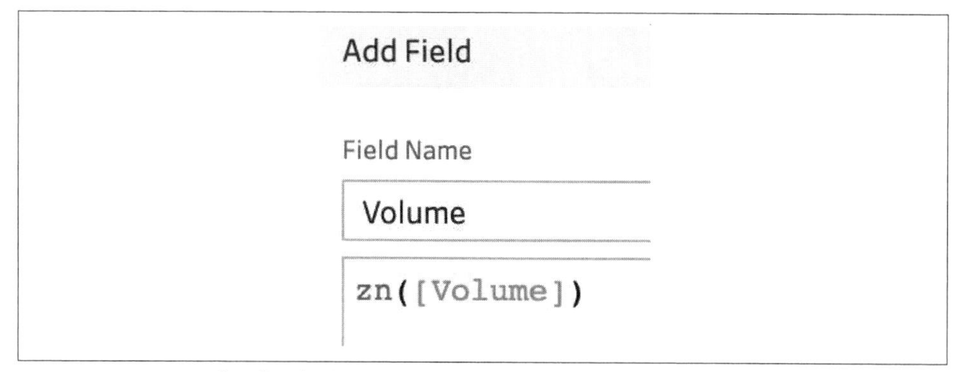

Figure 27-8. Zero if null calculator setup

In this example, Prep Builder will overwrite any null values with a zero instead. You can overwrite nulls in Prep Builder by double-clicking the null in the Profile pane and entering `0` (zero) instead. If the data field is non-numeric, you may wish to use `IFNULL()` instead, as it returns the value given if it finds a null within the data field (Figure 27-9).

Field Name

Name of Shape

ifnull([Name of Shape],'Unknown')

Figure 27-9. Using an `IFNULL()` function to return `'Unknown'` instead of a null

Merge

Not all nulls are supposed to be zeros—they might be due to data entry or system errors. This means that if you have a value within the data set that can replace those nulls, you can use a merge. The situation where this mostly occurs is during a union: if data field names do not exactly match, then the union will create separate columns for the mismatched fields.

Let's use the Preppin' Data 2020: Week 8 (*https://oreil.ly/y-VGs*) challenge to show the merge operation in action. This challenge starts with a wildcard union but results in two columns for the volumes (Figure 27-10).

Date	Type	Sales Volume	Volume
05/02/2020	BAR	462	*null*
06/02/2020	BAR	851	*null*
07/02/2020	BAR	214	*null*
08/02/2020	BAR	574	*null*
09/02/2020	BAR	246	*null*
10/02/2020	LIQUID	*null*	951
11/02/2020	LIQUID	*null*	183
12/02/2020	LIQUID	*null*	280
13/02/2020	LIQUID	*null*	734
14/02/2020	LIQUID	*null*	794

Figure 27-10. Data set where the Sales Volume and Volume fields need to be merged

There are a couple of easy ways to merge the columns in the Profile pane. First, drag one of the fields and drop it onto the other field you wish to merge it with. Second, select both columns by holding down the Ctrl key (Command on the Mac), and then click the ellipsis. Select Merge (Figure 27-11).

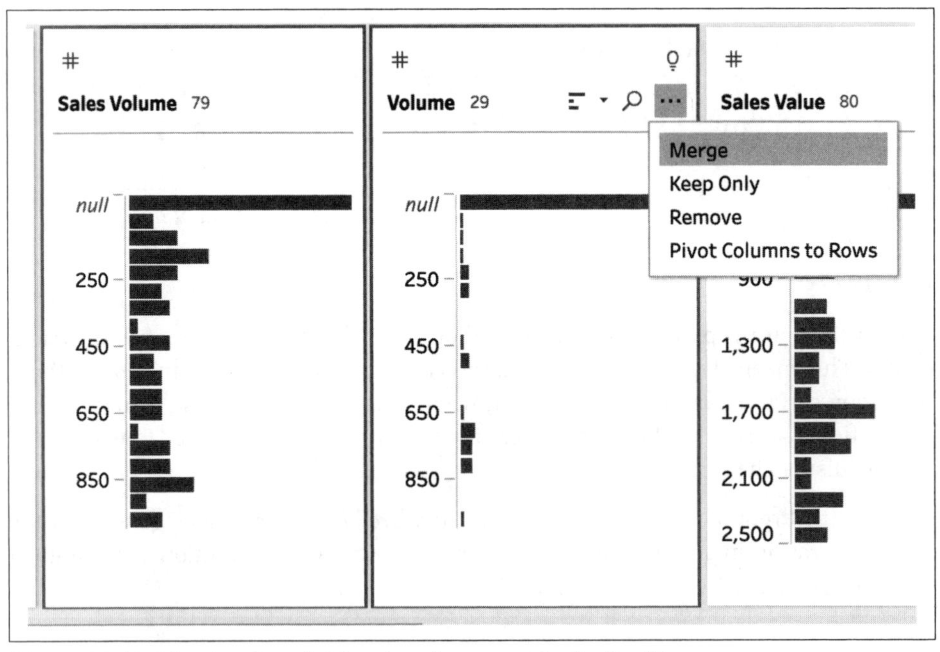

Figure 27-11. Merging data fields using the menus in the Profile pane

Now you should see those two fields merged together with the null values from one column replaced by the corresponding values from the other. The resulting data set is shown in Figure 27-12.

Type	Date	Sales Volume	Week	Sales Value
liquid	01/01/2020	862	1	1,817
liquid	02/01/2020	680	1	2,265
liquid	03/01/2020	894	1	1,757
liquid	04/01/2020	433	1	1,531
liquid	05/01/2020	650	1	1,728
bar	01/01/2020	998	1	1,979

Figure 27-12. Result of the merged data set

Nulls can be frustrating, but hopefully this chapter has helped you to understand the choices you have, when the nulls can remain, and when and how to remove or replace them.

Summary

Nulls are the absence of values in a data set. There are multiple situations where their presence is necessary for you to properly analyze the data. Where the nulls need to be removed or altered, however, there are lots of different techniques available to you. To choose the right one, you'll need to think through the impact of your actions on the subsequent analysis.

CHAPTER 28

Using Data Roles

As you saw in Chapter 26, Tableau Prep's Group and Replace functionality is pretty incredible with the rich, easy-to-use options that are built into the tool (Figure 28-1).

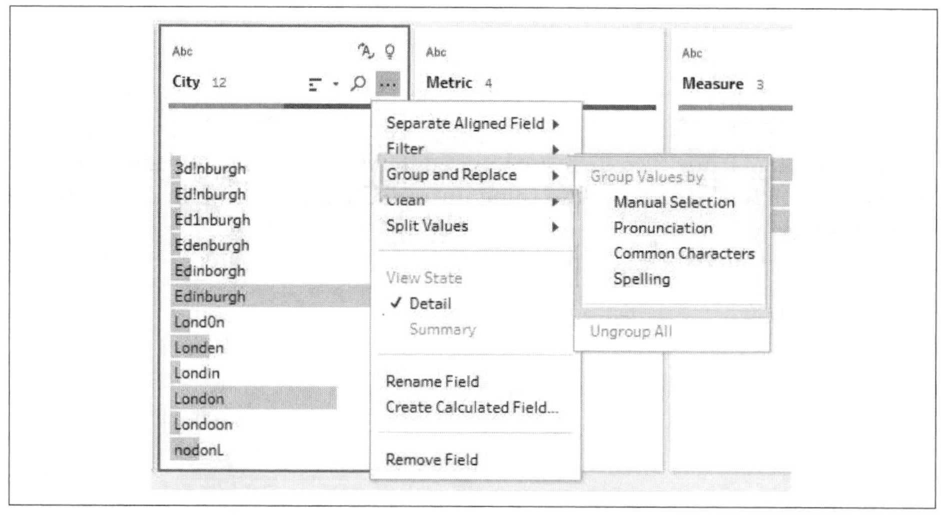

Figure 28-1. Using Group and Replace in Prep to clean string data

If you haven't had a chance to explore these, I recommend both revisiting Chapter 26 and attempting Preppin' Data 2019: Week 2 (*https://oreil.ly/rgvlA*), where you get to use these techniques on the City field. Do this before continuing to read this chapter if you want to avoid challenge spoilers!

The Tableau developers have added an extra level of validation to data cleaning with *data roles*.

For string data fields, you can now set a specific data role for Prep to test the data against. You can test geographic roles, email addresses, and URLs against Tableau's own list to see if they are valid.

How to Use Data Roles

When you click on the Data Type icon to change the data type, you will see a drop-down list of options (Figure 28-2). In this example, I'm assigning the City role.

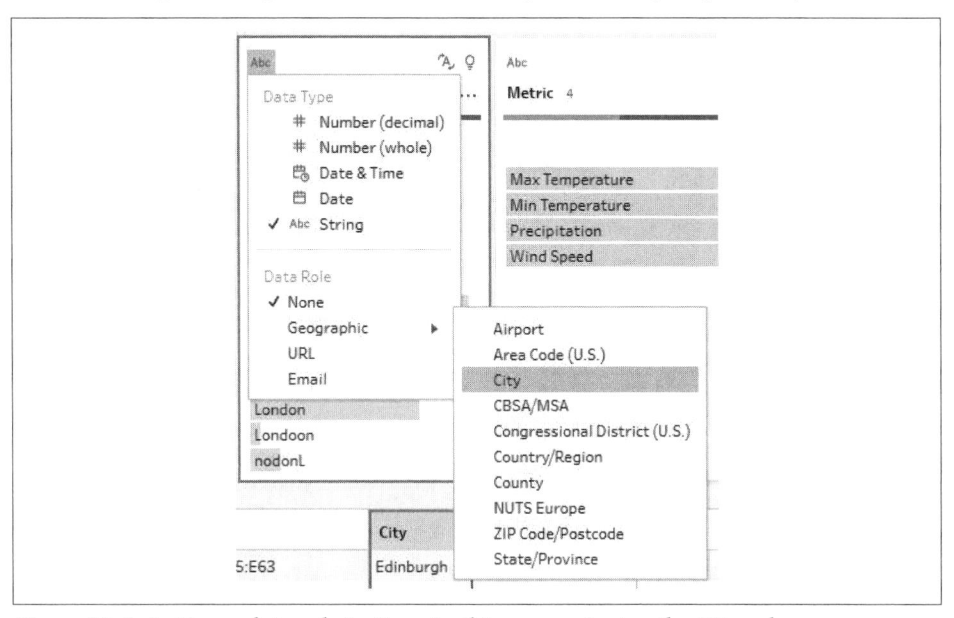

Figure 28-2. Setting a data role in Prep, in this case assigning the City role

By selecting City to compare the list of city names from the Week 2 exercise, you can see what Prep recognizes as a true city name and what it doesn't (Figure 28-3).

With this highlighting, you can easily work through all of the problems using Prep's fantastic Group and Replace functionality.

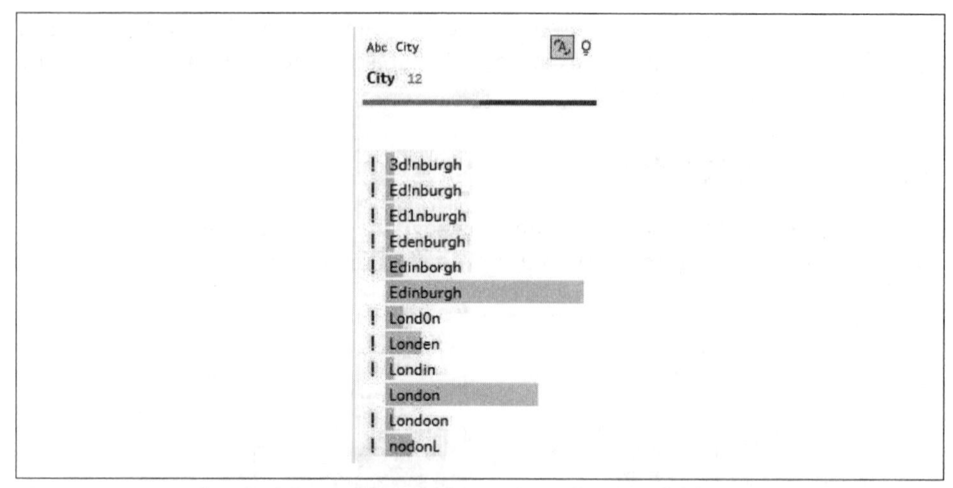

Figure 28-3. Results of applying the City data role to the City field

But what if you want to get rid of those errors? Well, by having an active data role, you actually get a few more options to play with (Figure 28-4).

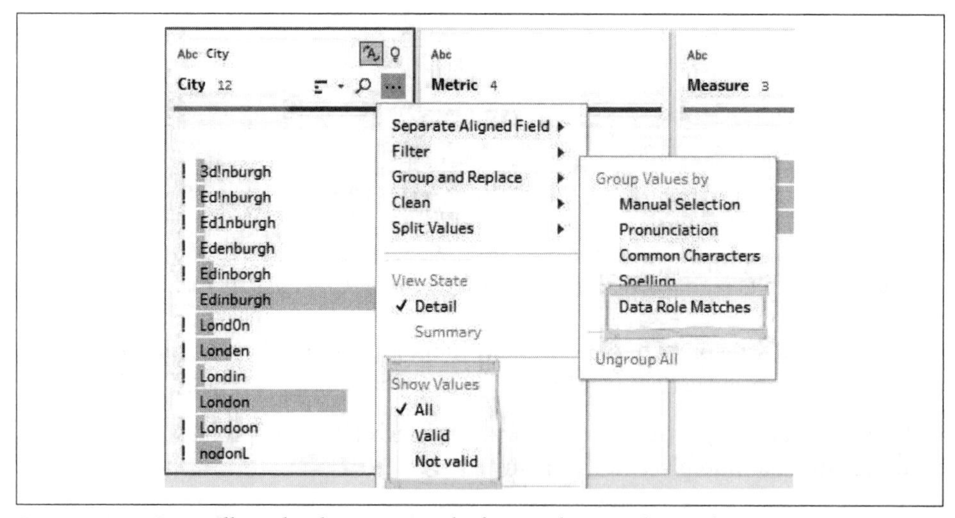

Figure 28-4. Controlling the data returned after applying a data role

You can select whether you want to see:

All
: Will show the results that meet the data role, and those that don't (which are indicated by an exclamation point)

Valid
: Will show only the results that meet the data role.

Not valid
: Will show only the results that don't meet the data role (so you can start cleaning them without being distracted by the valid results).

You can also group by the data role members, but I haven't found a use case where this has helped my data preparation yet. Maybe that will be a future Preppin' Data challenge.

Custom Data Roles

As of version 2019.3.1, not only can you use Tableau's data roles, but you can also create a custom data role from any string or integer data fields. The data role will comprise all of the values in the field. To create a custom data role, in a Clean step open the ellipsis menu in the data field you want to use (Figure 28-5).

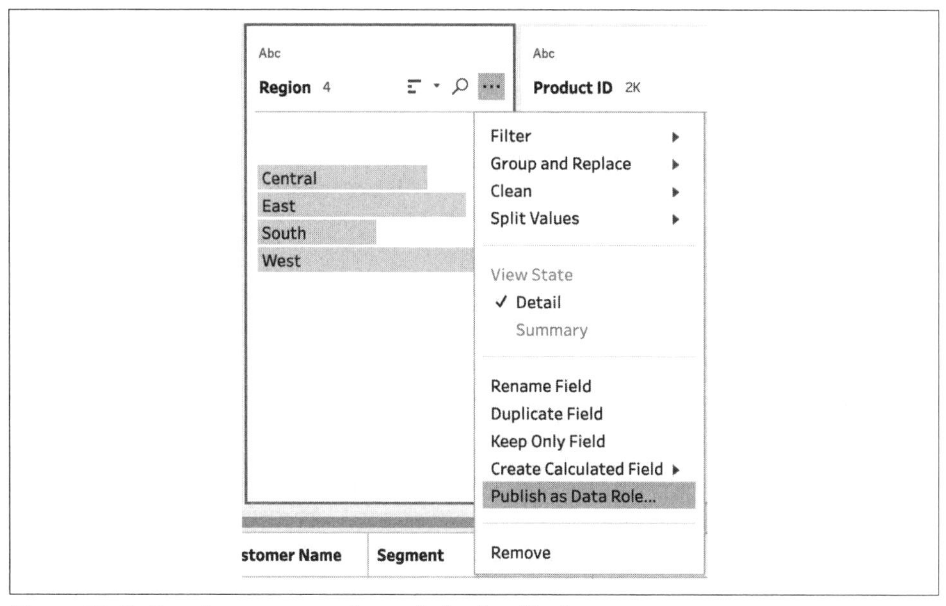

Figure 28-5. Creating a custom data role in the ellipsis menu

Custom data roles are held on Tableau Server or Tableau Online, so you'll need to publish them. Selecting Publish as Data Role will place an output for the data role into the Flow pane (Figure 28-6).

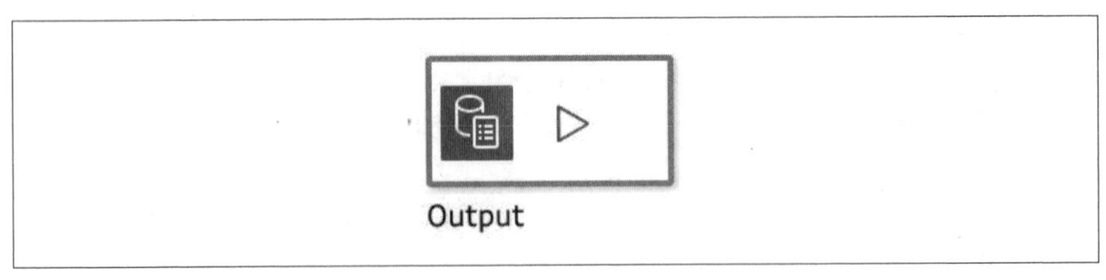

Figure 28-6. Data role output icon

The configuration of the output for the custom data role is similar to that for a data source being published to a Tableau Server or Online location. You must specify the server, site, and project to publish the data role to. Custom data roles can be published only to a single site on the Tableau Server or Tableau Online instance. The site also needs to be the same as the data set's output location if you are publishing the data source. In the configuration, you can give the custom data role a name as well as a description to help others understand it (Figure 28-7).

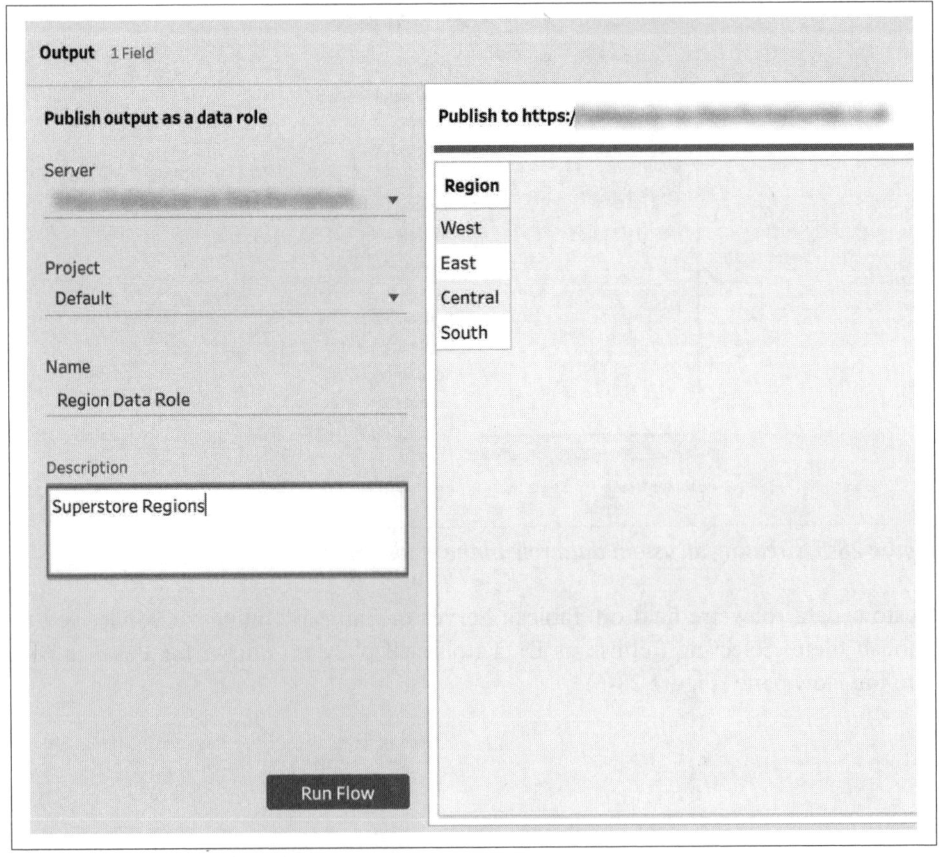

Figure 28-7. Custom data role output configuration pane

Clicking Run Flow publishes the custom data role. Your custom data role and others published can be found in the Explore section of the Tableau Server or Tableau Online instance (Figure 28-8).

Figure 28-8. The custom data role in Tableau Server

You can view the data role by clicking on its name. This reveals the values it includes and gives you the opportunity to edit the description (Figure 28-9).

Figure 28-9. Detailed view of the data role in Tableau Server

To be able to use the custom data role, you will need to be signed into the site on the Tableau Server or Online instance in which you published it. Once you are logged in, you can select the custom data role or any of the default data roles available from Prep Builder (Figure 28-10).

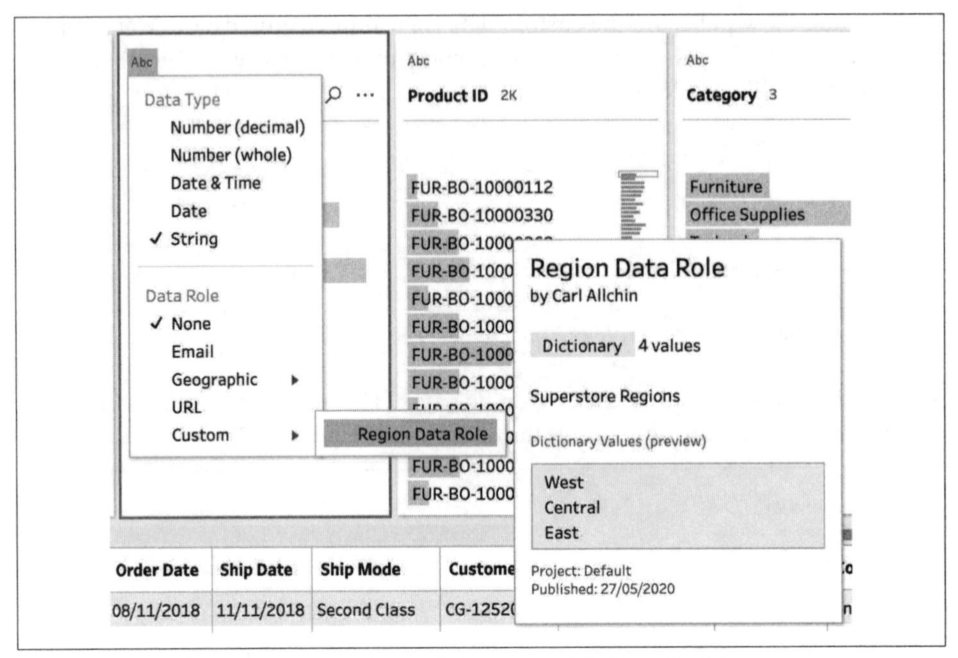

Figure 28-10. Selecting a custom data role for use within Prep Builder

If you change Central to North and apply the custom data role Region Data Role, Prep Builder highlights the values that do not match (Figure 28-11).

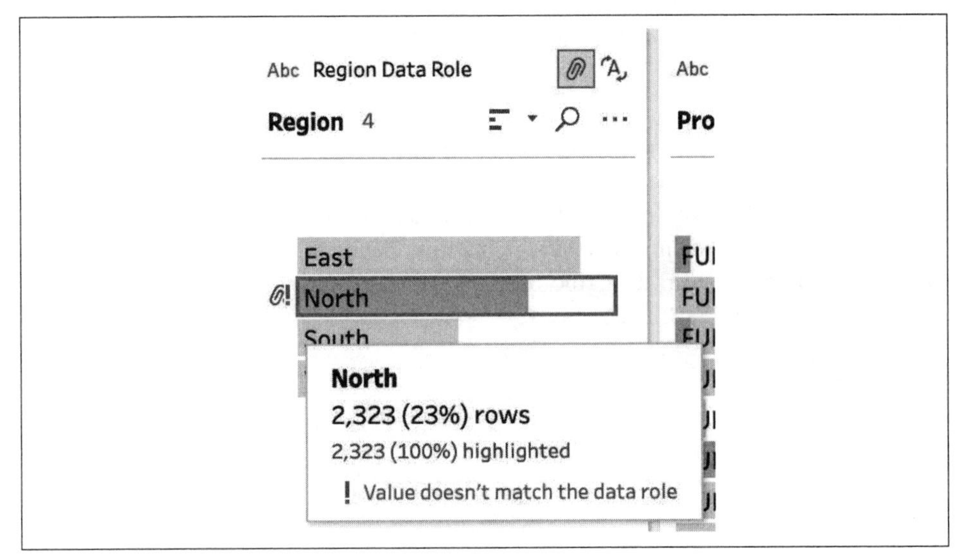

Figure 28-11. Values not matching the custom data role are highlighted

Custom Data Roles have the same options as the default data roles: Valid, Not valid, or All values (Figure 28-12).

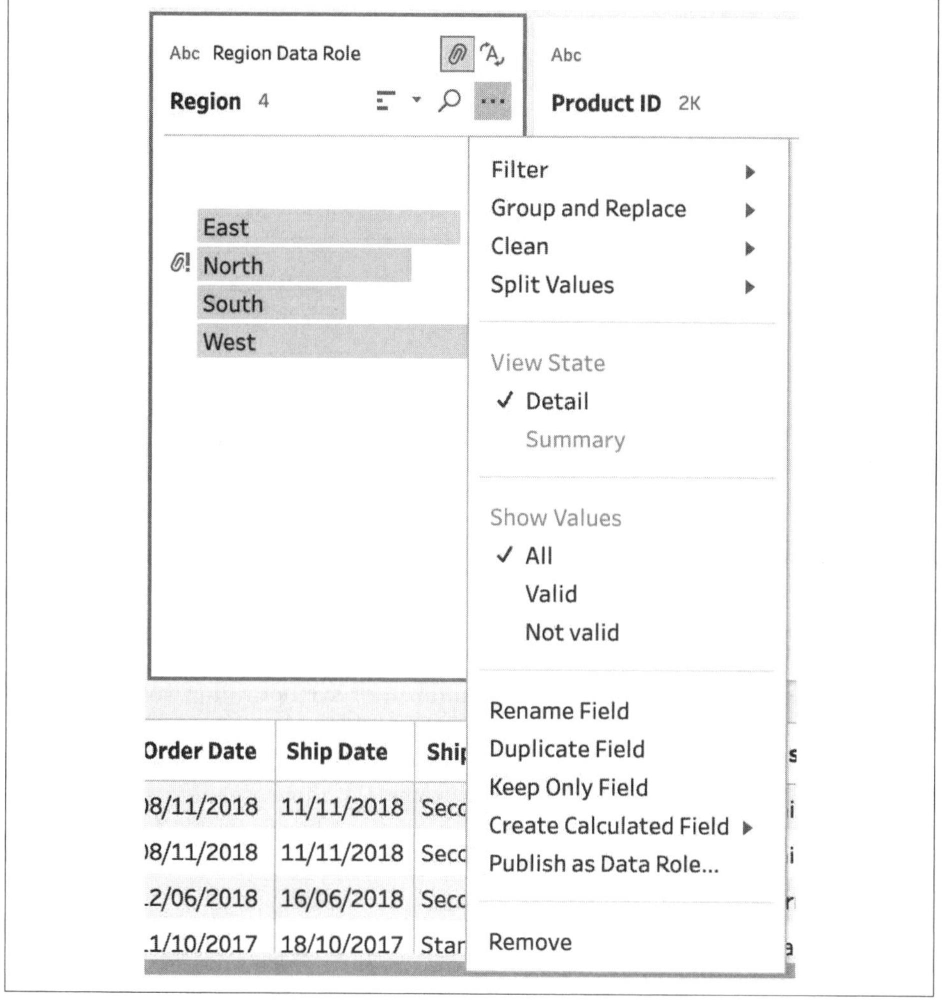

Figure 28-12. Options for returning data roles' matching values

Summary

Data roles in Prep Builder are a similar concept to Desktop's geographical roles but go further by giving you insight into whether data needs to be cleaned. By selecting the "Not valid" option, you can choose to filter or clean those values. Selecting "Valid" can give you the reassurance that the remaining data values meet set criteria. Custom data roles can allow an organization to match the values that are important to it, or those that are otherwise difficult to validate.

CHAPTER 29

Dealing with Unwanted Characters

When cleaning data, finding unexpected characters in your data fields can cause significant issues. Those issues can occur at multiple points during your data preparation: loading, using, and outputting the data. Therefore, this chapter is focused on building your understanding of what unwanted characters are, the problems they introduce, and how to remove them.

What Is an Unwanted Character?

An unwanted character is simply a letter, number, or symbol within your data field that you do not need or that introduces potential problems in your data output. Data software is often very precise about what it is processing, and rightly so; otherwise, it could easily produce erroneous output. For example, if there are different data types within a single data field, it affects whether a field can be aggregated logically. Let's look at the three main types of data fields that can be affected by unwanted characters when they're loaded into Prep Builder:

Numeric fields
: If a non-numeric character is loaded into a numeric field, the field will be imported as a string and thus can no longer be used in aggregations. For example, what should 10 + 1c3 equal? As you can see, that calculation isn't possible, and that is why a numeric field must contain only numeric values.

Dates
: If a non-numeric character is found in a field expecting only date values, the date value with the unwanted character will appear as a null because the date will be in an invalid format—for example, 30/0/4/2023.

Strings

Strings are very flexible data types and therefore generally won't cause an error if they're imported with unwanted characters. The only exception is when a character is used outside those permitted by the data source or Tableau Prep, such as certain characters from non-English alphabets.

The most common unwanted character in data preparation is the humble space. The space between characters is easy to spot, but that's not the case with leading or trailing spaces (i.e., a space at the start or end of a string). These spaces still count as characters in string functions like `LEFT()`, `RIGHT()`, `MID()`, and `SPLIT()`, so they can cause issues in most of the common string data preparation steps.

Issues Caused by Unwanted Characters

The challenge of unwanted characters isn't very different from many others in data preparation. However, where they do differ is in the fact that they present potentially hard-to-find, *individual* values to clean rather than entire data fields. Identifying those individual values with the unwanted characters can be a challenge, especially in a string field, which doesn't simply return a null value on input.

The main issue with unwanted characters is that they can increase the complexity of your data prep by preventing you from being able to apply a single rule to all values in a data field. For example, a numeric field with a hidden non-numeric character lurking in just one value cannot be simply aggregated, preventing many of the common steps from working normally. Fortunately, instead of trying to find the needle in the haystack, you can use some of Prep Builder's built-in functionality to assist with this task.

Figure 29-1 includes at least one unwanted character in each data field to demonstrate the potential effects. In the Date field, the letter `c` has replaced the day. In the Store field, an exclamation mark has replaced the `l` in `Clapham`. The Type field has two issues: an `8` instead of a `B` and a leading space.

Date	Store	Type	Sales
01/04/2020	Clapham	8ar	100
02/04/2020	Clapham	Bar	d50
0c/04/2020	Clapham	Bar	200
04/04/2020	C!apham	Bar	150

Figure 29-1. A data set containing unwanted characters

The data set has been saved as a comma-separated value (CSV) file. Tableau Prep Builder loads a CSV file's data fields as strings by default. This means you will always need to set the data types in Tableau Prep before using functions specific to types other than strings (Figure 29-2).

Sheet1 Fields selected: 4 of 4 Filter Values...

Select the fields to include in your flow, apply a filter o the flow pane.

☑	Type	Field Name	Original
☑	Abc	Date	Date
☑	Abc	Store	Store
☑	Abc	Type	Type
☑	Abc	Sales	Sales

Figure 29-2. The Input metadata grid of a CSV file showing only string data types

Changing the different fields into the data type required for analysis highlights the unwanted characters shown earlier in Figure 29-1. Date values that do not conform to the expected date format are converted to nulls (Figure 29-3).

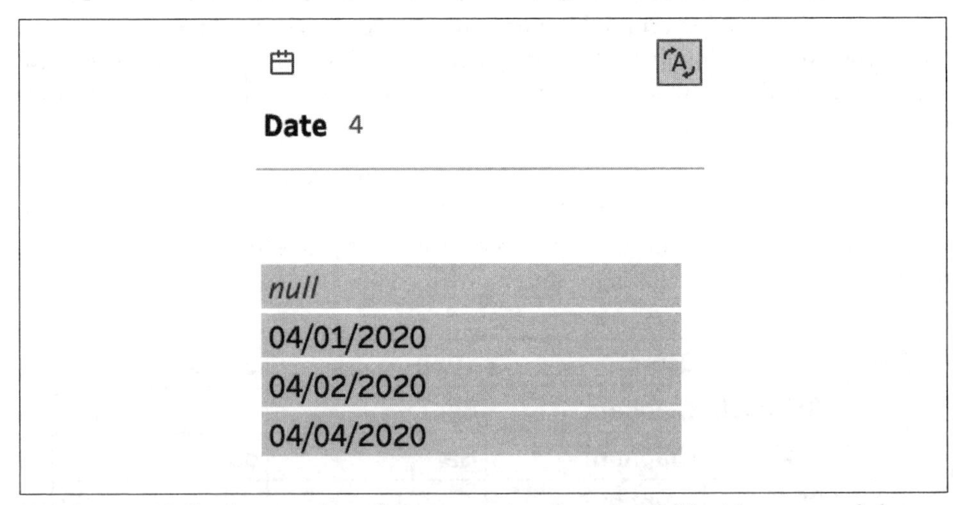

Figure 29-3. Null values resulting from converting the string field with unwanted characters to a date

The sales figures that are not solely numeric are also converted to nulls (Figure 29-4).

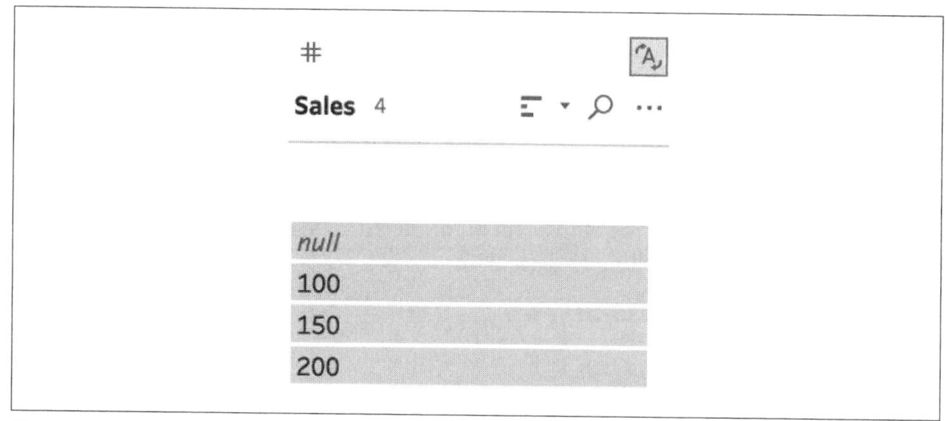

Figure 29-4. Sales records with non-numeric values are also converted to nulls

The string values with the exclamation point and number 8 remain as strings, but because they clearly differ they are split apart in the Profile pane (Figure 29-5). The `Bar` value with a leading space has been cleaned up naturally by Prep. It can do this for some data sources, but not all.

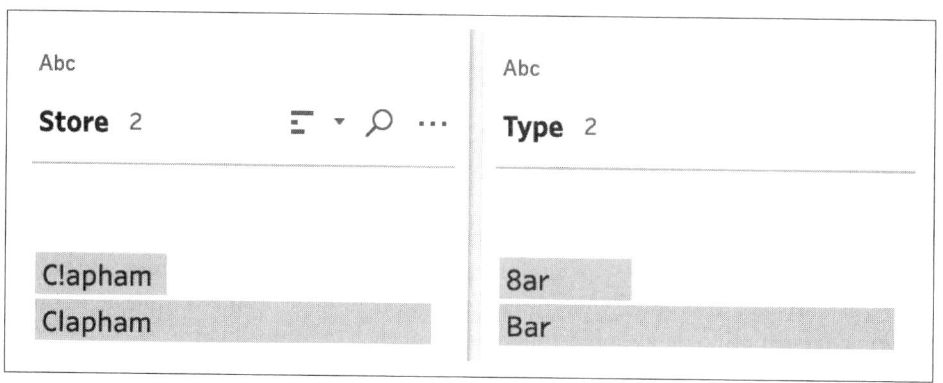

Figure 29-5. Unwanted characters shown in the Profile pane

Therefore, to make the data usable, you have to remove the unwanted characters before completing the data type conversion.

Removing Unwanted Characters

How you remove unwanted characters depends on the character and the data type the field should be.

Strings with Mistyped Characters

With free text entry, words and names can easily be mistyped by the person who entered the information. Often, the values with typos are close enough to other similar values that Prep Builder's Group and Replace functionality can correct the unwanted characters.

To use this technique, open the ellipsis menu of the string data field, select Group and Replace, and then select Spelling (Figure 29-6).

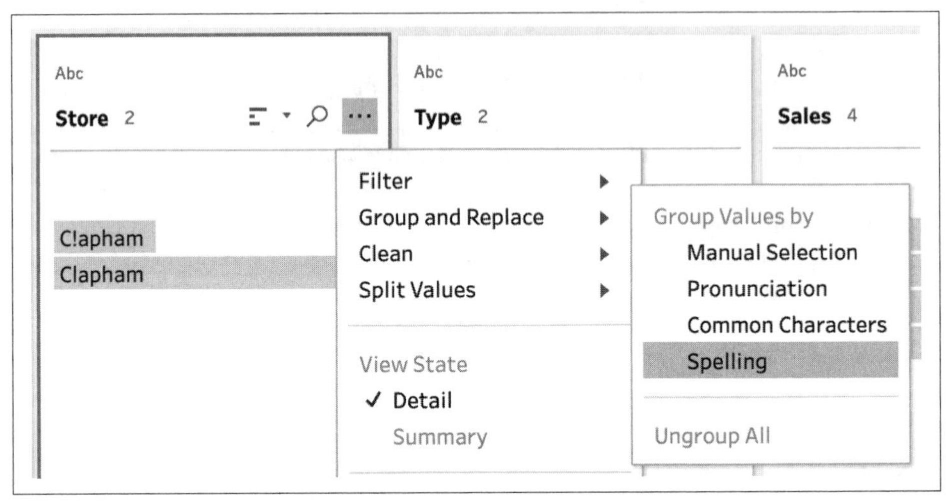

Figure 29-6. Group and Replace menu

This results in just a single value for Clapham, and the correct spelling is taken from the most common result. You can apply the same technique using numbers instead of letters, as in the 8ar example. After you apply the Spelling grouping technique to both columns, the resulting data set is a single name in each column (Figure 29-7).

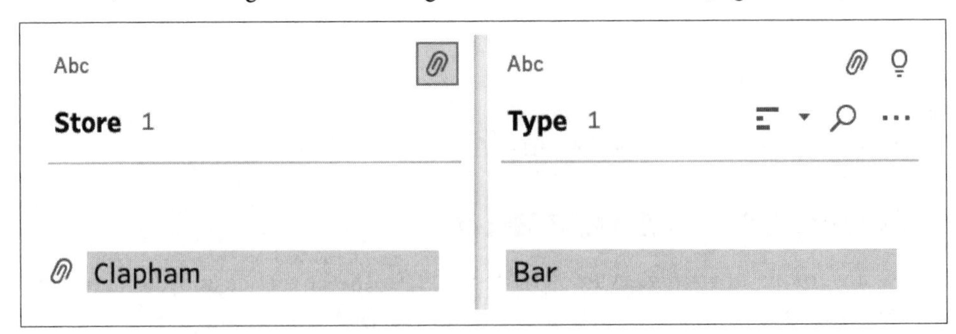

Figure 29-7. Result of Spelling grouping

Grouping techniques are covered in more detail in Chapter 26.

Numbers with Unwanted Characters

When you convert the Sales field to a numeric data type, the `d50` value results in a null due to the letter `d`. To resolve this, first you use Prep Builder's cleaning functionality in the ellipsis menu to remove letters from the data field (Figure 29-8).

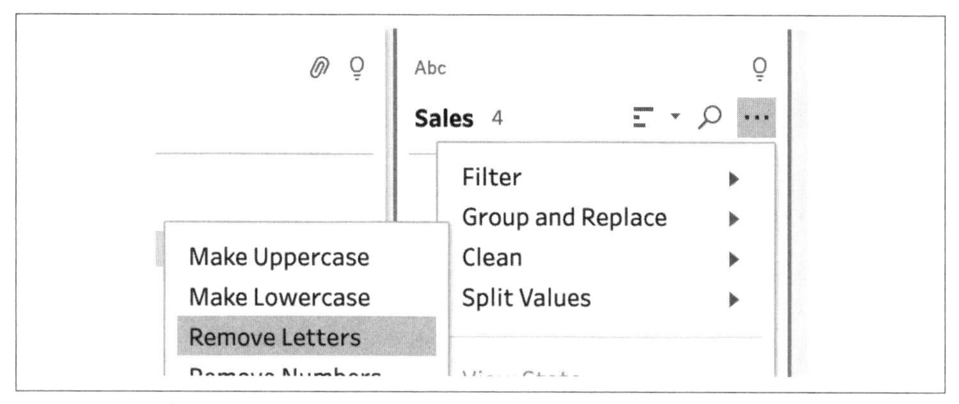

Figure 29-8. Clean menu Remove Letters option

Next, you convert the data field to a numeric data type (Figure 29-9).

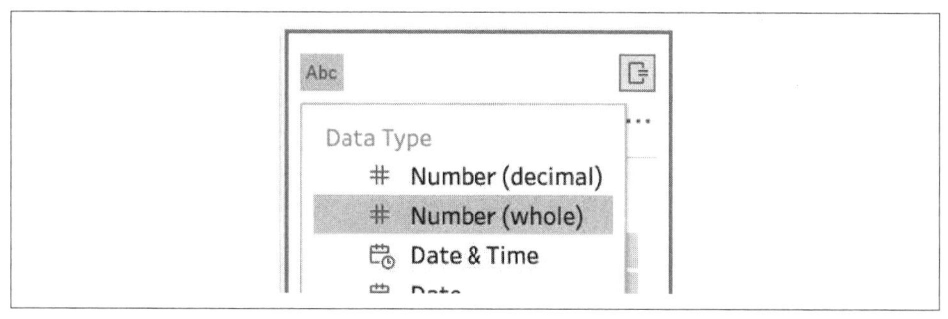

Figure 29-9. Converting a string field to an integer field

This results in a set of values that you want for the analysis (Figure 29-10).

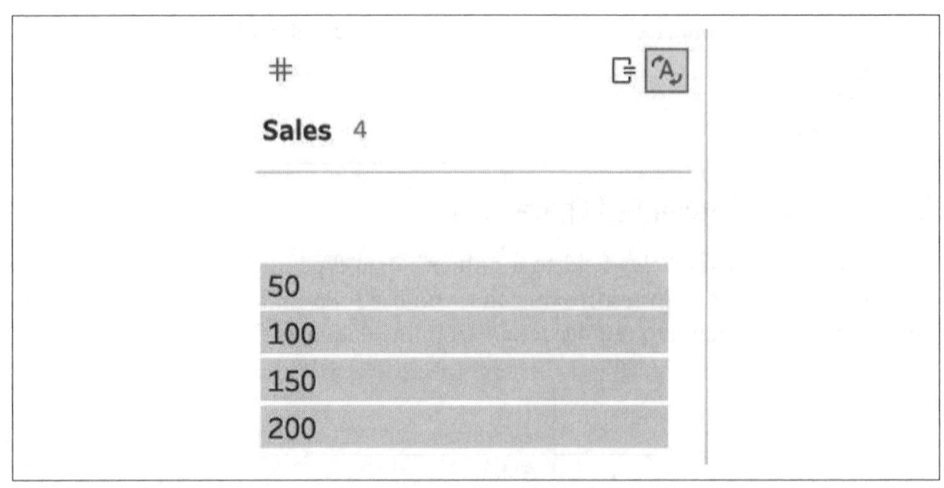

Figure 29-10. Result of type conversion after removing unwanted characters

Dates with Mistyped Characters

If there are additional unwanted characters, in many cases you can simply repeat the cleaning technique described in the preceding section to remove them. If the character is mistyped, however, it might require some human logic to determine what the actual value should be. You can search manually, but in large data sets this can take a significant amount of time. Instead, regular expressions (regexes), covered in Chapter 31, can help you find the unwanted characters (a letter in this case). The regex calculation shown in Figure 29-11 returns `True` when it finds the letter *c* in the Date field.

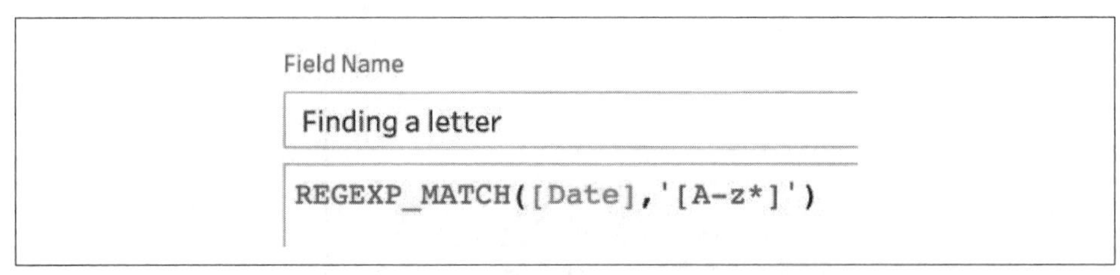

Figure 29-11. Regex calculation to identify values containing a letter character

Prep Builder clearly indicates when a letter appears somewhere within the string field. The Profile pane highlights the value(s) in question (Figure 29-12).

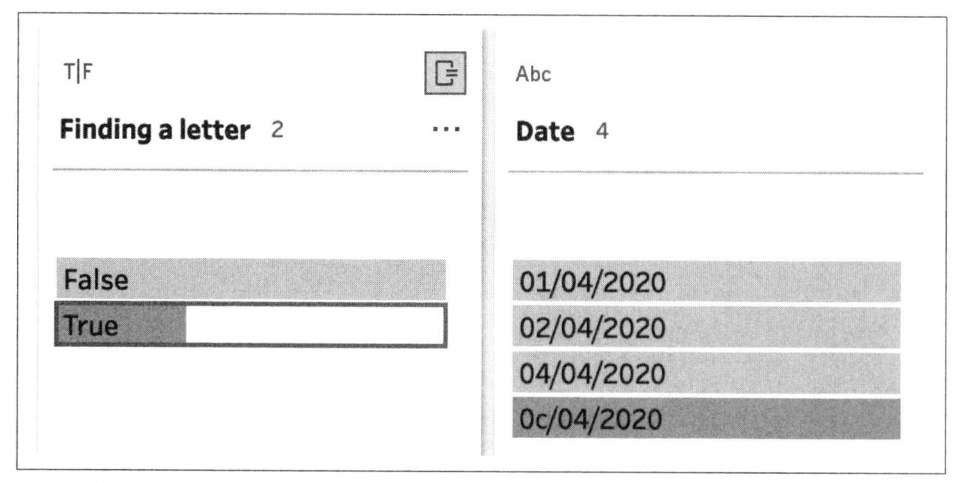

Figure 29-12. Letter characters highlighted in the Profile pane

You can then manually assess how to resolve the issue. In this example data set, we expect to have a record per date, so clearly we need to rename 0c to 03. To do this, simply double-click the record in the Profile pane and update it with the correct value (Figure 29-13).

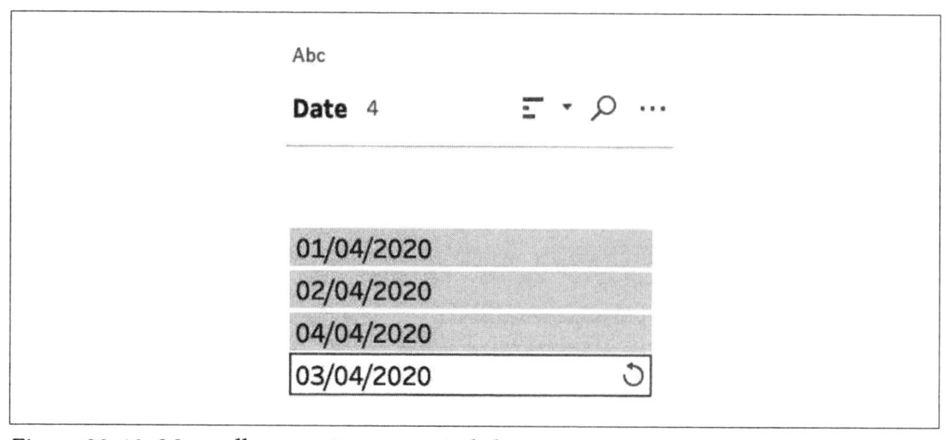

Figure 29-13. Manually removing unwanted characters

You can then change the data type to date without converting any values to null.

Summary

Unwanted characters can be quite challenging, but you can largely handle them with Prep Builder's built-in functionality. For some typos, however, you'll need to create more custom solutions to help identify and then manually correct the issue.

CHAPTER 30

Deduplicating

Understanding the level of granularity in your data is key to preparing it well for analysis. When investigating granularity, however, you might find some unclear answers. The cause of this lack of clarity is often duplications. This chapter will go through how to recognize duplicates in your data set and what you can do about them.

How to Identify Duplicates

Unless you are intentionally looking for duplicates, you are relying on someone "knowing the numbers" to inform you that something's wrong. Therefore, it is important that you actively try to prevent duplicates in your data set and know how to remove them if required. Removing duplicates makes aggregations easier because you can simply sum records to find totals, which in turn makes the resulting data set easier to analyze.

Let's look at an example where a system captures when orders come in to Chin & Beard Suds Co. When analyzing orders, we'd expect to have a data set where each order has its own row. When loading a data set into Prep Builder, you can easily determine:

- How many rows there are for each order (here represented as Case ID)
- Whether there is an even distribution of rows as shown in the Profile pane

By clicking on a single Case ID, you can see in the Data pane that there are multiple rows per ID, and different IDs have a different number of rows in the data set (Figure 30-1).

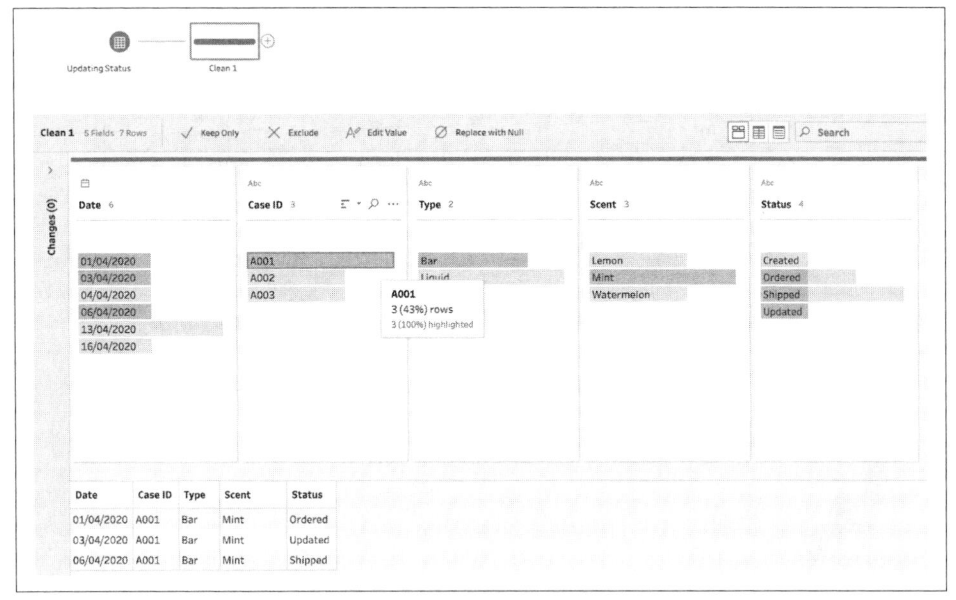

Figure 30-1. Using Case ID to identify the number of rows

Causes of Duplicates

Now that you know there is duplicate data in the example data set, let's look at some potential causes.

System Loads

As operational systems track order statuses, customer complaints, and inventory, they will capture operational state at different points. For each record, if the previous state is still stored and not overwritten by the new state, a duplicate can result.

In the example shown in Figure 30-2, the Status field updates over time as the company processes orders. Only a *distinct count* of Case ID would return the correct number of orders, as distinct counts count a value only once and ignore further instances of it.

Date	Case ID	Type	Scent	Status
01/04/2020	A001	Bar	Mint	Ordered
03/04/2020	A001	Bar	Mint	Updated
04/04/2020	A002	Liquid	Watermelon	Ordered
06/04/2020	A001	Bar	Mint	Shipped
13/04/2020	A003	Liquid	Lemon	Created
13/04/2020	A003	Liquid	Lemon	Shipped
16/04/2020	A002	Liquid	Watermelon	Shipped

Figure 30-2. Duplicates resulting from creating an additional row per status

If you require a distinct count, you can create one by using an Aggregate step.

The Aggregate step is covered in Chapter 15.

Row per Measure

Duplicates can also be caused where each row contains an individual measure. In the data set shown in Figure 30-3, you would have to use distinct counts on the Store and Type fields in order to count the number of stores selling each type of product. Distinct counts require significant computing resources, however, so this isn't a very efficient way to work when analyzing data.

Store	Type	Measure	Value
Clapham	Bar	Volume	302
Clapham	Bar	Value	10292
Clapham	Bar	Profit	1038
Lewisham	Bar	Volume	220
Lewisham	Bar	Value	9277
Lewisham	Bar	Profit	1405
Clapham	Liquid	Volume	472
Clapham	Liquid	Value	13021
Clapham	Liquid	Profit	2046
Lewisham	Liquid	Volume	204
Lewisham	Liquid	Value	10238
Lewisham	Liquid	Profit	297

Figure 30-3. Duplicates resulting from the data structure

Joins

Where joins do not have a perfect like-for-like join condition, the resulting data set can include multiple rows for every original single row. If you wanted to analyze the following data sets in terms of Sales (Figure 30-4) and Profits (Figure 30-5) at the same time, for example, there is no perfect way to join them that would result in only four rows of data.

Date	Type	Sales
01/04/2020	Bar	20493
01/04/2020	Liquid	12303
02/04/2020	Bar	24202
02/04/2020	Liquid	14039

Figure 30-4. Sales data set

Date	Type	Store	Profit
01/04/2020	Bar	Clapham	302
01/04/2020	Bar	Lewisham	283
01/04/2020	Bar	Wimbledon	198
01/04/2020	Liquid	Clapham	78
01/04/2020	Liquid	Lewisham	83
01/04/2020	Liquid	Wimbledon	40
02/04/2020	Bar	Clapham	305
02/04/2020	Bar	Lewisham	307
02/04/2020	Bar	Wimbledon	204
02/04/2020	Liquid	Clapham	89
02/04/2020	Liquid	Lewisham	94
02/04/2020	Liquid	Wimbledon	58

Figure 30-5. Profits data set

If Date and Type were matched as join conditions, the resulting table would be 12 rows long. This means the `Sales` value would be repeated multiple times. If the resulting Sales column were to be simply summed up, the profit would be overstated threefold.

A great tip from Jonathan Drummey is to use Aggregate steps on branches separate from the main flow of data. You can use these Aggregate steps to count the number of rows when adding all the key data fields into the Group By section and then compare this value to the number of rows shown at the top-left corner of each data set in the Profile pane. If those numbers differ, you'll know you have duplicates that need to be addressed.

How to Handle Duplicates

Once you know there are duplicates in your data set and understand the likely cause, you can use Prep Builder to unpick the duplicates, leaving just the records you want for your analysis. There are two principal approaches to this, and the one you choose will likely depend on the cause of the duplication.

Aggregating: Technique 1

Returning to the data set from Figure 30-2, say you want the latest status per Case ID. To remove the duplication, you need to retain the latest date per Case ID. To do this, use an Aggregate step, grouping by Case ID, and return just the maximum date (Figure 30-6).

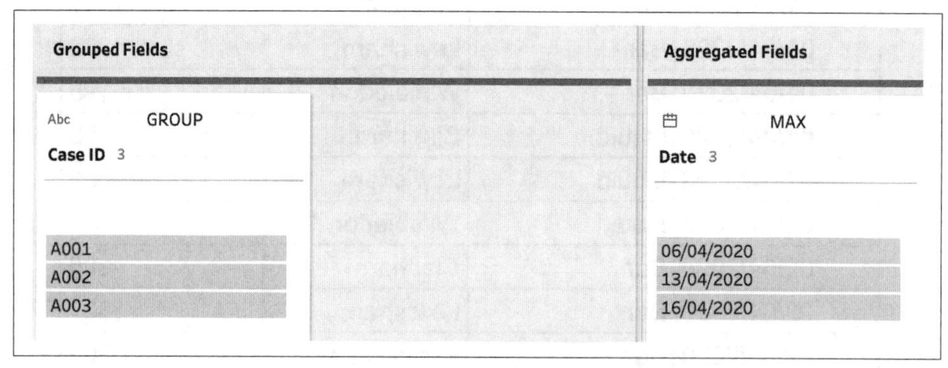

Figure 30-6. Using an Aggregate step to remove duplication

This step results in just three rows of data, one per case. The challenge here is that an Aggregate step in Prep Builder removes all columns that are not used either as a Group By or Aggregated field. To add the status to the latest Case ID row (and any other columns you wish to add back in), add a Join step and set the join conditions on Case ID = Case ID and Date = Date for an inner join (Figure 30-7).

If you haven't used joins before, read Chapter 16 before working through these examples.

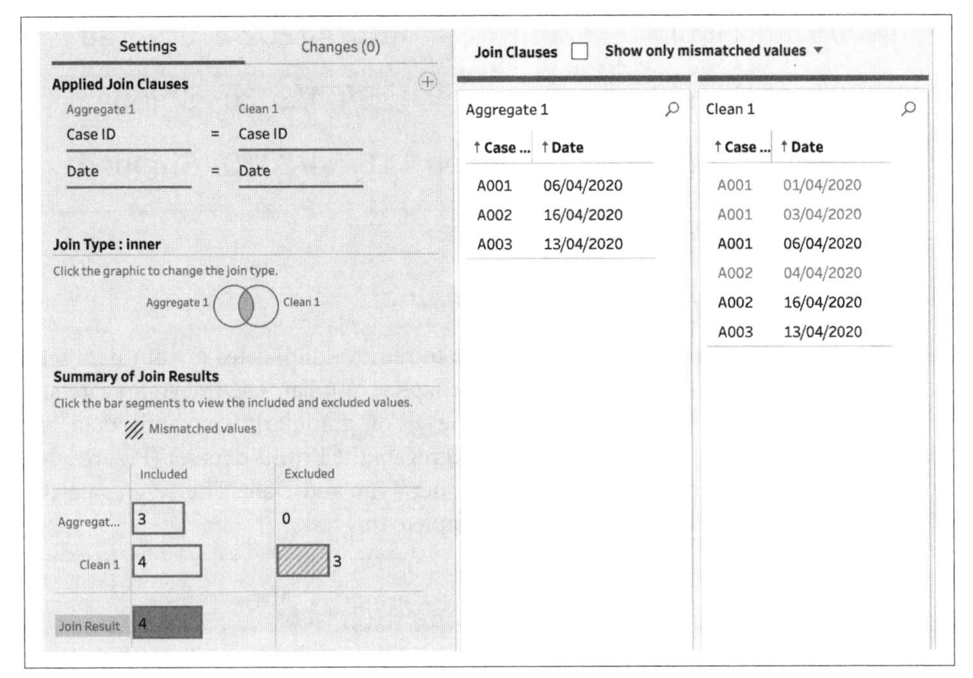

Figure 30-7. Join setup showing mismatched values

Figure 30-8 shows the resulting flow.

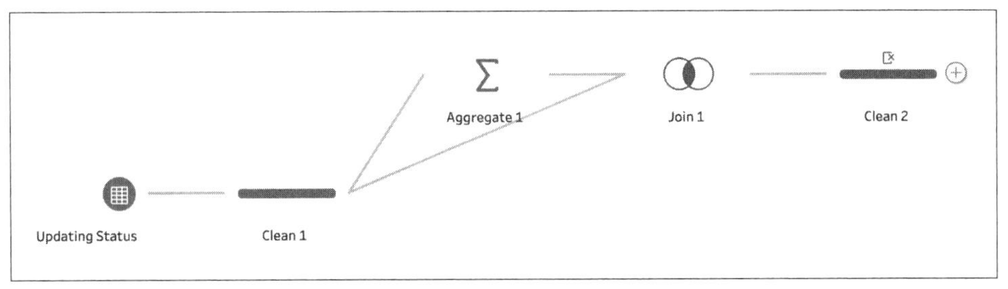

Figure 30-8. Flow using a join to remove duplication

This results in a *deduplicated* data set: you're left with all the data fields you need for your analysis after you remove the duplicated Case ID and Date field from one side of your join (Figure 30-9).

Case ID	Type	Scent	Date	Status
A001	Bar	Mint	06/04/2020	Shipped
A003	Liquid	Lemon	13/04/2020	Created
A003	Liquid	Lemon	13/04/2020	Shipped
A002	Liquid	Watermelon	16/04/2020	Shipped

Figure 30-9. Resulting data set after deduplication

Aggregating: Technique 2

The other way you can use an Aggregate step to remove duplicates in your data set is to help in the "Joins" section earlier in the chapter. You can use the Aggregate step before the Join step to create the same level of granularity in both data sets (Figure 30-4 and Figure 30-5). You need to aggregate the Profits data set (Figure 30-5) to a single row for each combination of Product Type and Date. The Aggregate step can be set up as shown in Figure 30-10 to complete this task.

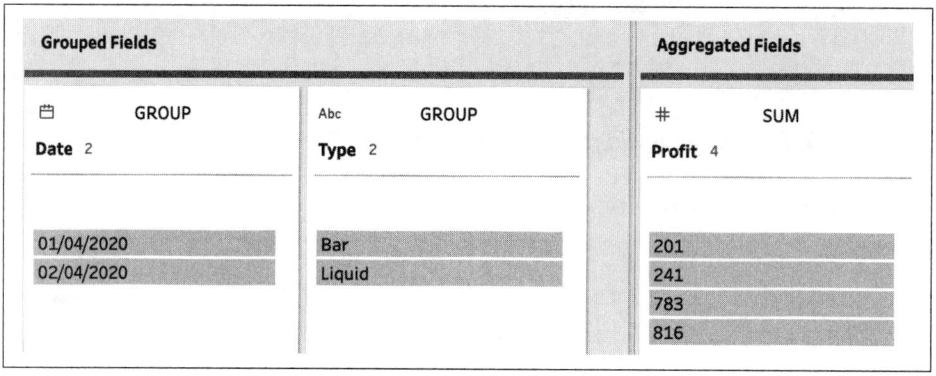

Grouped Fields

GROUP

Date 2

01/04/2020
02/04/2020

Abc GROUP

Type 2

Bar
Liquid

Aggregated Fields

SUM

Profit 4

201
241
783
816

Figure 30-10. Setup of Aggregate step to remove duplicates

Because the purpose of this join is to add the Profits column to the Sales table, you don't need to add the Store Name back in using a join as in the first Aggregate step technique. Figure 30-11 shows how the Join step is set up.

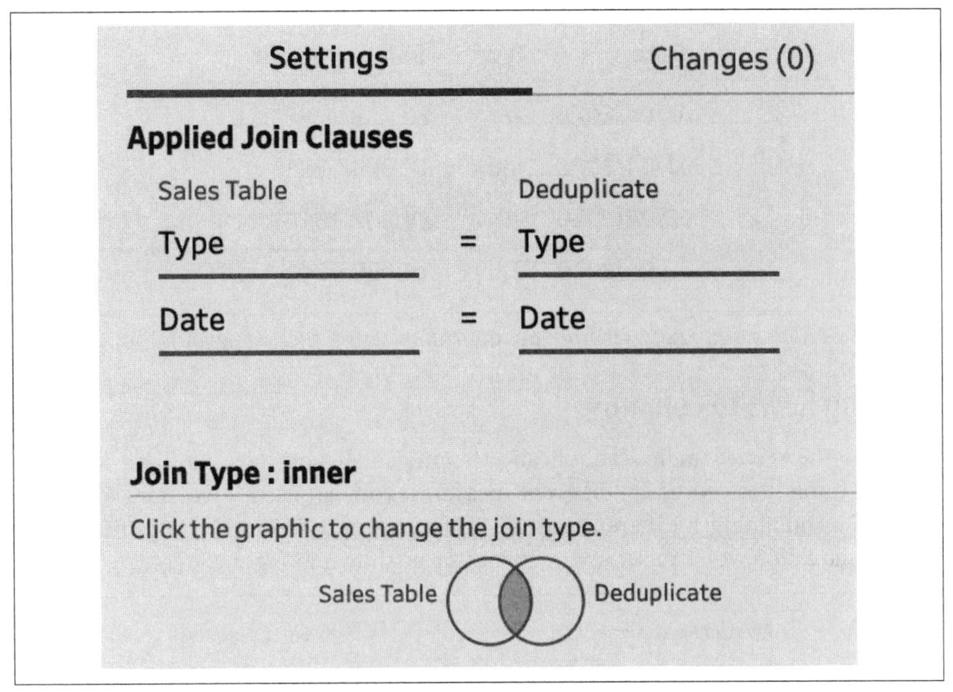

Figure 30-11. Join step setup to remove duplicates

After the duplicated fields (Type and Date) are removed because they are in both data sets going into the join, the overall flow for the deduplication looks like Figure 30-12.

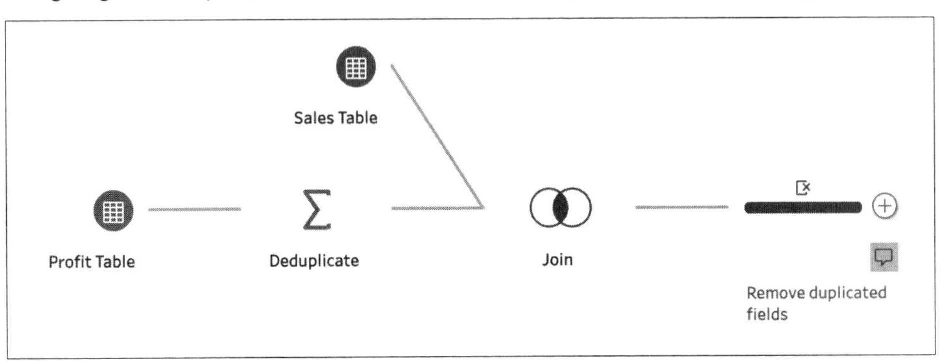

Figure 30-12. Flow using the join technique to remove duplicates

The resulting data set looks like Figure 30-13.

Date	Type	Sales	Profit
01/04/2020	Bar	20,493	783
01/04/2020	Liquid	12,303	201
02/04/2020	Bar	24,202	816
02/04/2020	Liquid	14,039	241

Figure 30-13. Resulting data set from deduplication techniques using joins

Pivoting Rows to Columns

Now for the second main deduplication approach. If you have multiple rows that contain duplicates caused by different measures, you can use a Pivot step to convert each row containing a different measure into a column instead. Using the data set from Figure 30-3, you can set up the Pivot step as shown in Figure 30-14.

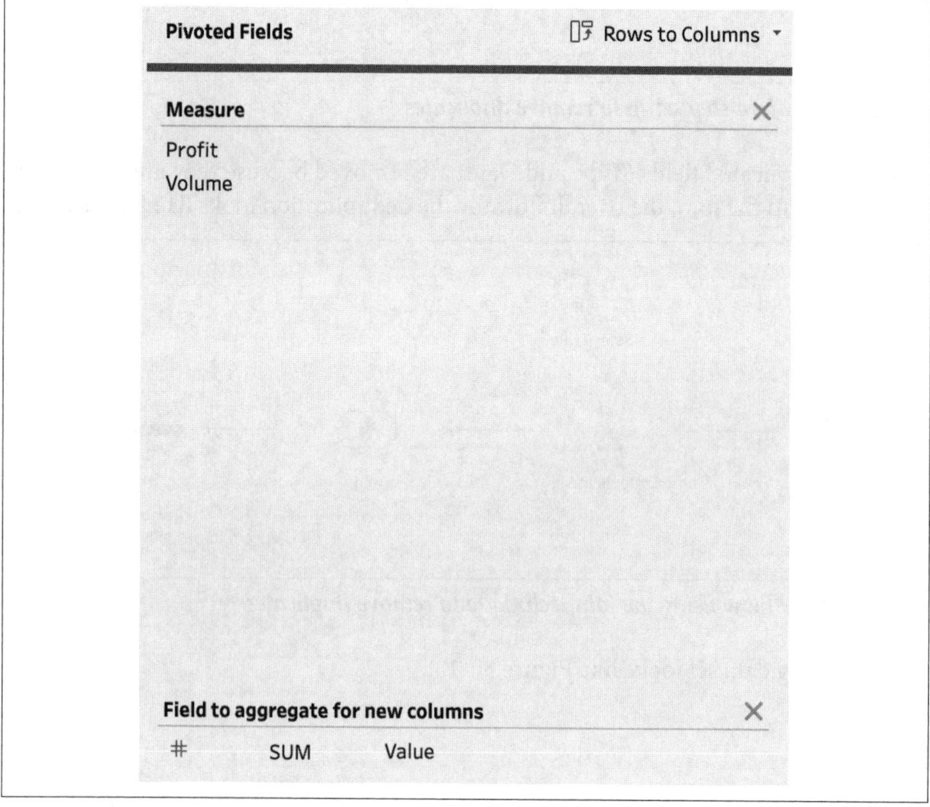

Figure 30-14. Rows to Columns Pivot step setup

There is no need to add any categorical/dimensional data back into the data set using a join step, as anything not used in the Rows to Columns pivot remains in the data set. The flow for this deduplication method is simple (Figure 30-15).

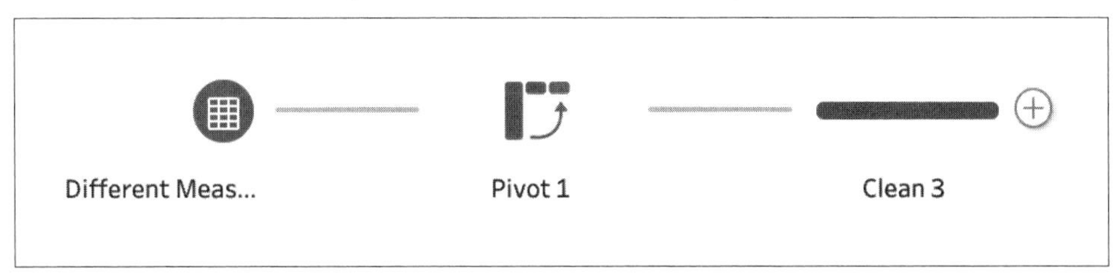

Figure 30-15. Flow using the Pivot step in Figure 30-14

The resulting data set looks like Figure 30-16.

Store	Type	Volume	Profit
Lewisham	Liquid	204	297
Clapham	Bar	302	1,038
Lewisham	Bar	220	1,405
Clapham	Liquid	472	2,046

Figure 30-16. Resulting data set from Rows to Columns pivot

Summary

Removing duplicates from data that you wish to analyze can save you from some challenging calculations and potentially some mistakes as well. The cause of the duplication will determine how you should approach the removal of the duplicate data, or whether it needs to be removed at all. As always, planning your preparation and being clear on what you require for the output is the key to choosing the correct steps.

CHAPTER 31
Using Regular Expressions

Data can be complicated; it's not all just adding values together or counting things. For example, string data fields can have a lot of complexity, but using them is often essential in forming your data sets and analysis. Regular expressions (or *regexes*) are a set of commands that make parsing difficult strings much more achievable. In this chapter, we will cover what regexes are exactly, how they can be used in Prep Builder, and how to use some common regex functions.

What Are Regular Expressions?

Regular expressions are calculations that search for a pattern of characters (e.g., the value `population`) in a string of characters (e.g., `"The population of the United Kingdom is 66,650,000 people"`). Regexes are much more flexible in the patterns they can identify compared to the other string functions we've seen like `LEFT()`, `MID()`, and `RIGHT()`.

One advantage of learning regexes is that they share a common set of commands across most data tools and therefore are highly transferable between data preparation tools. Another is how they can help you avoid building very complex string calculations using the simple string functions this book has covered thus far.

How to Use Regexes in Prep

There are three main ways to use regexes in Prep (Figure 31-1).

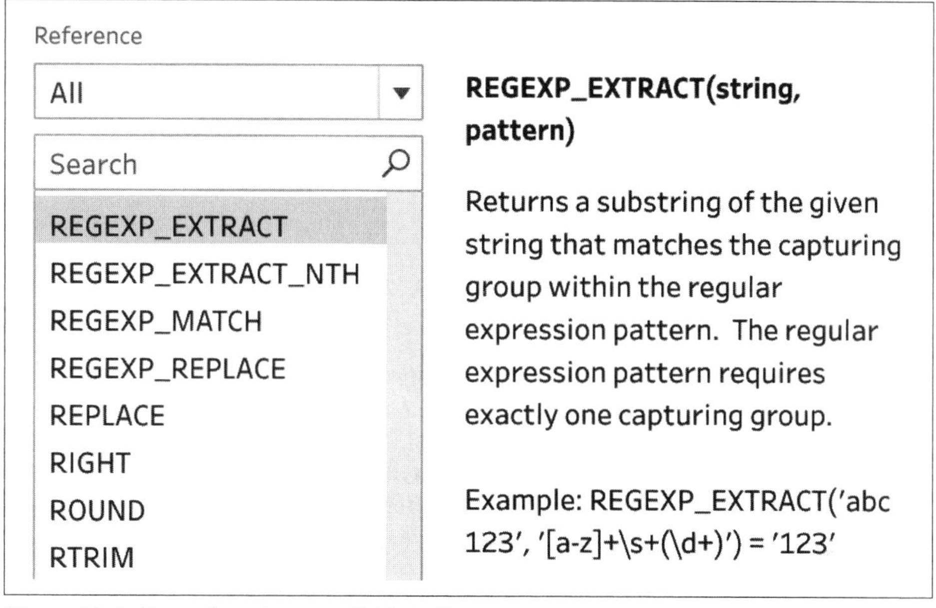

Figure 31-1. Regex functions available in Prep

To help you understand how to make use of regexes in Prep Builder, we'll go through each function before addressing specific examples.

REGEXP_EXTRACT() and REGEXP_EXTRACT_NTH()

`REGEXP_EXTRACT()` allows you to extract part of a string (a *substring*). The part returned is enclosed by brackets within the regular expression; this is called the *capturing group* and it appears next to the regex pattern inside parentheses.

```
REGEXP_EXTRACT([String Field],'regex pattern')
```

The `REGEXP_EXTRACT_NTH()` function returns the *n*th instance of the capturing group if it exists.

REGEXP_MATCH()

The `REGEXP_MATCH()` function returns a Boolean result of either `True` or `False` depending upon whether or not a given string matches the specified regex pattern. The entire string does not have to match the regex pattern to return a `True` result, only a substring.

```
REGEXP_MATCH([String Field],'regex pattern')
```

REGEXP_REPLACE()

The `REGEXP_REPLACE()` function specifies a part of a string field to replace as well as the value to replace it with.

```
REGEXP_REPLACE([String Field],'Substring to replace','Replacement value')
```

Regex Use Cases

The flexibility offered by regexes allows for a vast array of use cases. This section describes some of the most common scenarios.

Replacing Common Mistakes

Many string fields come from manually entered text. This can mean a data field contains repeated mistakes that will need to be removed before accurate analysis can occur. For example, if you have the times of different events captured but in different formats, you'll need to extract the relevant values from the strings before you can begin your analysis (Figure 31-2).

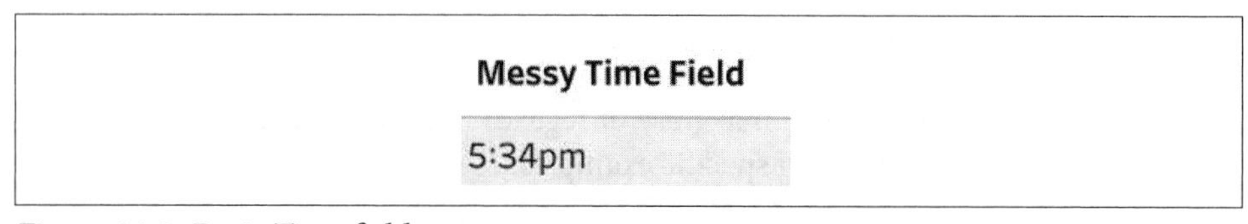

Figure 31-2. Basic Time field

In this instance, the time needs to be extracted from this string. Figure 31-3 shows how you can use `REGEXP_EXTRACT()` to achieve this.

Field Name

Clean Time

```
REGEXP_EXTRACT([Messy Time Field],'(\d+:\d+)\D')
```

Figure 31-3. Using REGEXP_EXTRACT() to extract the time

This returns the result you need (Figure 31-4).

Messy Time Field	Clean Time
5:34pm	5:34

Figure 31-4. Result of REGEXP_EXTRACT() calculation in Figure 31-3

Anonymizing Comments or Feedback

One of the biggest strengths of regexes is finding substrings in a much longer string that match a specified pattern. One case where this functionality can be especially useful is identifying a person's details and removing them from the data source. Let's use the sample output from a Chin & Beard Suds Co. feedback form shown in Figure 31-5.

Description
I've had a terrible experience and want a reply. Please email carl.allchin@preppindata.com
I'd love to get an alert when the product is available again. Please let me know at jeremy@preppindata.com
Just message me toni.feather@preppindata.com and I'll solve your transport issues
No I'm not sharing my email address. I just want you to charge me less!

Figure 31-5. Customer feedback to be anonymized

By identifying a common pattern in the text, you can specify that pattern in a regex to replace the text in question. For this example, we want to identify the pattern *name@domain.com*. We use the encoding `[\.\w]+@[\.\w]+` to anonymize any word characters before the @ symbol and any word characters after it. A *word character* is a letter, digit, or underscore. The space at the start and end of the email address indicates that the whole Description field isn't converted.

In our example, we will replace the email address with `[Hidden Details]` so it is easy to remove or leave in future analyses. Figure 31-6 shows the calculation to do this within Prep Builder.

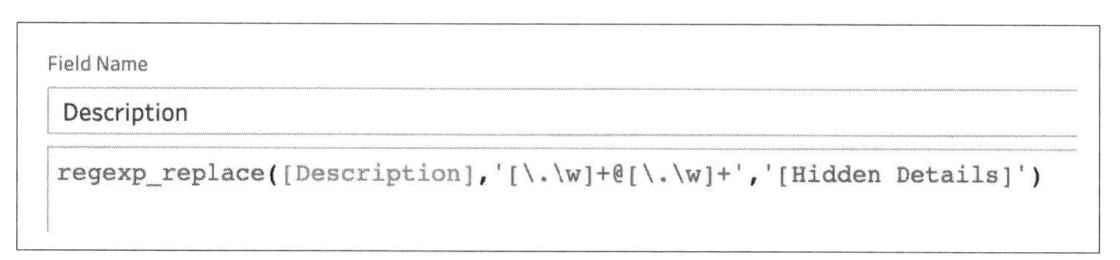
Field Name
Description
regexp_replace([Description],'[\.\w]+@[\.\w]+','[Hidden Details]')

Figure 31-6. Using REGEXP_REPLACE() to anonymize email addresses

The resulting data set is shown in Figure 31-7.

Description
I've had a terrible experience and want a reply. Please email [Hidden Details]
I'd love to get an alert when the product is available again. Please let me know at [Hidden Details]
Just message me [Hidden Details] and I'll solve your transport issues
No I'm not sharing my email address. I just want you to charge me less!

Figure 31-7. Resulting data set from the REGEX_REPLACE() example

Common Regex Commands

In order to be able to use regex functions, you will need to be able to specify the patterns you're looking for using regex commands. In regexes, case sensitivity depends on the data source Prep Builder is connected to, so you will need to take care when entering string names. Table 31-1 lists the key regex commands you'll use.

Table 31-1. Common regex commands

\d	Any digit 0–9
\d+	One or more digits
\a-z	Any lowercase letter
\A-Z	Any uppercase letter
\D	Any nondigit
\s	Whitespace
\w	Any word character
^	Start of a string
$	End of a string
( )	Capture anything within the parentheses, as in REGEXP_EXTRACT() function
{ }	Exactly a number of the given substring; set up by adding a number between the brackets (i.e., {3})

For testing, you can use Prep or get instant feedback on websites built for testing regexes (e.g., http://regex101.com).

Summary

Regular expressions allow you to clean complex string fields in a much more simplistic and reusable way. Although there is more to learn with regexes compared to the other string functions used to clean data, the investment of time is worth it, as they make it much easier to prepare your data for analysis.

CHAPTER 32

Completing Advanced Joins

Joining is one of the more powerful techniques for preparing your data for analysis. Adding another data set to your existing data can either give you more context about why something is happening or answer more granular questions. In Chapter 16 we covered using join types and join conditions to get to the data you want in a straightforward way. However, as you'll see in this chapter, there are many use cases for joins where the logic is more complex.

Multiple Join Conditions

Frequently, the process of linking two data sets will require more than a simple, single join condition. When using multiple join conditions, keep in mind that they behave as an AND rather than an OR statement. In other words, for the data record being assessed to be output by the inner section of a Join step, every join condition must be met.

In the example challenge, Preppin' Data 2020: Week 8 (*https://oreil.ly/m_WoZ*), joining the Actual Volume and Value data set (Figure 32-1) to the Profit data set (Figure 32-2) requires multiple join conditions.

Type	Week	Sales Value	Sales Volume
BAR	2020_3	9,889	3,305
BAR	2020_6	12,254	3,530
BAR	2020_8	12,237	4,197
BAR	2020_5	10,140	3,868
BAR	2020_7	7,752	4,467
LIQUID	2020_1	9,098	3,519
LIQUID	2020_5	13,192	2,366
LIQUID	2020_8	12,089	3,055
LIQUID	2020_6	8,998	3,347
LIQUID	2020_2	9,531	5,149
LIQUID	2020_7	9,315	4,307
BAR	2020_2	10,271	3,333
LIQUID	2020_3	9,187	3,537
LIQUID	2020_4	11,729	3,438
BAR	2020_4	13,490	3,620
BAR	2020_1	7,015	2,824

Figure 32-1. Actual Volume and Value data set

Week	Type	Profit Min Sales Volume	Profit Min Sales Value
2020_1	BAR	2,000	7,500
2020_2	BAR	3,000	10,000
2020_3	BAR	3,000	10,000
2020_4	BAR	3,000	10,000
2020_5	BAR	3,000	10,000
2020_6	BAR	3,000	10,000
2020_7	BAR	3,000	10,000
2020_8	BAR	3,000	10,000
2020_1	LIQUID	3,000	8,000
2020_2	LIQUID	3,500	11,000
2020_3	LIQUID	3,500	11,000
2020_4	LIQUID	3,500	11,000
2020_5	LIQUID	3,500	11,000
2020_6	LIQUID	3,500	11,000
2020_7	LIQUID	3,500	11,000
2020_8	LIQUID	3,500	11,000

Figure 32-2. Profit data set

Each data set has only 16 rows of data, but with just one join condition (Type), the output of the join is 128 rows (Figure 32-3).

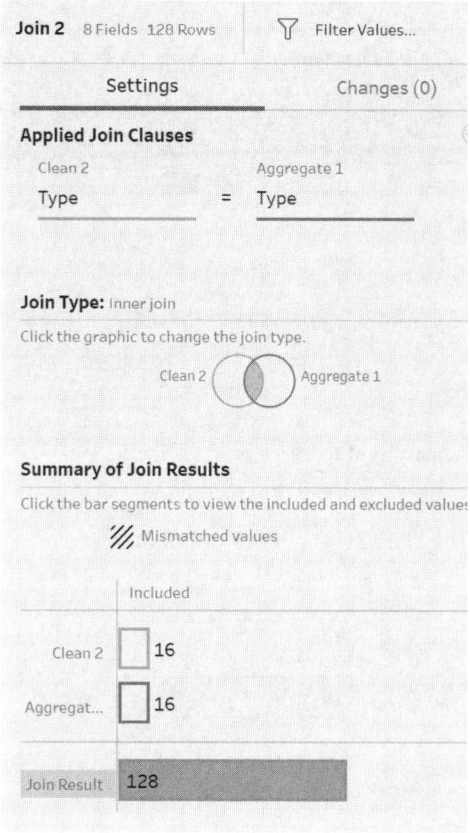

Figure 32-3. Duplication of rows caused by a join

Let's look at what caused this duplication. Each data set has eight rows for Type, as each type of soap (bar and liquid) is sold in each of the eight weeks recorded. Therefore, all the Bar records in one set are joined to the Bar records in the other set (also eight), so 8 × 8 gives us 64 rows for each type and we have two different types, so 64 × 2 = 128 rows resulting from the join.

By adding a second join condition to match the rows of data on a one-to-one basis—in this case setting Week = Week as a condition—we get the desired result (Figure 32-4).

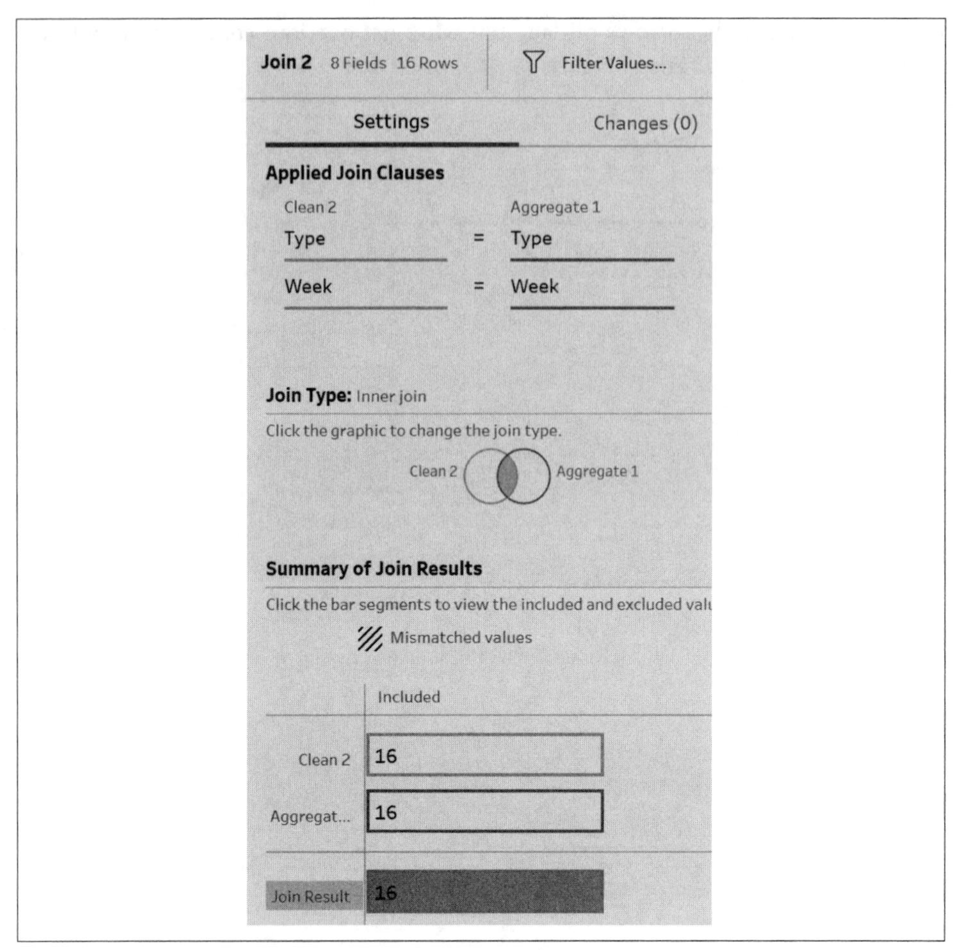

Figure 32-4. Setting multiple join conditions to remove the duplication shown in Figure 32-3

To add the second condition, click the plus sign in the circle in the top-right corner of the Join configuration pane.

Join Conditions Other Than Equals

In many cases, when thinking about how to join two data sets together, we look for common fields in the two data sets. We also often think about finding things that are equal to each other. However, there are many situations where using a condition other than equals can be very powerful. Let's look again at the examples from the Preppin' Data 2020: Week 8 challenge.

Filtering with a Join

There are a number of weeks' worth of sales (Figure 32-1) that need to be compared to the values and volumes that would make the company profitable (Figure 32-2). The Profit data set has one record for each week to assess. We want to include only those weeks that exceed the Profit Minimum values.

This is where join conditions that return only values that exceed values taken from another data set can come in particularly handy. Remember, the less data being returned, the less processing time required in data preparation.

In this example, the required join condition is that the actual values and volumes recorded exceed the profit minimum expectations. You set this using two join conditions that are *greater than or equal to*, as shown in Figure 32-5.

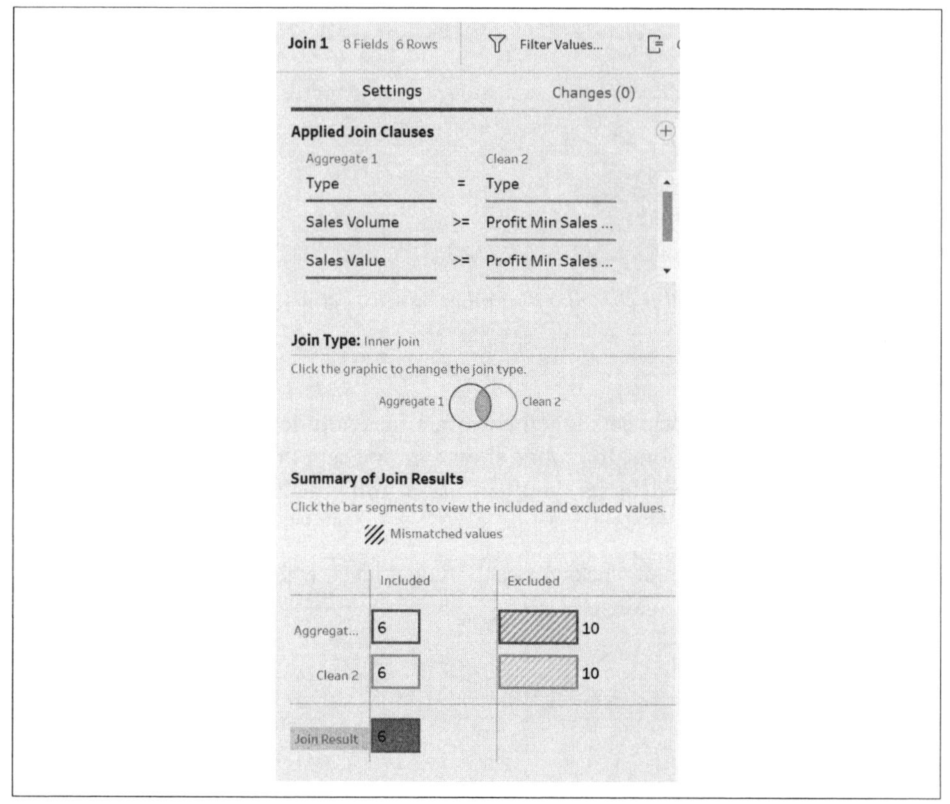

Figure 32-5. Setting up greater than or equal to join conditions to act as a filter

There are actually four join conditions in use here. Both Type and Week from each data set should be equal, but to return only records that beat the minimum level of profitability, Sales Volume has to be equal to or greater than (=>) Profit Min Sales Volume, and Sales Value has to be equal to or greater than Profit Min Sales Value. All four of these conditions must be met for the records to be output from the Join step. In this case, we can see that several records haven't met this requirement (Figure 32-6).

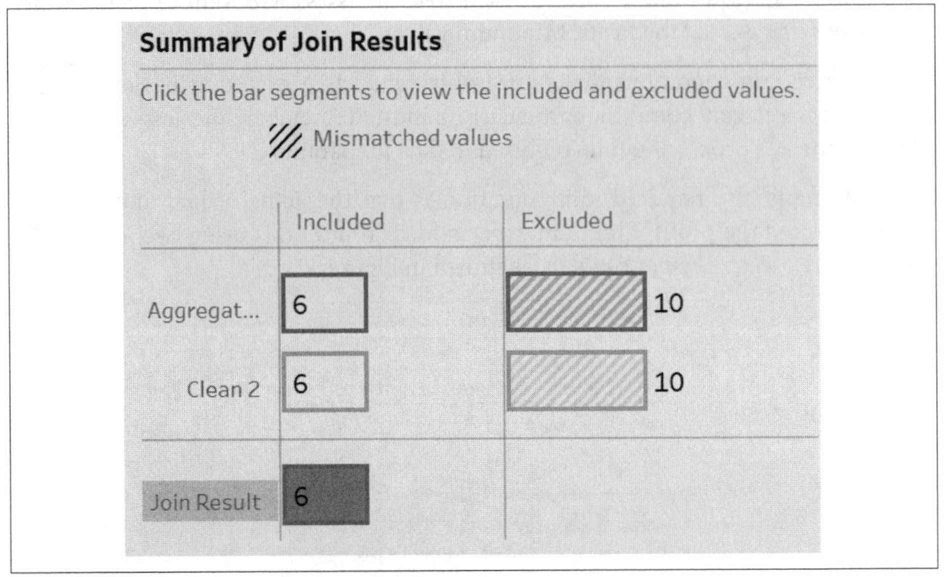

Figure 32-6. The Join result is showing a number of rows being excluded

Joining by a Range

When joining different data sets together, sometimes you don't have an exact match to join on. In these situations, if a range shown in one set of data corresponds to column values in the other data set, you can use a non-equals join condition to join these together.

Figure 32-7 shows a range of weeks for each budget level, which we need to assess the actual value and volumes against.

Range of Weeks	Pivot1 Values	Type	Measure
1-2	1,875	Bar_01893	Budget Volume
1-2	5,500	Bar_01893	Budget Value
1-2	2,225	91374__Liquid	Budget Volume
1-2	6,050	91374__Liquid	Budget Value
3-5	2,225	Bar_01893	Budget Volume
3-5	8,075	Bar_01893	Budget Value
3-5	2,750	91374__Liquid	Budget Volume

Figure 32-7. Week ranges within a data set

To make the range available as a minimum and maximum value to test the actual week against, you need to split the value in the Range of Weeks column (Figure 32-8).

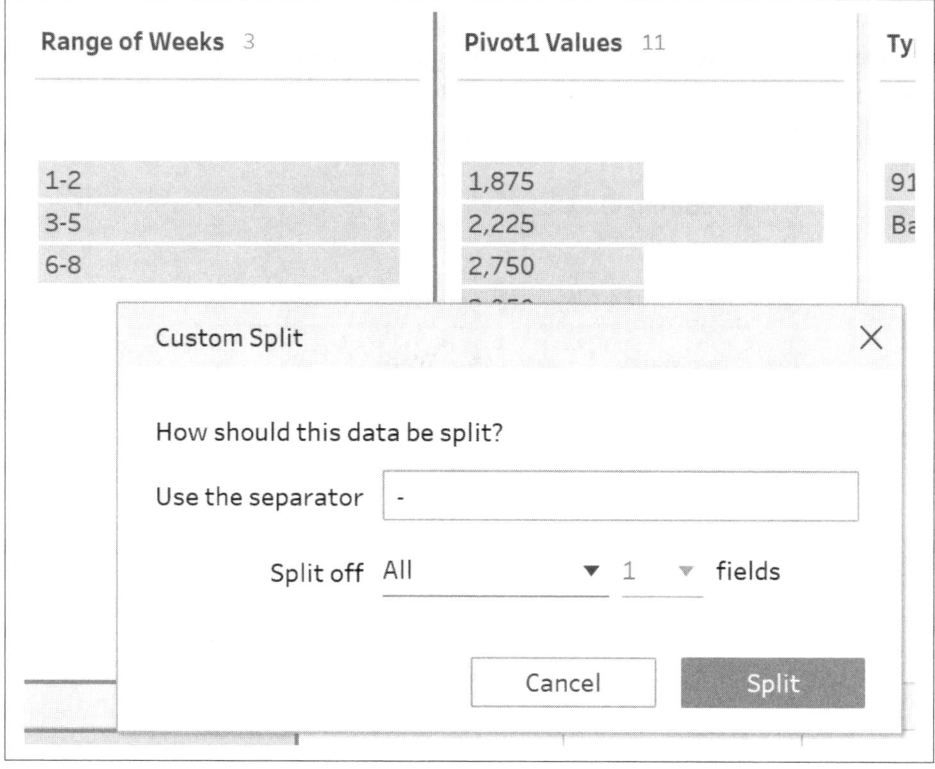

Figure 32-8. Split setup

This results in a data set with two additional columns: Start Week and End Week (of the range), as shown in Figure 32-9.

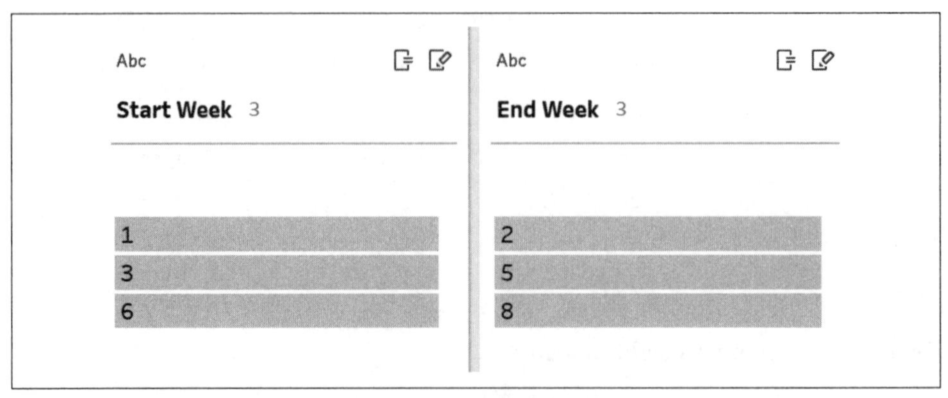

Figure 32-9. Result of splitting the Range of Weeks field

These fields can then be used to set the range for the actual week in the original data set. Using the minimum week of the range as the less than or equal to (<=) condition and the maximum week of the range as the greater than or equal to (>=) condition allows you to add the relevant range records to the Actual Volume and Value data set (Figure 32-10).

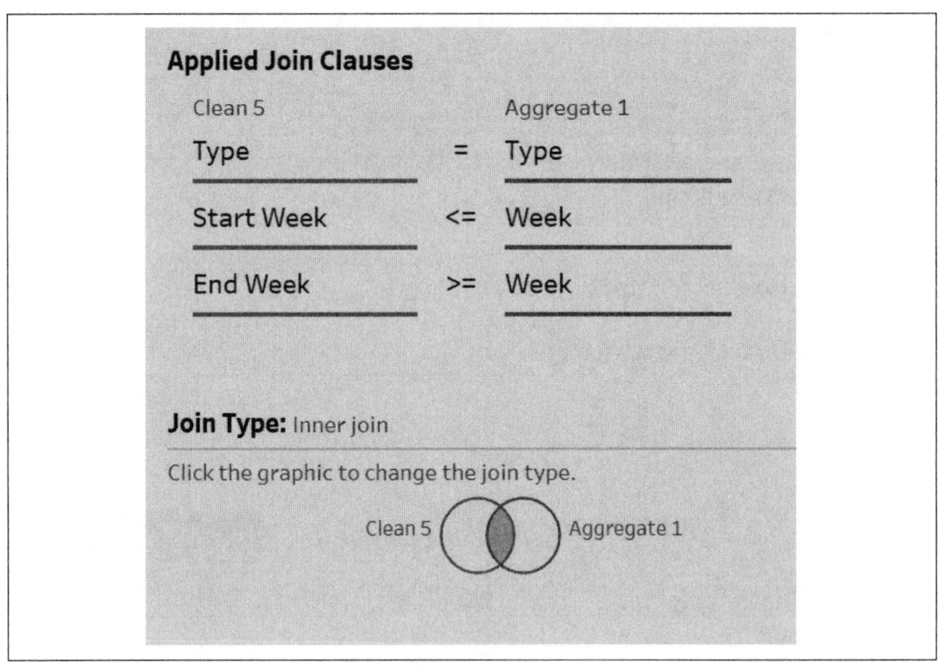

Figure 32-10. Multiple join conditions and operators

OR Statements

As discussed earlier in this chapter, multiple join conditions behave as an AND statement, but what happens if you don't want all the records to meet all the join conditions you need to assess? In this example, we're assessing the Profit data set (Figure 32-2) against the Actual Volume and Value set (Figure 32-1). The output required from the join is whether either actual volume *or* value is below the budget for that week. You can mimic the behavior of OR statements through joins, but you actually need to do two joins and then union those results back together to get a complete set of records that meet your requirements.

Because the join conditions are set up very similarly, I will just show one of the two. For the other, just switch Volume for Value. Type = Type is a join condition, but it's offscreen in Figure 32-11.

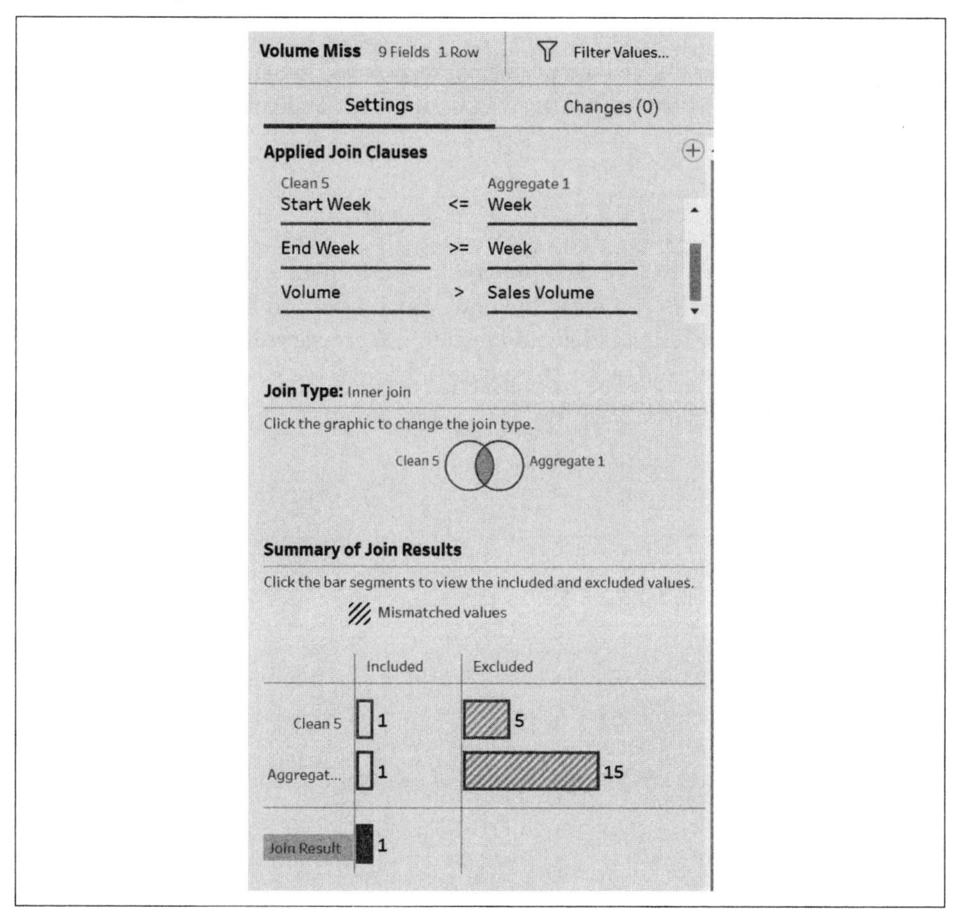

Figure 32-11. Join condition setup for the OR example

Then, you can simply union the output of these two joins—that is, stack the results on top of each other (Figure 32-12).

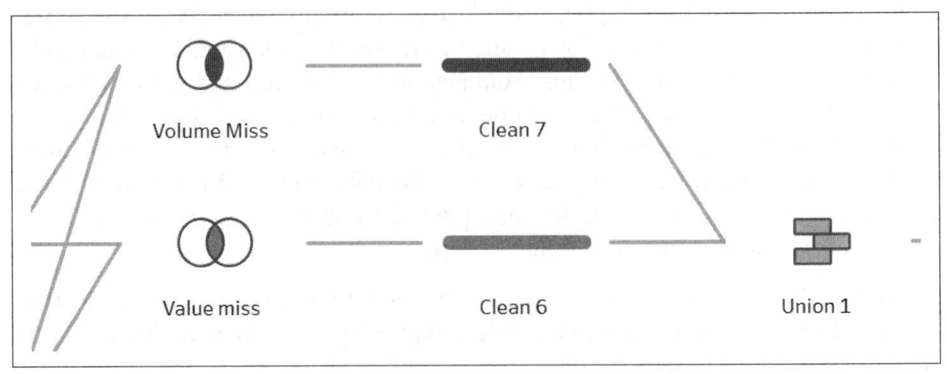

Figure 32-12. Unioning results of multiple joins

The ability to replicate OR statement behavior gives you much greater flexibility in your analysis. You don't have to work around join types and conditions to be able to return the data you need.

Summary

Joins are a fundamental part of data preparation, and there are some advanced join techniques that can make complex situations much easier to handle. Whether it is using multiple join conditions to avoid creating unnecessary duplicates or handling OR statements, being able to tackle these more advanced use cases can save your data set's end users hours of workarounds.

CHAPTER 33

Creating Level of Detail Calculations

As you've seen, joining data sets together is a very common task in data preparation. But there is one use of joins that needs highlighting due to its slightly different use—a technique called *appending*. Within Tableau's suite of products, appending is often used in Level of Detail (LOD) calculations, even if you don't realize you are doing it.

This chapter will use LOD calculations to show the benefits of appending values to your data set for easier analysis. LOD calculations can be tricky to learn, so this chapter will walk you through what the LOD calculation is actually doing to the data set to cement your understanding.

What Is Appending?

Appending is the addition of a data field, or fields, onto an existing data set. This sounds very similar to joining, but we use the term *append* when adding a field that contains a constant, or set of constants, rather than the more row-to-row relationship that is found within joins.

Exploring Appending Through LOD Calculations

As Prep Builder is designed to prepare data for Tableau Desktop and Server, many of you likely will be familiar with Level of Detail calculations from using them within Desktop. If not, let's have a quick recap.

When to Use an LOD Calculation

Tableau sets a measure's level of aggregation by the granular dimension, or combination of dimensions, used within the view of Tableau Desktop. In the example in Figure 33-1, using data from Tableau's sample data set Superstore, we'll use an LOD calculation to set the total sales for Category—no matter the view makeup in Desktop—so the percentage of sales within each category can be calculated for a subcategory.

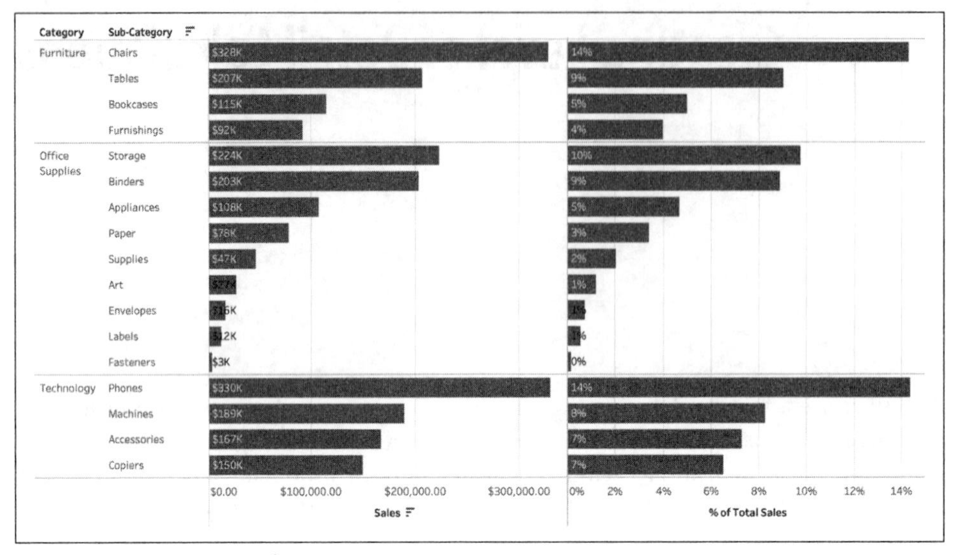

Figure 33-1. Superstore data set

What is being calculated is the percentage of total sales across all categories. We could determine the subcategory percentage of categorical sales with a Table Calculation as well, but it's not necessary for the data set we're trying to return (Figure 33-2).

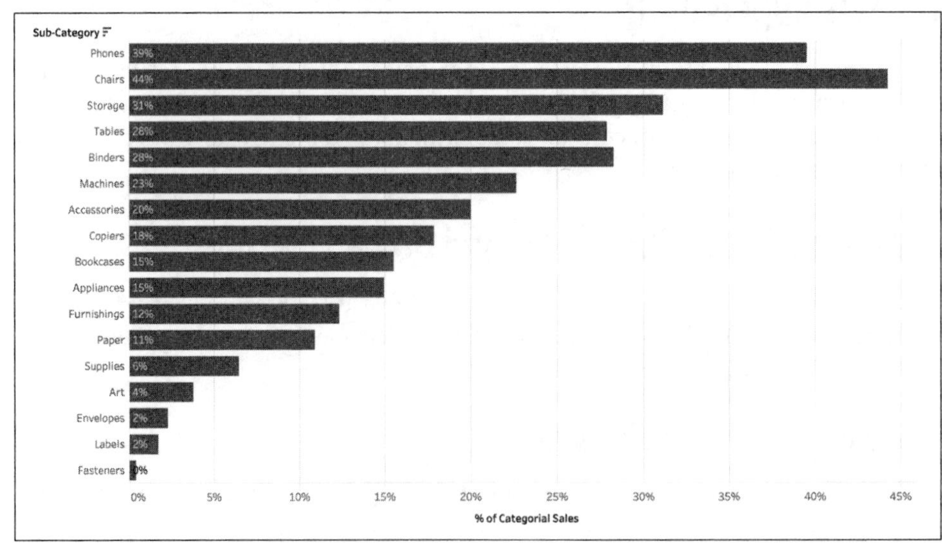

Figure 33-2. Resulting data set to be formed

How to Write an LOD Calculation in Prep Builder

As of version 2020.1.3, you can create LOD calculations in Prep Builder. As mentioned in Chapter 18, an LOD calculation sets a hierarchical level at which an aggregation will be processed. This is applied to just one measure per calculation, compared to the Aggregate step, where the whole data set's granularity is changed to the specified hierarchical level.

You have two different options in Prep Builder to create an LOD calculation: Custom Calculation or Fixed LOD. Both are accessed from a data field's ellipsis menu in the Clean step through the Create Calculated Field submenu (Figure 33-3).

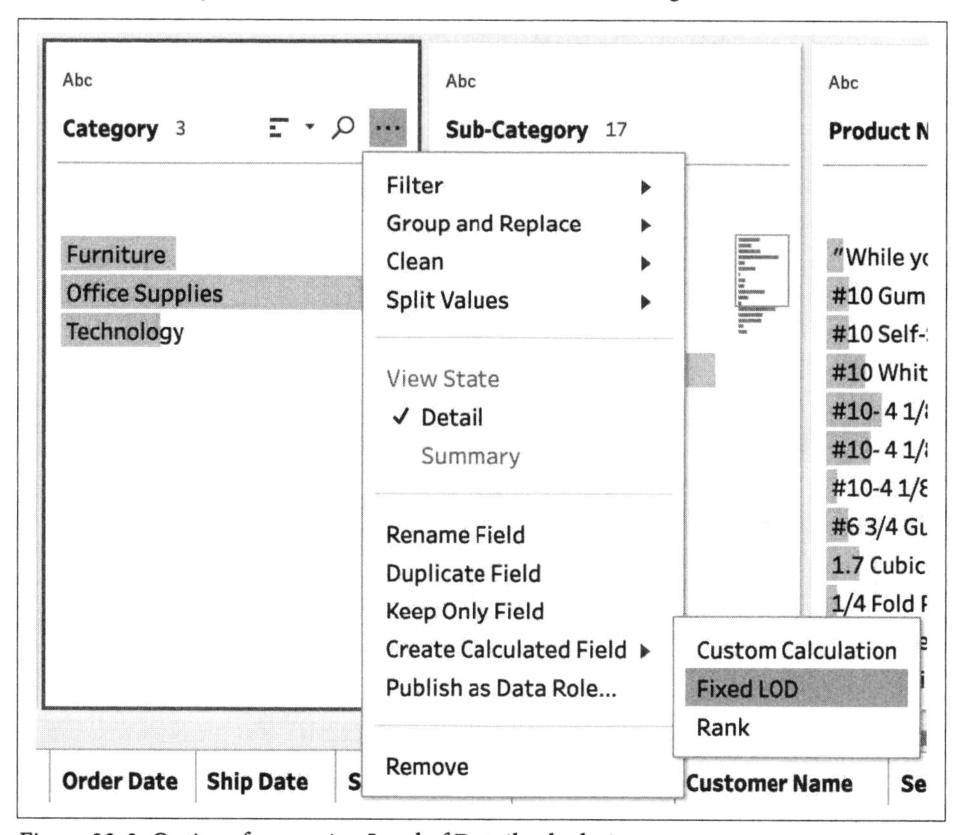

Figure 33-3. Options for creating Level of Detail calculations

Selecting Fixed LOD opens the Visual Calculation Editor, which was shown in Chapter 18. So, this chapter will focus instead on the other option: creating an LOD calculation in the Custom Calculation Editor. The syntax for LOD calculations is:

```
{ fixed [optional categorical field(s)] : aggregation ( measure ) }
```

To build the LOD calculation needed to output the data set in Figure 33-2, first we'll create the field Categorical Sales (Figure 33-4).

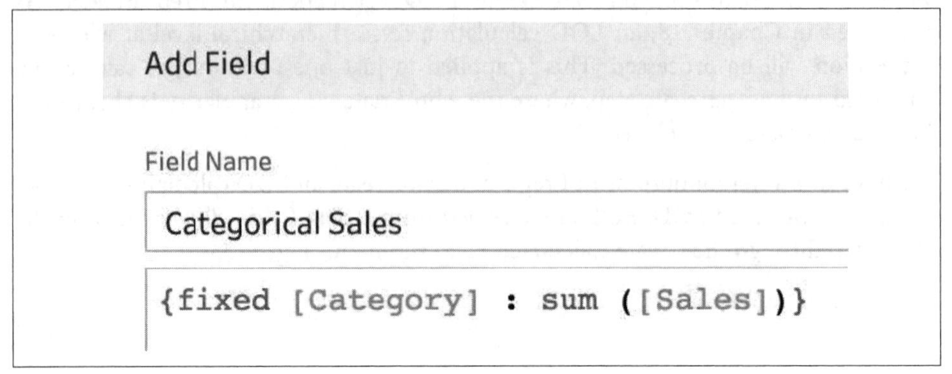

Figure 33-4. Categorical Sales LOD calculation

This calculation has set a constant sales total for each category. The Profile pane is a good place to check for the effects of the calculation (Figure 33-5).

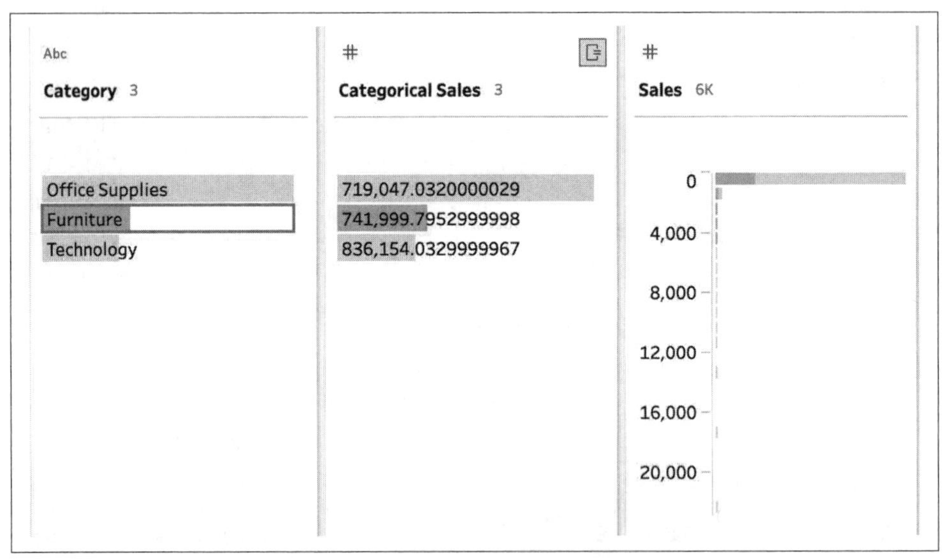

Figure 33-5. Results from the Categorical Sales LOD calculation

By selecting a category in the Profile pane, you can see the values in other columns containing that category. In this case, Furniture has a constant Categorical Sales value, but there are still a range of Sales values within the selected rows. Prep Builder is *appending* the Categorical Sales value to the existing data set, as you can see on a per-row level in the Data pane (Figure 33-6).

Category	Categorical Sales	Sales
Furniture	741,999.7952999998	261.96
Furniture	741,999.7952999998	731.9399999999999
Furniture	741,999.7952999998	957.5775
Furniture	741,999.7952999998	48.86
Furniture	741,999.7952999998	1,706.1840000000004
Furniture	741,999.7952999998	71.37199999999999
Furniture	741,999.7952999998	1,044.63

Figure 33-6. Results from the Categorical Sales LOD calculation in the Data pane

Due to the hierarchy of the Superstore data set, a product can belong to only one category. This data set is ready for use within Tableau Desktop. If you want a quick answer for your analysis, you can select Preview in Desktop by right-clicking on any step, and Tableau Desktop will launch with a Hyper file of that data set already connected (Figure 33-7).

Figure 33-7. Preview in Tableau Desktop option in Prep Builder

To create the set shown earlier in Figure 33-2, divide the Sales value by Categorical Sales data in Desktop (Figure 33-8).

% of Categorical Sales

```
[Sales] / [Categorical Sales]
```

Figure 33-8. Calculating % of Categorical Sales in Desktop

This calculation determines the percentage contribution by each row of data, which is then summed up by the dimension(s) used in the view in Desktop; in this case, that's Sub-Category, although other members of the product hierarchy could be used instead (Figure 33-9).

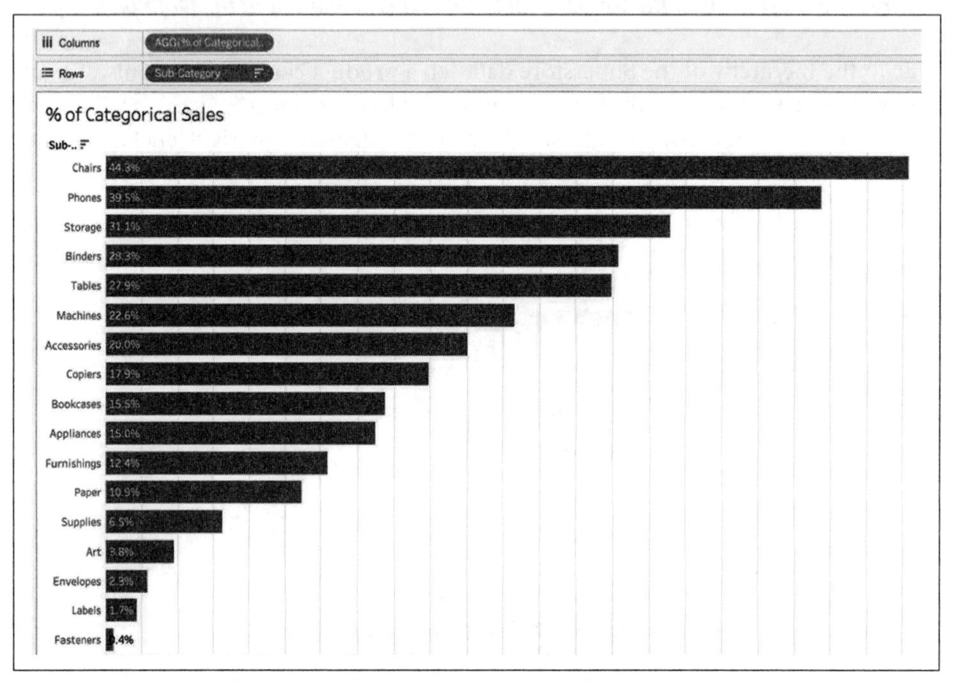

Figure 33-9. The % of Categorical Sales analysis in Desktop

By appending the Categorical Sales value to the data set, you save your end users from having to create complex calculations themselves. However, these calculations (as well as any data fields) must be used correctly to avoid incorrect or misleading analysis.

What a Level of Detail Calculation Is Doing

To help describe what the LOD calculation is doing, let's walk through appending a value to a data set. To make it easier to see what is happening within Prep Builder, we'll use the same calculation as before, just with the data set aggregated to the Sub-Category level (Figure 33-10).

Abc Category	Abc Sub-Cat...	# Sales
Office Supplies	Binders	203,413
Office Supplies	Storage	223,844
Technology	Accessories	167,380
Furniture	Furnishings	91,705
Furniture	Tables	206,966
Furniture	Bookcases	114,880
Technology	Phones	330,007
Office Supplies	Art	27,119
Office Supplies	Paper	78,479
Technology	Machines	189,239
Office Supplies	Appliances	107,532
Office Supplies	Envelopes	16,476
Furniture	Chairs	328,449
Office Supplies	Supplies	46,674
Office Supplies	Fasteners	3,024
Technology	Copiers	149,528
Office Supplies	Labels	12,486

Figure 33-10. Superstore data set aggregated to the Sub-Category level

Step 1: Calculate the categorical sales

Using an Aggregate step, group by Category and sum the Sales values (Figure 33-11). This will result in each category having a single Categorical Sales value. Using an Aggregate step removes the other data in the set.

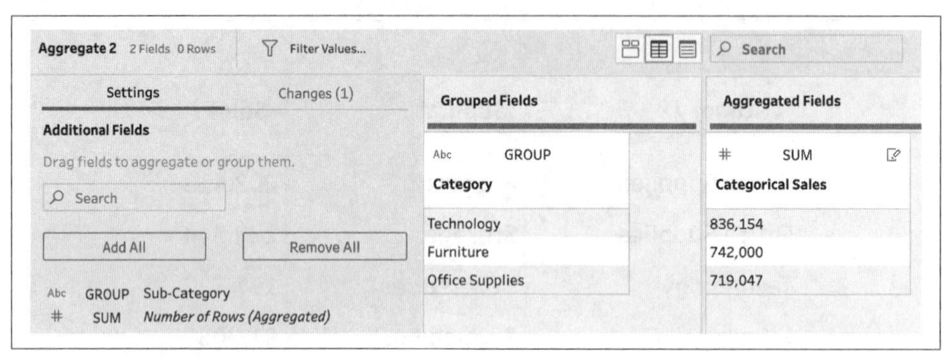

Figure 33-11. Using an Aggregate step to calculate Categorical Sales

Step 2: Join the aggregated results back to the original data set

Completing a *self-join*—joining an earlier stage of the data preparation flow to the output of the Aggregate step—adds the original data at a Sub-Category level back into the flow (Figure 33-12).

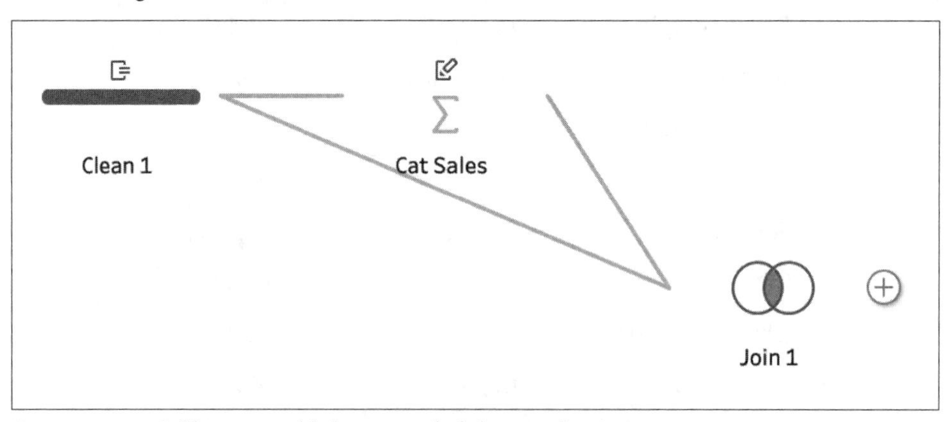

Figure 33-12. Self-join to add the appended data to the Aggregate step output

Using an inner join with a join condition of Category = Category then adds Categorical Sales back to the original data set (Figure 33-13). With an inner join, you should delete the matching field, now called Category-1, as Prep doesn't allow duplicate field names.

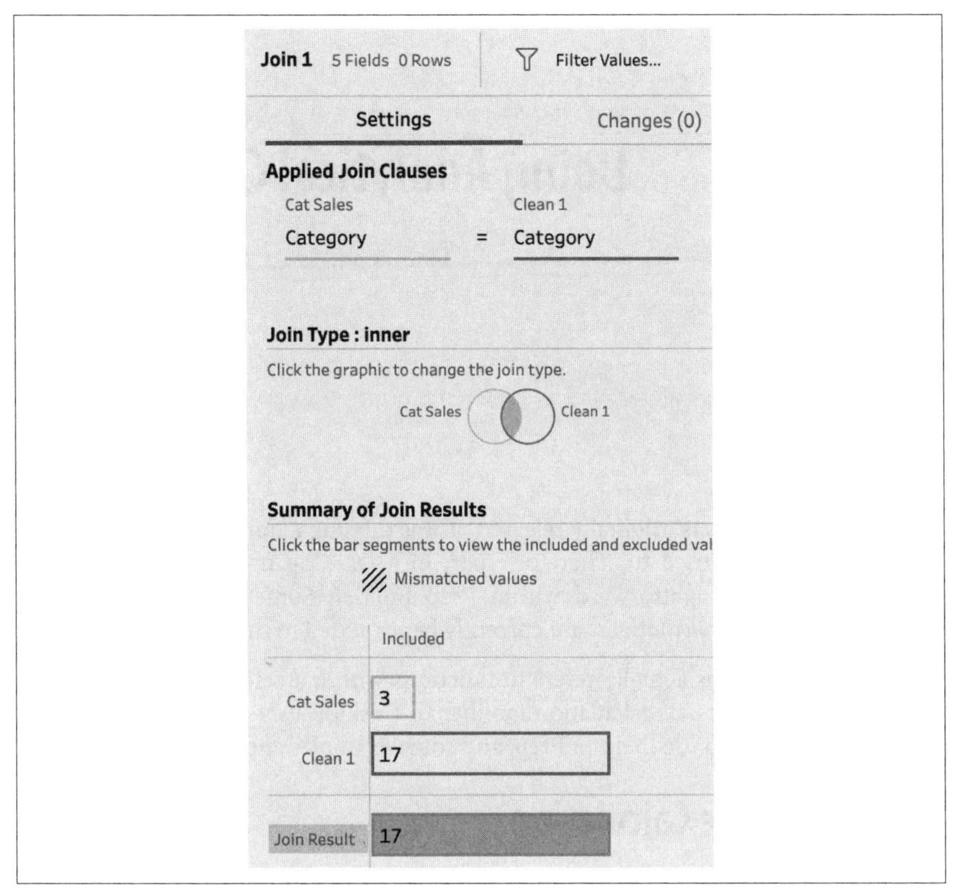

Figure 33-13. Configuring the self-join

Summary

Level of Detail (LOD) calculations are very useful when analyzing data but can be difficult for end users of the data set to form themselves. LOD calculations work by aggregating a measure and then appending the result to the data set. The LOD calculations formed in Prep Builder can be used in further calculations or exported for use within Tableau Desktop to simplify the analytical process for users.

CHAPTER 34

Doing Analytical Calculations

When Prep Builder was originally released, it was missing a number of functions that Desktop users were used to. Therefore, part of a Desktop user's normal analytical process was not straightforward within Prep Builder. Some of these capabilities, known as *analytical calculations*, are currently being added to the tool.

Analytical calculations actually refers to functions within a calculation. In this chapter, we'll look at a type of calculation familiar to Desktop users called *Table Calculations*, how and why to use them in Prep, and some examples and use cases.

What Is a Table Calculation?

Table Calculations in Tableau Desktop are highly flexible and easy to use with a couple of clicks. Table Calculations allow you to handle secondary aggregations, like rank or running total, that would not otherwise be possible in Tableau.

The downside of Table Calculations is exactly the same as what makes them great: their flexibility. A Table Calculation can be easily reconfigured to work on whatever dimensions are within the view. Therefore, whenever a user wants to use the calculation, they need to understand how to reconfigure it. With complicated views, if the user doesn't reconfigure the calculation correctly, this can impact both the performance of the view and the accuracy of the data output. Taking this work on during the data preparation stage can help mitigate these risks.

For example, let's use Tableau's Superstore data set to cover a use case assessing the ranking of each region based on its sales. In Tableau Desktop, you'd need to include both the Date and Sales numbers in the view as well as breaking down quarterly sales into Region rows. Figure 34-1 shows what that view might look like in Desktop.

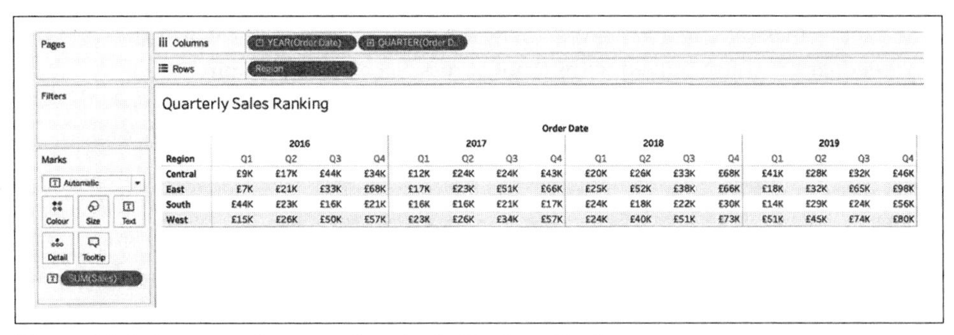

Figure 34-1. Superstore's quarterly sales in Tableau Desktop

To rank those regional sales in each quarter in Tableau Desktop, we'd use a Table Calculation. Table Calculations are always applied to continuous or measure fields in the view, as they make use of the results shown in the view—in this case, the sum of sales for each region and quarter. To create the rank value, right-click on the `sum(Sales)` data field, select Quick Table Calculation, and then select Rank in the submenu (Figure 34-2).

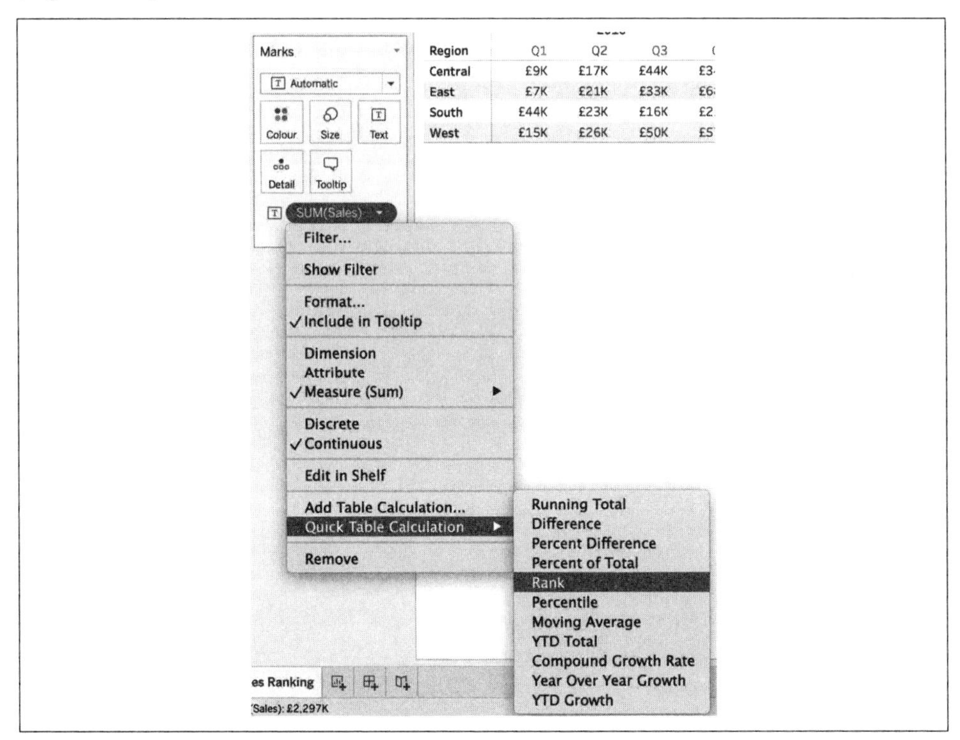

Figure 34-2. Table Calculation menu in Tableau Desktop

The rank values returned are not actually what we need, as the calculation is returning the quarterly sales ranked within, not across, each region (Figure 34-3).

Quarterly Sales Ranking

	Order Date															
	2016				2017				2018				2019			
Region	Q1	Q2	Q3	Q4	Q1	Q2	Q3	Q4	Q1	Q2	Q3	Q4	Q1	Q2	Q3	Q4
Central	16	14	3	6	15	12	11	4	13	10	7	1	5	9	8	2
East	16	13	9	2	15	12	7	3	11	6	8	4	14	10	5	1
South	2	7	15	10	13	14	9	12	5	11	8	3	16	4	6	1
West	16	13	8	4	15	12	11	5	14	10	7	3	6	9	2	1

Figure 34-3. The default results from applying a Rank Table Calculation

We wanted to see how the regions rank against each other per quarter, not how each region ranks against its past performance. To change this, we need to edit the Table Calculation. Right-clicking on `sum(Sales)`, which now has a delta symbol on the right-hand side, opens a menu with the option Edit Tableau Calculation. Selecting this option displays the screen in Figure 34-4.

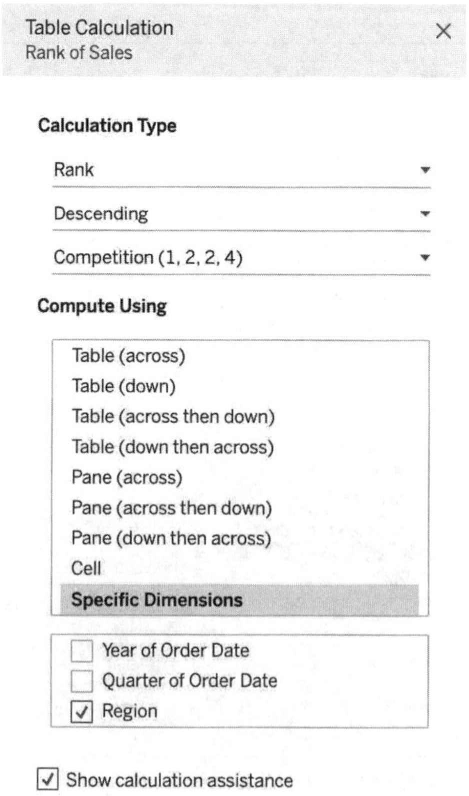

Table Calculation
Rank of Sales

Calculation Type

Rank

Descending

Competition (1, 2, 2, 4)

Compute Using

Table (across)
Table (down)
Table (across then down)
Table (down then across)
Pane (across)
Pane (across then down)
Pane (down then across)
Cell
Specific Dimensions

☐ Year of Order Date
☐ Quarter of Order Date
☑ Region

☑ Show calculation assistance

Figure 34-4. The Table Calculation Editor

To reflect how we want to process the Rank calculation, in the Table Calculation Editor, we need to change the Specific Dimensions Tableau Desktop is using for the calculation. The rank of the selected dimensions will be calculated for every combination of the dimensions that are unselected. Figure 34-5 shows the results of the updated calculation.

Quarterly Sales Ranking

	Order Date															
	2016				2017				2018				2019			
Region	Q1	Q2	Q3	Q4	Q1	Q2	Q3	Q4	Q1	Q2	Q3	Q4	Q1	Q2	Q3	Q4
Central	3	4	2	3	4	2	3	3	4	3	3	2	2	4	3	4
East	4	3	3	1	2	3	1	1	1	1	2	3	3	2	2	1
South	1	2	4	4	3	4	4	4	3	4	4	4	4	3	4	3
West	2	1	1	2	1	1	2	2	2	2	1	1	1	1	1	2

Figure 34-5. Updated results for the Rank Table Calculation

Table Calculations are extremely useful because they update as additional dimensions are brought into the view. However, as you've seen, creating them in Tableau Desktop is a multistep process and also requires some background knowledge to be able to make them work as required. So, as mentioned earlier, precalculating this metric in Prep Builder can save users from having to do it themselves, although they will lose some of the flexibility of updating the calculation as other dimensions are brought into the view.

Applying Table Calculation Logic in Prep Builder

To save Desktop end users the trouble, you can create the calculation in the data preparation stage, which can also benefit from the addition of analytical calculations. You've seen that the analytical process can either begin or be completed in Prep Builder, so having these functions already there can make analysis easier.

The syntax for analytical calculations differs from that of other functions found in Tableau Prep:

```
{ PARTITION [Dimension] : {ORDERBY [Measure] : ANALYTICAL_CALC( ) }}
```

Let's explore each option listed in the Analytic Reference menu within the Calculation Editor (Figure 34-6).

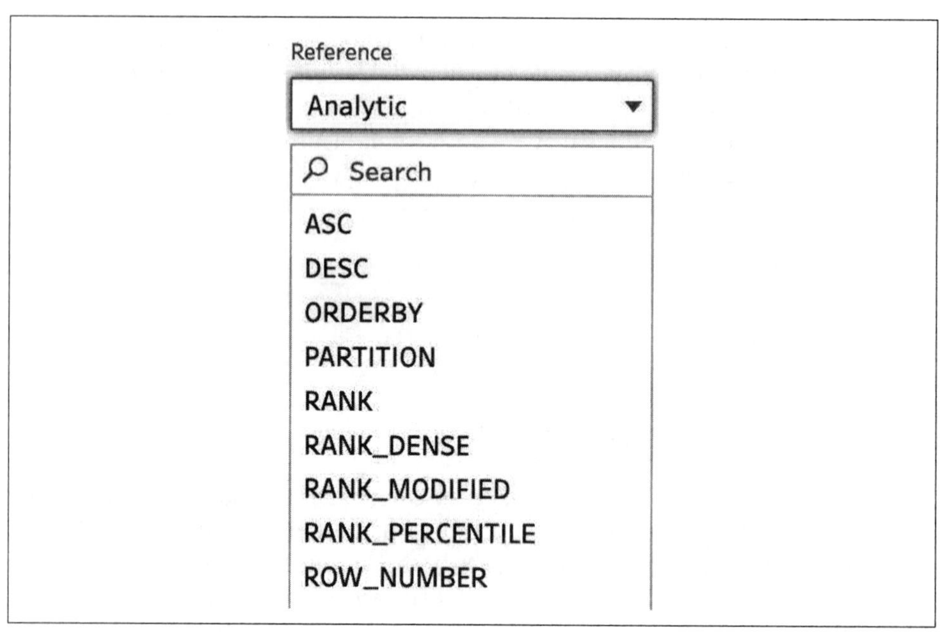

Figure 34-6. The analytical calculation options

Keywords

The Analytic Reference menu includes the following keywords:

`PARTITION`
: Like partitions in Table Calculations, this option allows you to restart an analytical calculation if required (for example, ranking each subcategory's sales but restarting the rank for each category).

`ORDERBY` *and* `ASC`/`DESC`
: This dictates the order in which the analytical calculation will be processed. The `ORDERBY` keyword sorts the data, and `ASC` and `DESC` specify ascending or descending, respectively.

Analytical Calculations

The analytical calculations are:

`RANK()`
: This function returns the standard competition rank (for each rank, we'll use the example values `3`, `7`, `7`, `10`, so this would return 1, 2, 2, 4).

`RANK_DENSE()`
: This function returns the dense rank (1, 2, 2, 3).

`RANK_MODIFIED()`
: This option returns the modified competition rank (1, 3, 3, 4).

`RANK_PERCENTILE()`
: This option returns the percentile rank (0.25, 0.75, 0.75, 1).

`ROW_NUMBER()`
: This function returns a row number.

Unlike Desktop, `RANK()` in Prep Builder does take into account nulls. The analytical calculation functions fall into a subset of Tableau calculations that do not require anything within their parentheses, along with `NOW()`, `TODAY()`, and `PI()`.

Now, with this introduction to the analytical calculations behind us, let's look more closely at how they are used in practice.

The ranking functions

Within Desktop, the ranking functions are used for many purposes, such as determining the best and worst performers in a data set, and setting limits on data sets (e.g., the Top 100).

To simplify this example, let's look at total sales for each subcategory (Figure 34-7).

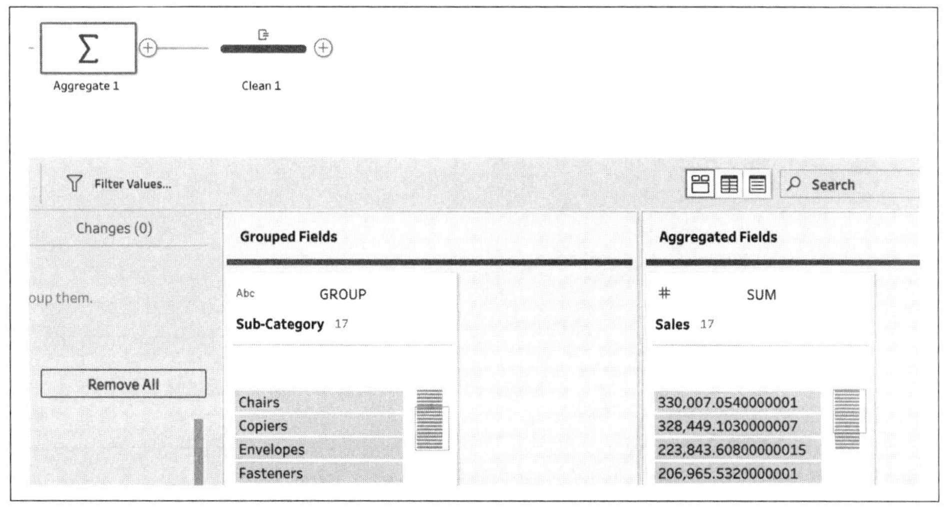

Figure 34-7. Calculating total sales in an Aggregate step

After the sales aggregation, let's rename the Sales field to Sub-Category Sales (Figure 34-8).

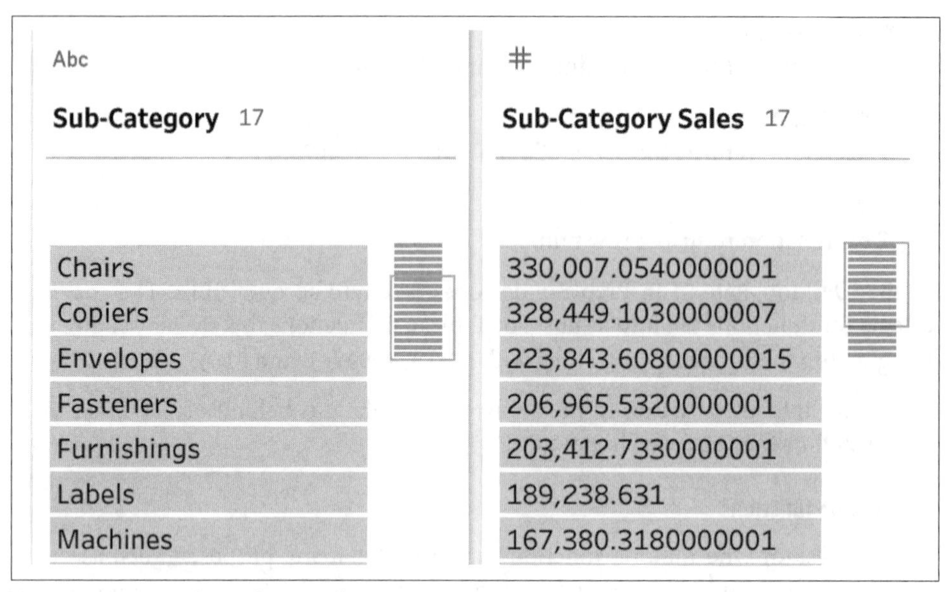

Figure 34-8. Resulting data set from Figure 34-2

To form a rank calculation, you must specify how to rank the values. In this case, setting the ranking along Sub-Category Sales using the `ORDERBY()` keyword will rank the values from 1 upward based on the size of the values from largest to smallest (Figure 34-9).

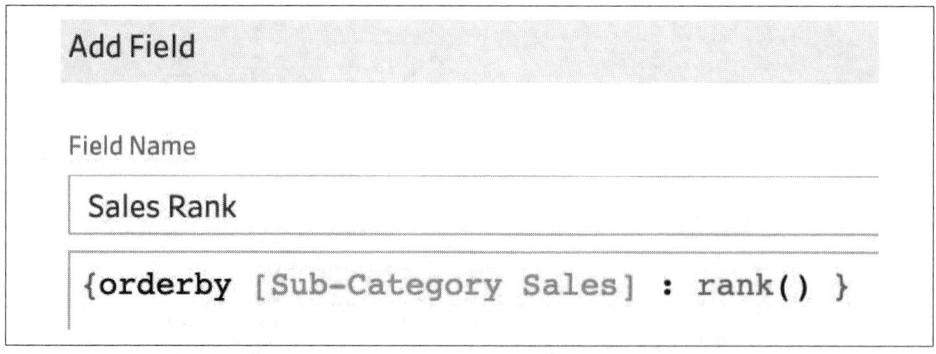

Figure 34-9. Creating a rank based on the Sub-Category Sales field

The result of the calculation is 1 through 17, as there are 17 Sub-Category Sales values (Figure 34-10).

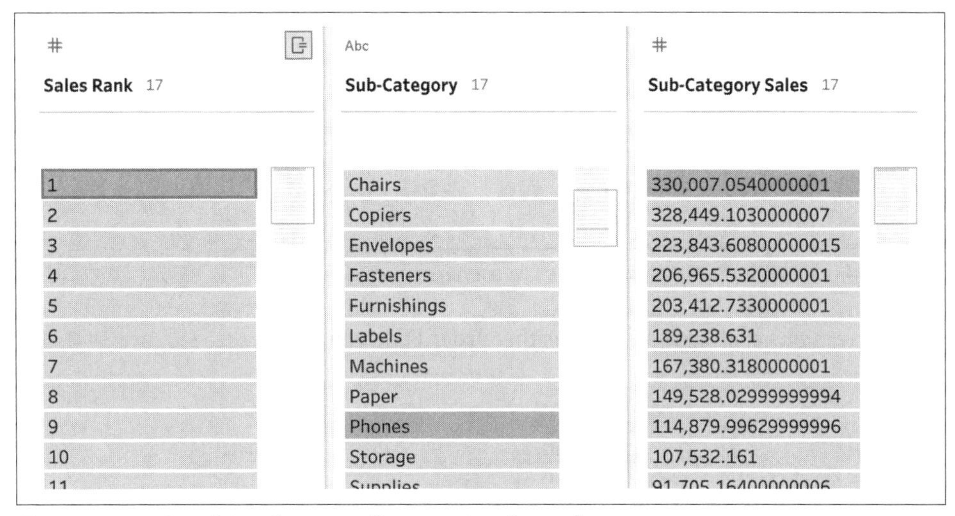

Figure 34-10. Resulting data set after creating the rank

The default ordering of the calculation is from largest to smallest, or *descending*. You can set this option by including `DESC` in the `ORDERBY` calculation. To reverse the rank, you'd include `ASC` (ascending) instead (Figure 34-11).

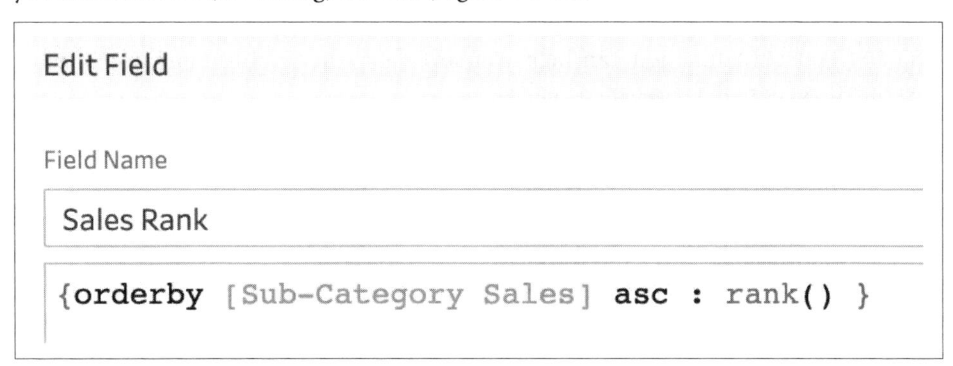

Figure 34-11. Creating an ascending rank

The ROW_NUMBER() function

The `Row_Number()` function allows you to assign a value to each row based on a direction specified in the `ORDERBY` keyword. The default order is descending (largest to smallest, or latest to earliest), so you'd need to add an `asc` to make the row number increase instead (Figure 34-12).

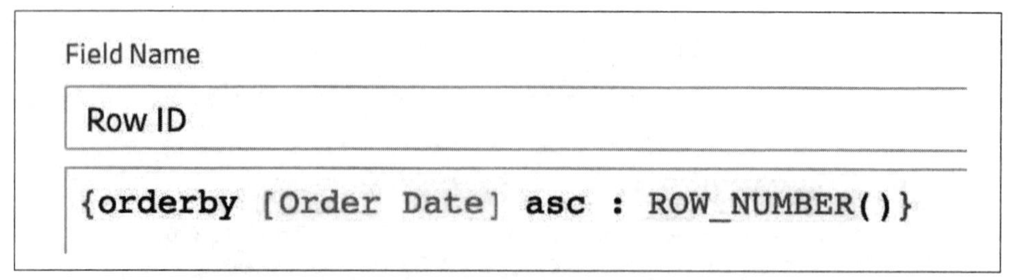

Figure 34-12. Setting up a row number to increase over time

This increments each row number as the Order Date values increase (Figure 34-13).

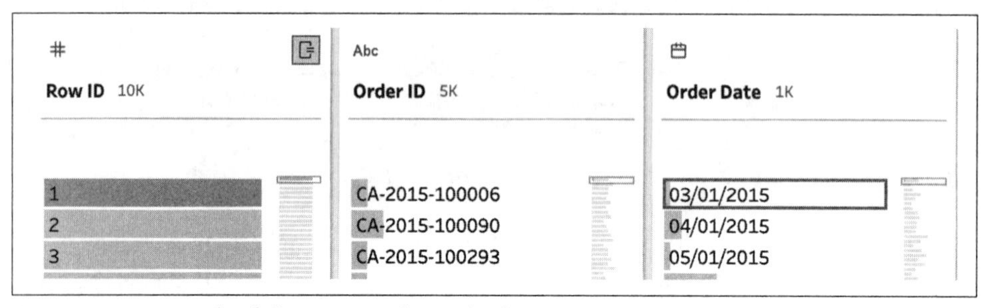

Figure 34-13. Result of the row number calculation

The `ROW_NUMBER()` function can help you sort data when doing more complicated cleaning tasks to make it easier to analyze.

Use Cases

In Prep, there are numerous use cases that can benefit from analytical calculations. This section will cover just a few examples.

Filtering for the Top N

In earlier versions of Prep Builder, creating "Top *N*" filters was quite the challenge. Ranking data involved a complicated self-join with tricky join conditions. Now, the rank functionality allows you to easily compile "Top *N*" lists by filtering. Let's return to the Sub-Category Sales rank from earlier and filter it down to just the Top 5 Sub-Categories by Sales (Figure 34-14).

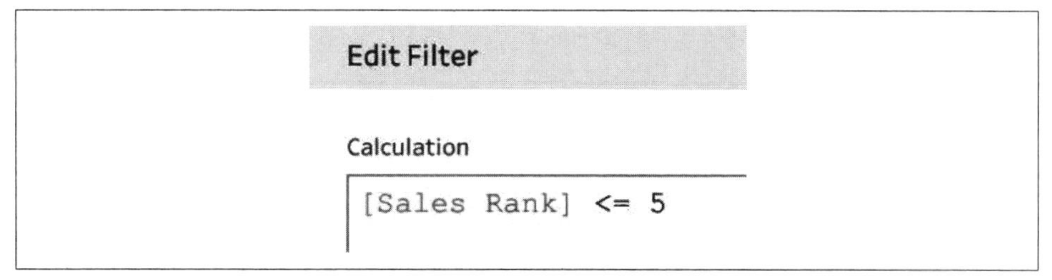

Figure 34-14. Calculating a Top 5 rank

This returns just the Top 5 values (Figure 34-15).

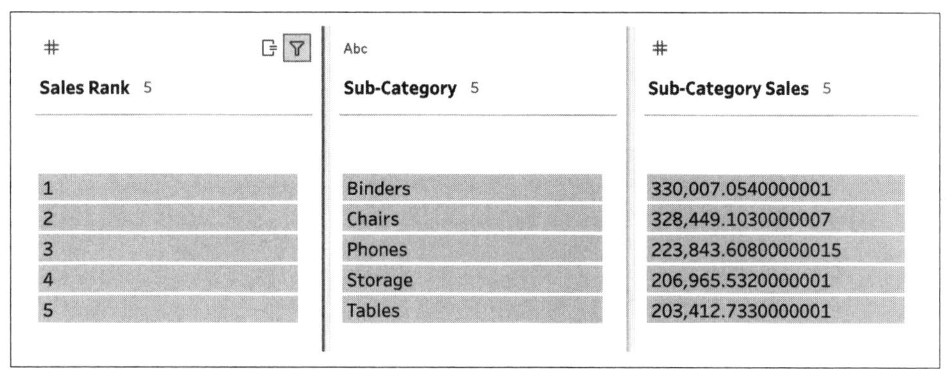

Figure 34-15. Resulting data set after applying the Top 5 filter

Filtering Out a Percentage of Data

Using a similar technique as the Top 5 ranking, you can set how much of a data set you wish to return (based on a value in the data). This is a useful technique if the volume of data, in this case Sub-Categories, will increase over time, but you still wish to return the same-size sample (Figure 34-16).

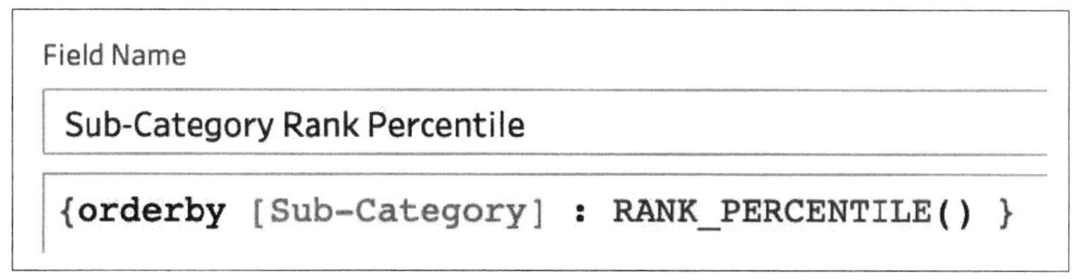

Figure 34-16. Calculating a rank to return a percentile

With this rank calculation, the values returned are between 0 (largest value—in this case, highest sales) and 1 (lowest). The resulting Rank Percentile field can be used in a filter to return the highest 50% of subcategories (Figure 34-17).

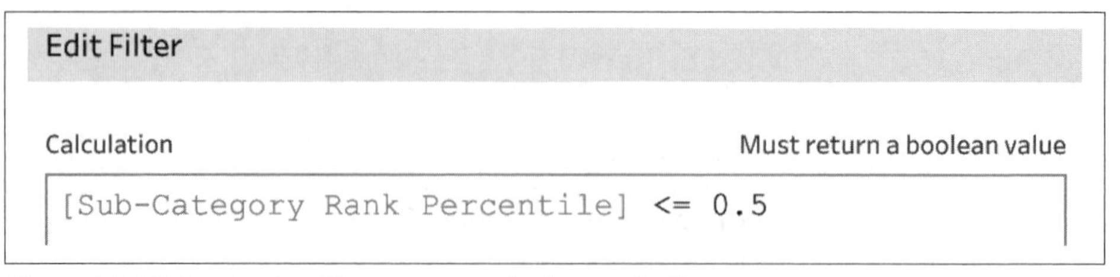

Figure 34-17. Setting the filter to return the largest half

For the 17 subcategories in the Superstore data set, this filter returns 9 subcategories (Figure 34-18).

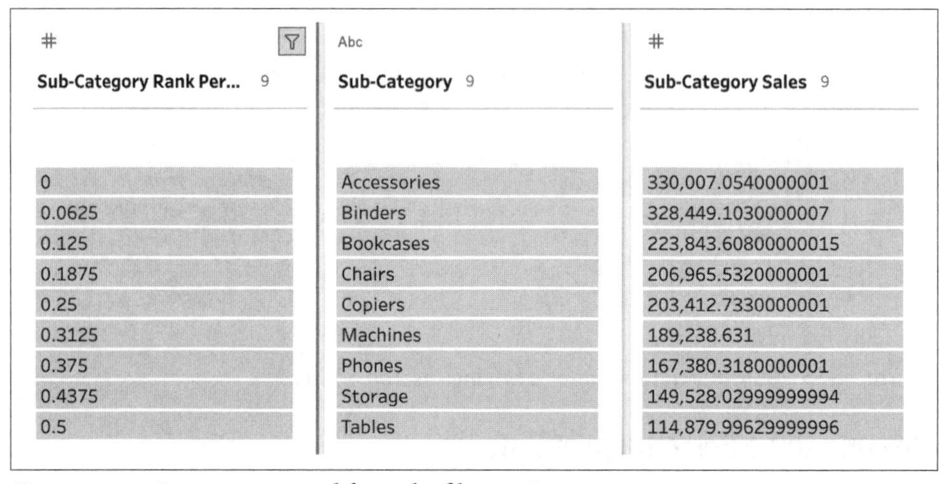

Sub-Category Rank Per... 9	Sub-Category 9	Sub-Category Sales 9
0	Accessories	330,007.0540000001
0.0625	Binders	328,449.1030000007
0.125	Bookcases	223,843.60800000015
0.1875	Chairs	206,965.5320000001
0.25	Copiers	203,412.7330000001
0.3125	Machines	189,238.631
0.375	Phones	167,380.3180000001
0.4375	Storage	149,528.02999999994
0.5	Tables	114,879.99629999996

Figure 34-18. Data set returned from the filter in Figure 34-17

The `RANK_PERCENTILE()` calculation doesn't have to be used just as a filter. You could also use it to set groupings in mailing lists based on the open rate or frequency of customers ordering.

Summary

Analytical calculations enable not just easier data preparation but also deeper analysis before the data is output to a visualization tool. Not everyone has the luxury of super-fast computing power, so any work you can save the visualization software from having to process will help end users find the answers to their questions more quickly. Analytical calculations are not just for making the visualization work easier, however. Both the ranking and row number functions allow you to build more advanced cleaning logic and, therefore, to tackle harder challenges in Prep without having to call on ad hoc programming scripts.

PART VI

Beyond the Basics

CHAPTER 35

Breaking Down Complex Data Preparation Challenges

Previous chapters have discussed techniques for determining the changes required to prepare a data set for analysis, albeit at a relatively simple level. What about those situations where the path isn't straightforward; how do you approach the problem then? Complex challenges can include building solutions requiring multiple steps, inputting data sources using multiple join conditions, or having to complete many reshaping steps throughout the data prep process. This chapter will cover this exact scenario by taking on one of the most complicated challenges Preppin' Data has covered to date: 2020: Week 3 (*https://oreil.ly/Q3naF*). The aim of this challenge, created by Jonathan Allenby (my fellow Dr. Prepper), is to turn the National Basketball Association (NBA) game results into the detailed standings you commonly see on websites or in newspapers.

The Challenge

This challenge involves taking the results and building the full conference league tables, including rankings, wins, and losses; recent performance; and even winning streaks (Figure 35-1). This really is a tough challenge with many facets; hence, it's a great example to use to show you how to break down complex problems into smaller, more manageable pieces.

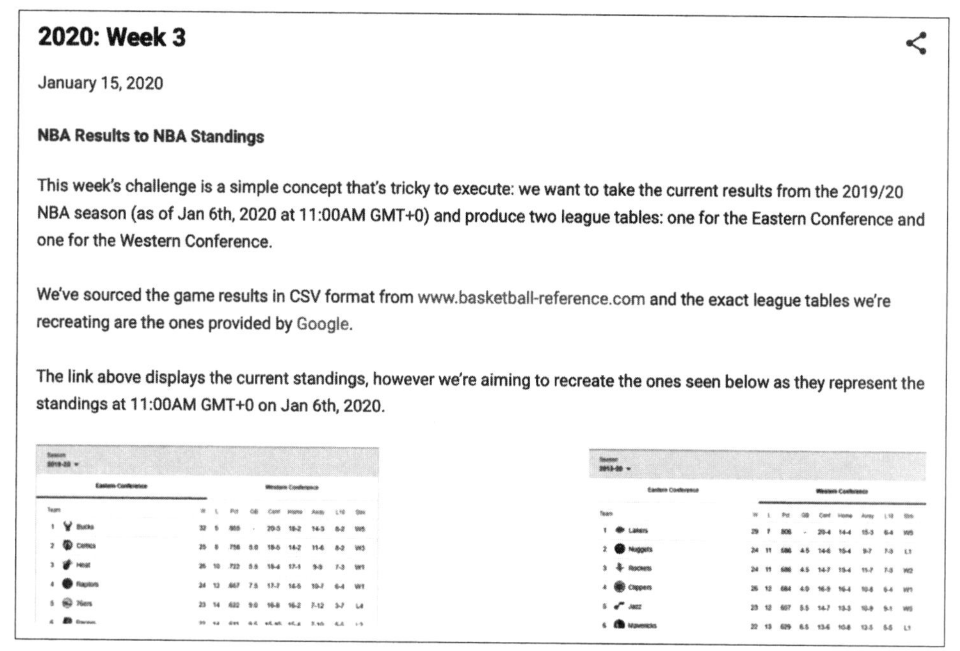

2020: Week 3

January 15, 2020

NBA Results to NBA Standings

This week's challenge is a simple concept that's tricky to execute: we want to take the current results from the 2019/20 NBA season (as of Jan 6th, 2020 at 11:00AM GMT+0) and produce two league tables: one for the Eastern Conference and one for the Western Conference.

We've sourced the game results in CSV format from www.basketball-reference.com and the exact league tables we're recreating are the ones provided by Google.

The link above displays the current standings, however we're aiming to recreate the ones seen below as they represent the standings at 11:00AM GMT+0 on Jan 6th, 2020.

Figure 35-1. Challenge post for Preppin' Data 2020: Week 3

As covered in Chapter 3, developing an understanding of the input and output will help you to get an overview of the task. On large, complex challenges, planning becomes especially important to ensure you are working toward the desired results.

Where to Begin

Figure 35-2 shows the first few rows of one of the monthly inputs for the challenge, demonstrating the structure of the data set.

Date	Start (ET)	Visitor/Neutral	PTS	Home/Neuti	PTS			Attend.	Notes
Tue Oct 22 2019	8:00p	New Orleans Pelicans	122	Toronto Rap	130	Box Score	OT	20787	
Tue Oct 22 2019	10:30p	Los Angeles Lakers	102	Los Angeles	112	Box Score		19068	
Wed Oct 23 2019	7:00p	Chicago Bulls	125	Charlotte Hc	126	Box Score		15424	
Wed Oct 23 2019	7:00p	Detroit Pistons	119	Indiana Pac	110	Box Score		17923	
Wed Oct 23 2019	7:00p	Cleveland Cavaliers	85	Orlando Ma	94	Box Score		18846	
Wed Oct 23 2019	7:30p	Minnesota Timberwolves	127	Brooklyn Ne	126	Box Score	OT	17732	
Wed Oct 23 2019	7:30p	Memphis Grizzlies	101	Miami Heat	120	Box Score		19600	
Wed Oct 23 2019	7:30p	Boston Celtics	93	Philadelphi	107	Box Score		20422	
Wed Oct 23 2019	8:30p	Washington Wizards	100	Dallas Mave	108	Box Score		19816	

Figure 35-2. Sample input data set for NBA challenge

To understand where to begin, it's important to think about what is required from the data set. As this is a Preppin' Data challenge, this has already been detailed for you (Figure 35-3).

Rank	Team	W	L	Pct	Conf	Home	Away	L10	Streak
1	Los Angeles Lakers	28	7	0.8	20-4	13-4	15-3	6-4	W4
3	Denver Nuggets	24	11	0.686	14-6	15-4	9-7	7-3	L1
3	Houston Rockets	24	11	0.686	14-7	13-4	11-7	7-3	W2
4	Los Angeles Clippers	25	12	0.676	16-9	15-4	10-8	5-5	L1
5	Utah Jazz	23	12	0.657	14-7	13-3	10-9	9-1	W5
6	Dallas Mavericks	22	13	0.629	13-6	10-8	12-5	5-5	L1
7	Oklahoma City Thunder	20	15	0.571	14-11	12-6	8-9	9-1	W5
8	Portland Trail Blazers	15	21	0.417	9-15	8-9	7-12	5-5	W1
9	San Antonio Spurs	14	20	0.412	9-12	10-9	4-11	5-5	L2
10	Phoenix Suns	14	21	0.4	9-17	7-12	7-9	3-7	W1
11	Memphis Grizzlies	14	22	0.389	9-14	7-12	7-10	5-5	W1
12	Minnesota Timberwolves	13	21	0.382	6-16	5-11	8-10	3-7	W1
13	Sacramento Kings	13	23	0.361	9-14	7-10	6-13	1-9	L1
14	New Orleans Pelicans	12	24	0.333	10-15	6-11	6-13	6-4	W1
15	Golden State Warriors	9	28	0.243	7-20	6-12	3-16	4-6	L4

Figure 35-3. Desired output for the NBA challenge

Here's my initial scope for this challenge (Figure 35-4).

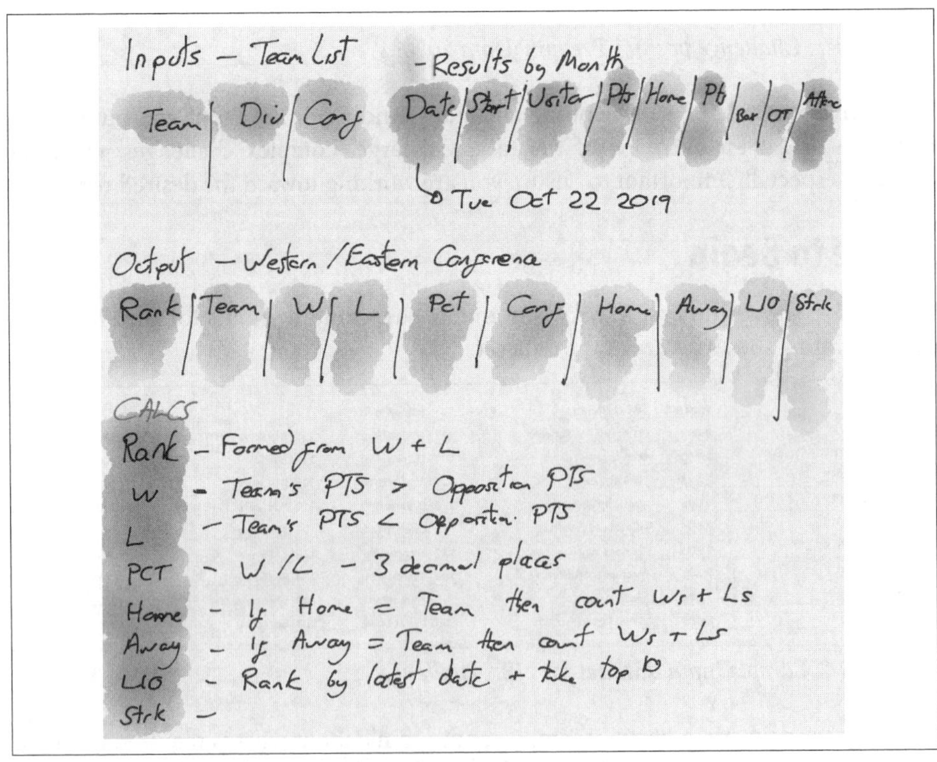

Figure 35-4. Sketched plan to solve the challenge

Let's revisit some of the steps covered in Chapter 3 and how they apply to this challenge:

1. Map out your inputs. What does each input contain? What dimensions and measures are there in the data? Are any data fields incorrectly formatted? What is the data's level of granularity?
   - In this case, only the Date field seems to potentially need cleaning.
2. Map out your outputs. How many output files will be required? What format will you need certain fields to be? Think about the granularity of the data required.
   - In this case, each team will have a single row within the output. This means there will be a lot of aggregation required to take the game results to the level we need.
3. Understand the gaps. What fields are missing within the data, and how do we add them? This will give us a list of fields that we need to create through either pivoting, joins, or calculations.

At this stage, you don't have to solve all the issues that might arise between loading the input data and generating the output data set for analysis. As discussed in Chapter 3, you might not spot all of the challenges in the data set at this point, but as you work with the data, they will emerge.

Logical Steps

Breaking the challenge down into individual chunks makes it a lot easier to work out how best to solve it. If you skip this step, the challenge may seem insurmountable. Creating the calculations you know you'll need is a good first step, as they will help:

- Guide next steps toward finding a solution
- Determine an order in which to proceed

Let's take wins as an example in this challenge. You have individual game results, and you need to determine who the winner is. This is easier said than done, though, because for each win there is a loss as well. This means two rows are required for each game: one to record the winner, and one for the loser. To ensure we capture every team's games, first we'll use the Team List data source and join all game results to that twice—once to match the home team and again to match the away team to the original Team List team names (Figure 35-5).

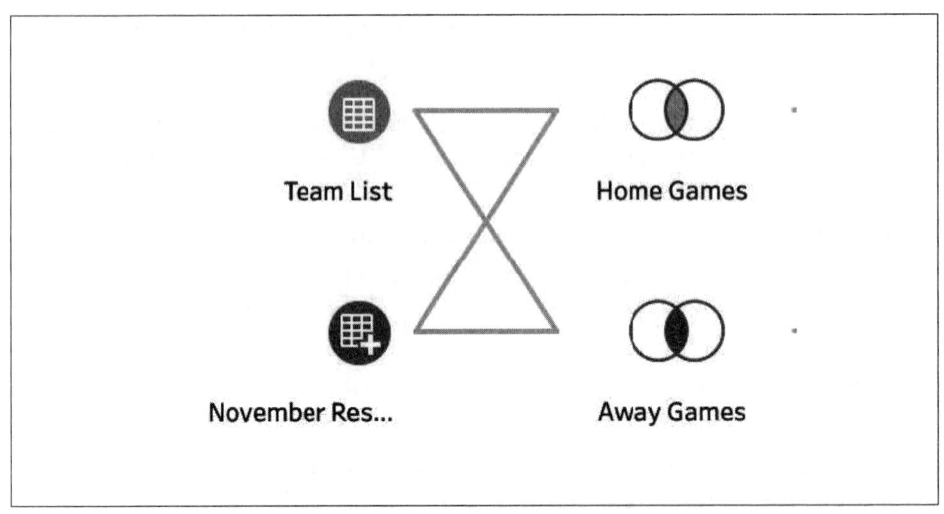

Figure 35-5. Join setup to create one set of results per team for home and away games

Using two calculations then allows us to calculate each team's points (Figure 35-6) and their opposition's points (Figure 35-7) for home games.

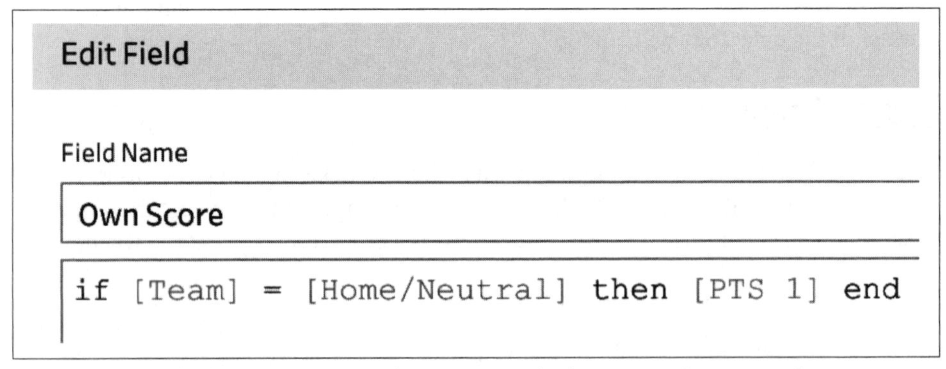

Figure 35-6. Calculating Own Score by returning the home team's points value

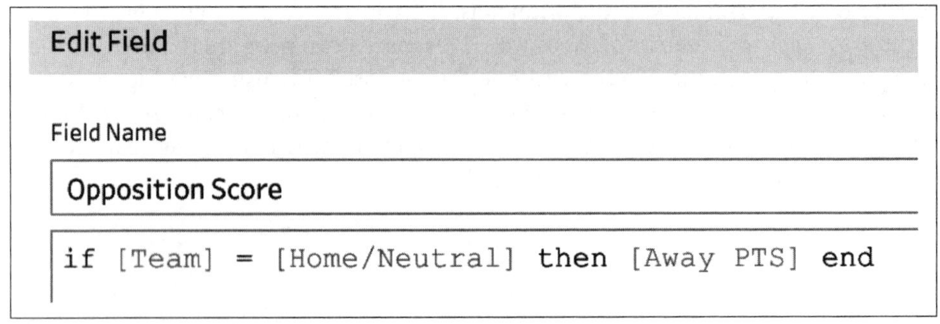

Figure 35-7. Calculating Opposition Score by returning the away team's points value

These calculations result in the data set shown in Figure 35-8.

Opposition Score	Own Score	Team	Divison	Conference	Visitor/Neutral	Away PTS	Home/Neutral	PTS 1	Date
116	123	Brooklyn Nets	Atlantic	Eastern	Houston Rockets	116	Brooklyn Nets	123	Fri Nov 1 2019
95	102	Indiana Pacers	Central	Eastern	Cleveland Cavaliers	95	Indiana Pacers	102	Fri Nov 1 2019
123	91	Orlando Magic	Southeast	Eastern	Milwaukee Bucks	123	Orlando Magic	91	Fri Nov 1 2019
102	104	Boston Celtics	Atlantic	Eastern	New York Knicks	102	Boston Celtics	104	Fri Nov 1 2019

Figure 35-8. Result of Own/Opposition Score calculations

We can then repeat these calculations, but this time testing whether the Team List team name is the away team. Figure 35-7 showed how to calculate the away team's score where the focus team plays away. Figure 35-9 shows how to calculate the opposition's score.

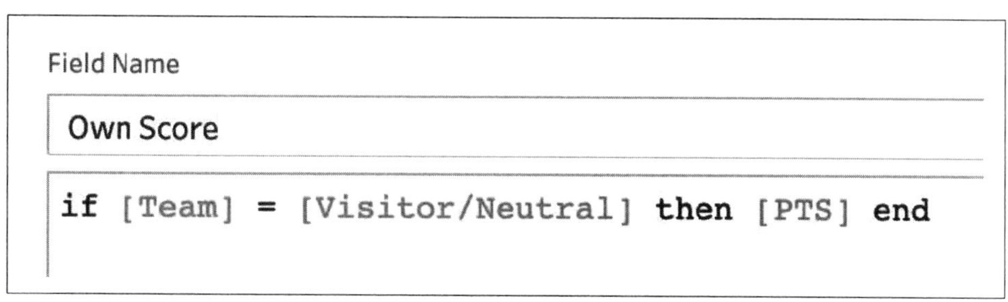
Field Name
Own Score
if [Team] = [Visitor/Neutral] then [PTS] end

Figure 35-9. Calculating the away team's Own Score from the second Join step in Figure 35-5

We can then assess these calculations for who won or lost the game (Figure 35-10). We can take the same approach for games in which the team played away and union the results together.

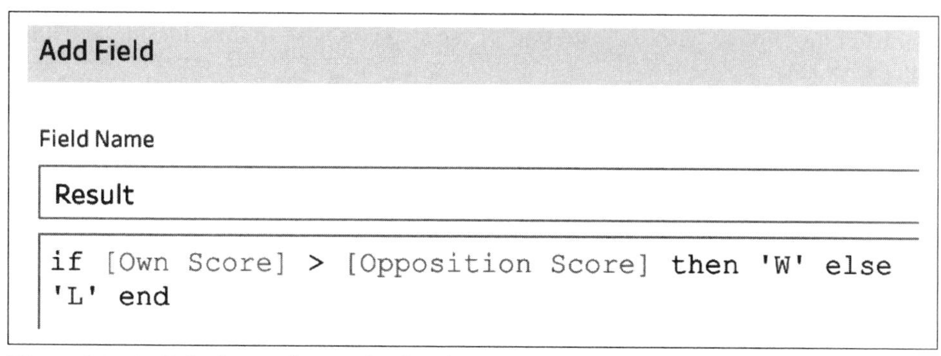
Add Field
Field Name
Result
if [Own Score] > [Opposition Score] then 'W' else
'L' end

Figure 35-10. Calculating the result of each game

A sample of the resulting data set is shown in Figure 35-11.

Result	Opposition Score	Own Score	Team	Divison	Conference	Visitor/Neutral	Away PTS	Home/Neutral	PTS 1	Date
W	116	123	Brooklyn Nets	Atlantic	Eastern	Houston Rockets	116	Brooklyn Nets	123	Fri Nov 1 2019
W	95	102	Indiana Pacers	Central	Eastern	Cleveland Cavaliers	95	Indiana Pacers	102	Fri Nov 1 2019
L	123	91	Orlando Magic	Southeast	Eastern	Milwaukee Bucks	123	Orlando Magic	91	Fri Nov 1 2019
W	102	104	Boston Celtics	Atlantic	Eastern	New York Knicks	102	Boston Celtics	104	Fri Nov 1 2019

Figure 35-11. Sample of the resulting data set to this point

We can then union together the two flows to create one large data set to record each team's full set of games in the season (Figure 35-12).

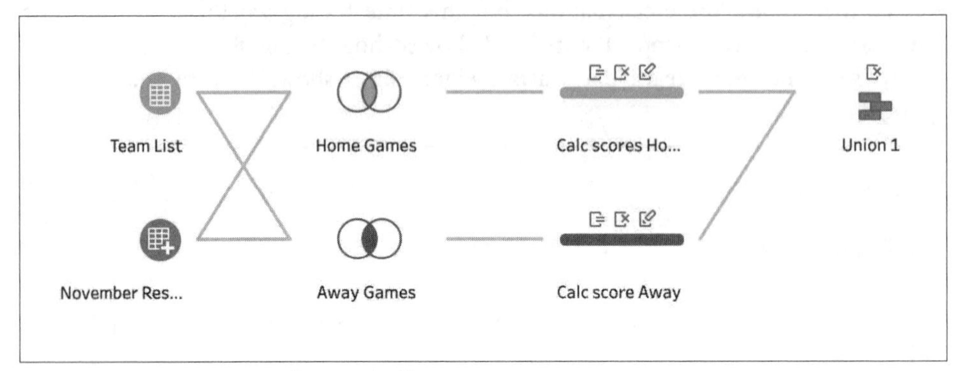

Figure 35-12. First stage of NBA challenge

Making Changes

You can repeat this approach to ensure you are tackling each of those calculations in turn. As you determine a solution for each subchallenge, you may need to change the order of the steps, or copy and paste entire sections. This is easy enough. By right-clicking on the linking line between two steps, you can delete it and then drag the step from the "pre-step" to the Add part of the step you wish to link it to in your flow (Figure 35-13).

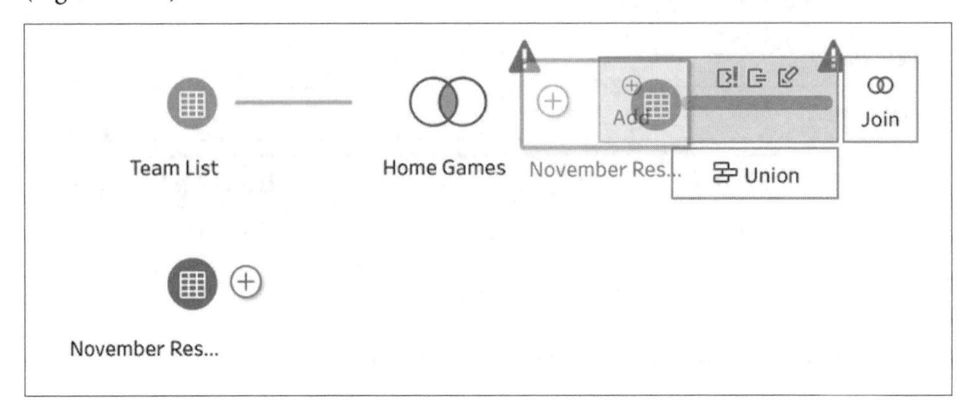

Figure 35-13. Reconnecting the steps

Be Ready to Iterate

Often only by working with the data will you arrive at the solution. Otherwise, it can be difficult to imagine how exactly the data will behave during each step and as you use the data when conducting your analysis. When analyzing the data, you may determine that you need to iterate further or remove certain steps. This is a good thing—it is all part of the data prep learning experience, and fortunately, with Prep Builder, you can make these changes quickly.

For example, when creating the Win (W), Loses (L), Home Results (Home), and Away Results (Away) columns, I knew they would involve similar calculations but wasn't sure in which order I would handle them. These columns represent the total wins and losses as well as the record of wins and losses achieved in home and away games. Figure 35-14 shows the flow I used to approach this task.

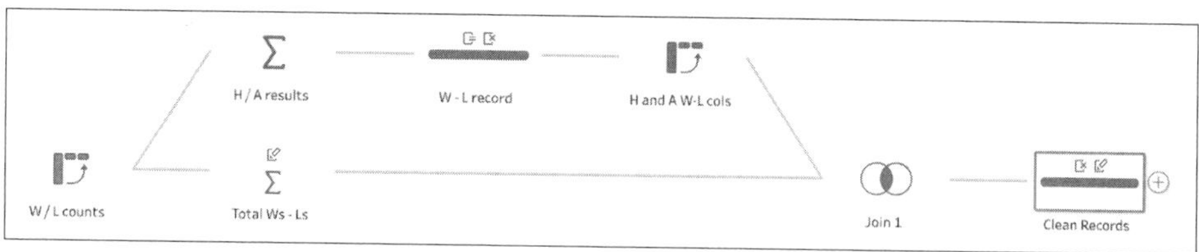

Figure 35-14. Flow to create Wins, Losses, Home Results, and Away Results columns

The primary data point I needed for each team was whether the team won or lost a game. I already had captured whether the team at the focus of the game had scored more points than the opposition. If they had scored more, I returned a column of W; otherwise, I returned a column of L. This logic was correct but wasn't ideal for aggregating to calculate all the relevant totals of wins and losses. Therefore, I pivoted this column to create a single column for both.

To make the next counts simple, I created a simple calculation of 1 to add to each win or loss column depending on the result (Figure 35-15).

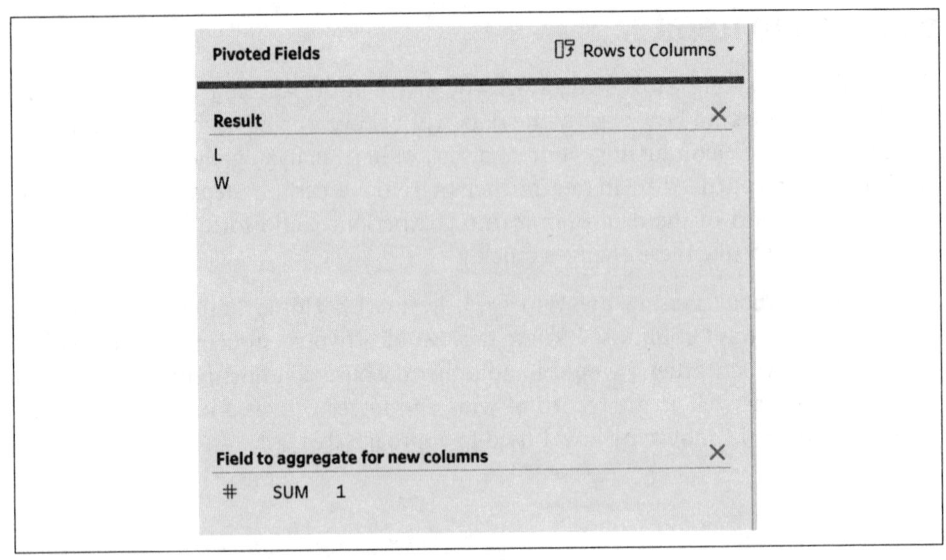

Figure 35-15. Pivot to calculate Win/Loss counts

Knowing that the next steps would require an Aggregate step, I realized I'd need to split the process into two streams at this point. The Aggregate step returns only the aggregated values and the Group By data fields (for a refresher on the Aggregate step in Prep, read Chapter 15). Even with very careful planning, it is unlikely that I would have predicted adding another branch to the flow at this point. Prep is a fantastically agile tool when your logic starts to diverge from your initial plans. Because I needed to aggregate at an overall level as well as to split out the home and away records, I needed two separate Aggregate steps (Figure 35-16).

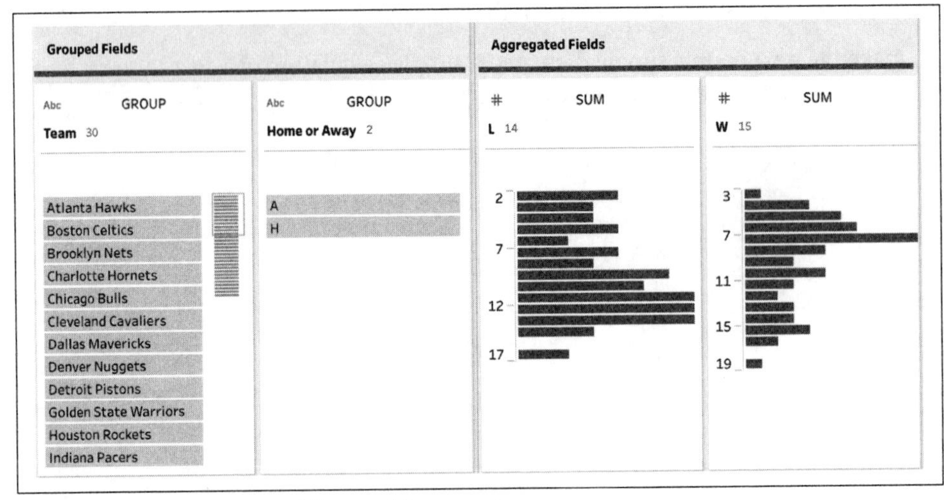

Figure 35-16. Aggregating Wins and Losses columns

These two flows were eventually joined back together to form one overall view of the team's records at the end of the flow shown in Figure 35-14. Using the Profile pane allows you to see which fields are required and which are not and remove them through the ellipsis menu. Figure 35-17 shows the resulting data set.

Team	Record	Home Record	Away Record
Philadelphia 76ers	10-6	3-6	7-0
New York Knicks	3-11	1-5	2-6
Milwaukee Bucks	15-1	8-1	7-0
Portland Trail Blazers	5-10	2-7	3-3

Figure 35-17. Resulting data set from this subsection of the flow

To practice what this chapter has covered, why not have a go at the challenge yourself? I won't spoil the whole challenge here and show every step, but this should get you started.

Summary

Being able to iterate your approach is a powerful data preparation skill, as often there are cleaning issues you didn't spot in your initial approach or your first solution doesn't work as intended. Enjoying the challenge of data preparation comes down to being able to focus on problem-solving. This is much easier if you break big challenges down into more manageable chunks.

CHAPTER 36

Handling Free Text

Free isn't always a good thing. In data, free text is a perfect example of this. However, there are lots of benefits to be gained from the data entered in free text fields.

What Is Free Text?

Free text is the string-based data resulting from answers people type into systems and forms. Surveys, social media posts, and operational system notes can all have free-text data points. The resulting data is normally stored within one column, with one answer per cell. By definition, free text means the answer could be anything, and this is what you get—absolutely anything. From expletives to slang, the words you will find in the data may be a challenge to interpret, but free text is the best way to collect the true voice of your customer/employee.

The free-text field is likely to contain long, rambling sentences that cannot be analyzed simply. Most entries are entirely unique, so it's unlikely that you'll be able to conduct any meaningful analysis by simply counting the frequency of each instance. The analytical value is in the individual words/ID numbers within the submitted sentences and paragraphs.

Why Is Free Text Useful?

Capturing simple survey results as the percentage of people who feel strongly for or against something won't give you much insight into where your business could improve. Hearing the "voice" of your customers in their own words will tell you more than even the best analysts in the world would be able to.

Free-text forms give you the flexibility to capture anything the form user wants to share, making for extremely useful data. However, that flexibility also makes the data harder to analyze. This has led organizations to take two different approaches to free-text analysis, each with pros and cons:

- Some organizations task third-party companies or junior employees with reading each comment and categorizing it. This method actually restricts the value of free-text comments, as the form should have forced users to make a choice in the first place so there's no ambiguity in their feedback.
- Paying third parties to conduct this work is expensive. Therefore, lots of organizations have removed the free-text entries altogether to make it easier to measure user responses in their analysis. By restricting feedback to a set of predefined choices, this approach removes the flexibility for form users to share their valuable insights.

Free text often fills the gaps in poor or developing systems. Organizations struggle with key operational system changes as either the system's complexity or the cost to make the changes becomes too significant. Free text allows them to capture and store the data required to make the system functional, but from a data analysis perspective, it also makes the data difficult to analyze on a large scale.

How to Analyze Free Text in Tableau

Despite its downsides, free text is extremely valuable and shouldn't be eradicated from forms and systems. It allows for a "catchall" response when the rest of the system or form doesn't provide another way for users to adequately share their feedback. Writing unbiased surveys is a very difficult skill to develop, for example, so a free-text box can capture information that wouldn't be gathered otherwise. This might lead to product innovations or highlight service gaps you wouldn't find without costly customer interviews.

Finding these nuggets of information, however, can be challenging for the following reasons:

- Each response is likely to be unique if you allow more than four words from a respondent.
- Finding a particular word is easy through a `CONTAINS()` function, but you need to know what to look for and hope respondents have spelled the word correctly. The whole point of free text is that your users might be telling you something that you don't know.
- There will be lots of words you need to get rid of, which can take a lot of work.

Therefore, the steps you need to take to analyze free text are:

1. Split sentences into multiple columns so each word is stored separately (this can create a very wide data set).
2. Pivot these columns into a single column to make cleaning a lot easier.
3. Clean the data by removing punctuation, spaces, and hashtags, and making case consistent across the data set (I recommend making all letters lowercase).
4. Join a data set of common words (*the*, *or*, *they*, etc.), as they will not add any value to your analysis, and remove any that appear in both data sets. Lists of the most common words are available through a Google search.
5. Use the Group and Replace options to add back in the original free text (if you've removed it throughout the flow) so you have context for the words in your final analysis.

Using the Tweets input data set from Preppin' Data 2019: Week 5 (*https://oreil.ly/QXPXT*), now we'll walk through these five steps in Prep Builder to make our free-text analysis easier.

Split the Strings

Breaking apart the longer strings into individual words is essential when you are working with free text. Prep Builder's Automatic Split option will often do the hard work for you by identifying the common separator within the text field and creating a new column each time it finds it (Figure 36-1). Remember there is a 150-split limit.

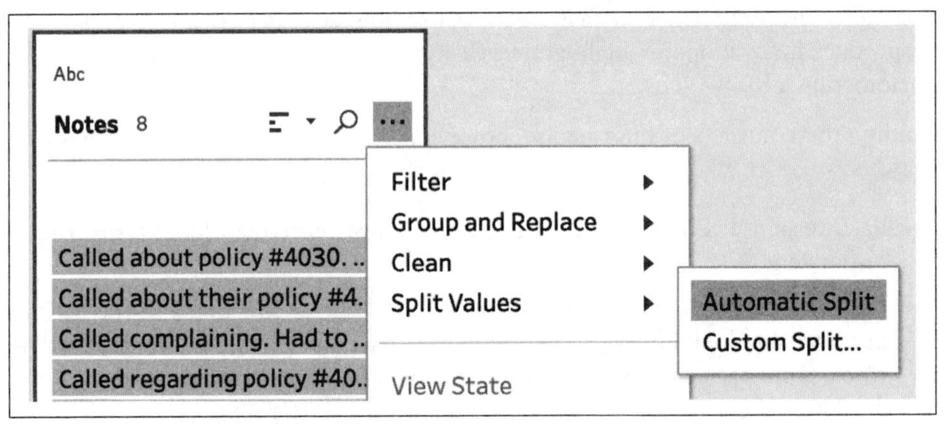

Figure 36-1. Automatic Split option

The new columns will have the same name as the existing field but with a "- Split *nn*" suffix (where *nn* is the split number) for each new split result. For example, in

Figure 36-2, there are 28 spaces in the longest Notes field; therefore, 29 new fields are created to make sure all words are returned.

Notes - Split 25	Notes - Split 26	Notes - Split 27	Notes - Split 28	Notes - Split 29
up.	Added	to	account	#3001

Figure 36-2. Part of the results from a split

Pivot Columns to Rows

As you can see, there are a number of blank values returned for the records that do not have that many spaces. Therefore, our second step, pivoting columns to rows, serves two purposes:

- Putting all the words in one column so you can complete the cleaning operations described earlier in step 3.
- Removing blank cells to reduce the data set (often considerably) and improving the processing time for the rest of your data preparation flow.

With potentially tens, if not hundreds, of columns to pivot, dragging each one into the pivot tool can take a long time. The Prep team has accounted for this, including the "Use wildcard search to pivot" option to make this process a lot easier (Figure 36-3).

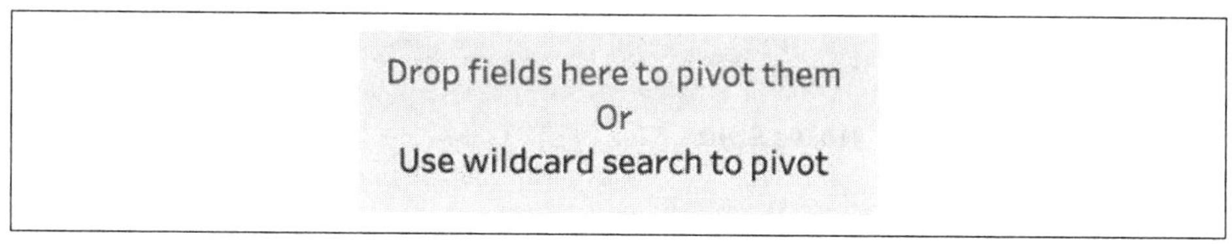

Figure 36-3. Wildcard search for Columns to Rows pivot

The wildcard search should not use the original field name (unless you have split multiple fields), as you want just the separated words. This way, using the "- Split" string as the search term will not add the original field but will instead pick up all the results of the split (Figure 36-4).

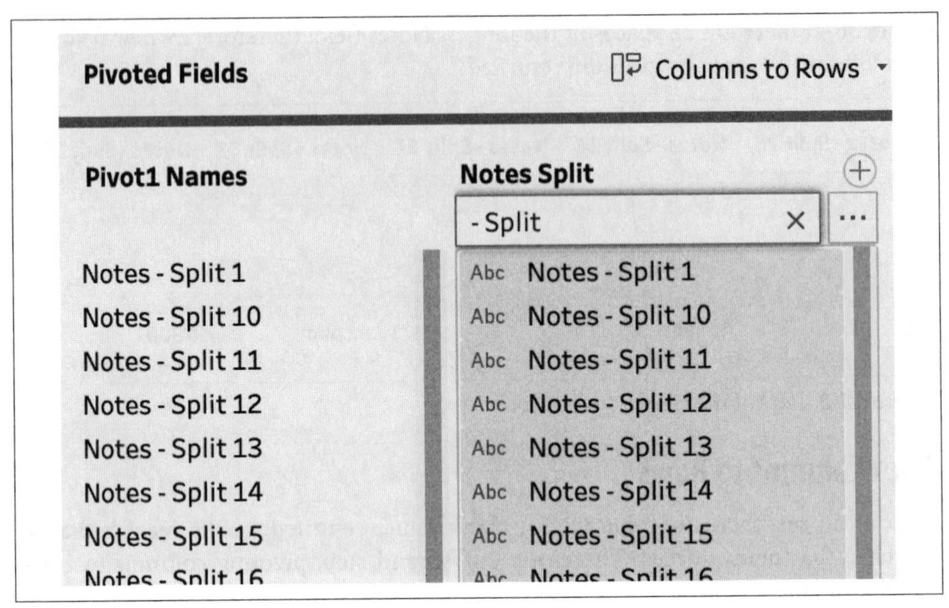

Figure 36-4. Setting up a wildcard search to pivot results from a split

After this step, it's very easy to remove the nulls from the data set—simply right-click on the nulls or blank values in the Profile pane and select Exclude (Figure 36-5).

See Chapter 23 for other filtering considerations.

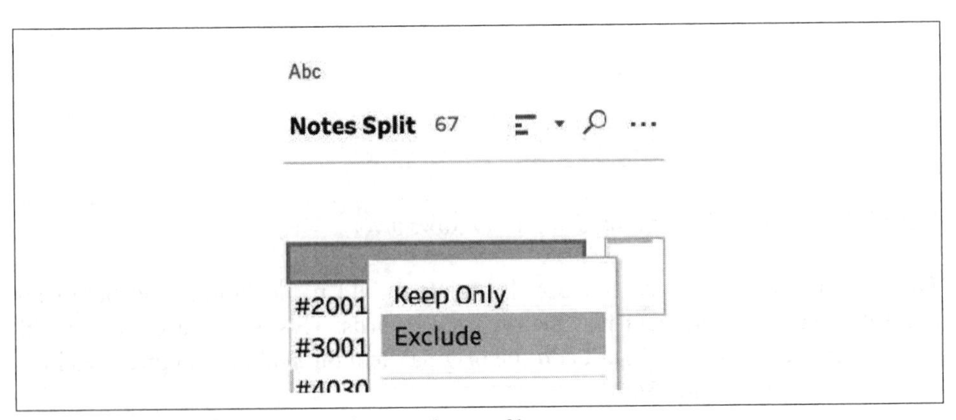

Figure 36-5. Excluding blank values in the Profile pane

Clean Cases and Punctuation

String-based data in Tableau Desktop is case-sensitive, so `'String'` and `'string'` would be counted as two different records. Prep Builder has a range of options to help you match as many values as possible through just a few clicks. In your data field that you want to make lowercase, click on the ellipsis menu, select Clean, and then select Make Lowercase (Figure 36-6).

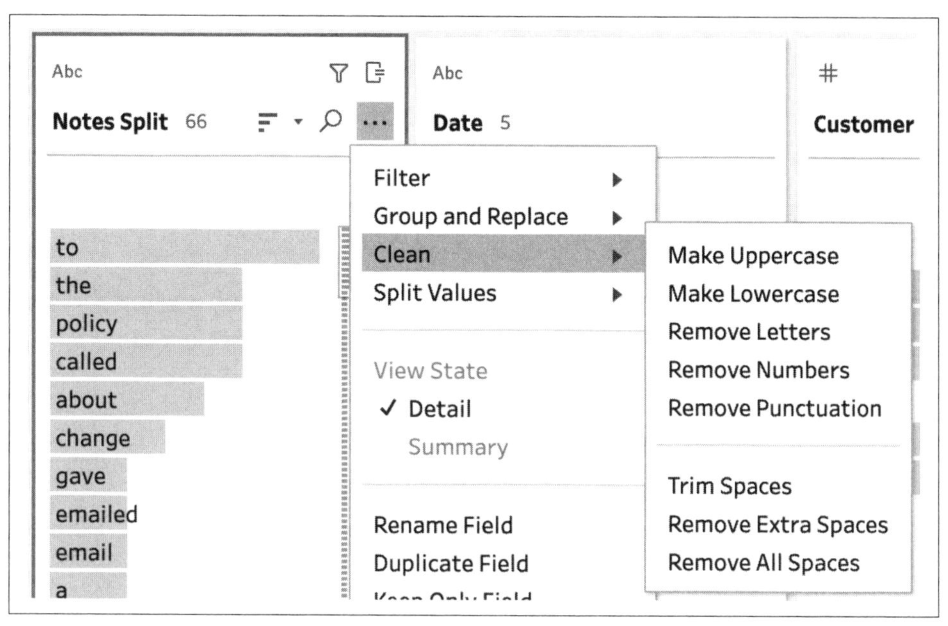

Figure 36-6. Cleaning values using the data field's ellipsis menu in the Profile pane

Getting rid of unnecessary punctuation is just as important as matching cases. `'String.'` is not the same as `'String'`, for example. Removing any extra spaces or other punctuation will ensure you develop the best analysis possible.

Use a Join to Remove Common Words

The results are already useful, but to save yourself a lot of time sifting through words that won't add much insight, you can remove the most common words (in the specified language) from the free-text field. For the 2019: Week 5 challenge example, 5 of the top 10 common words—*to, the, about, gave, a*—would give us no insight at all (Figure 36-7).

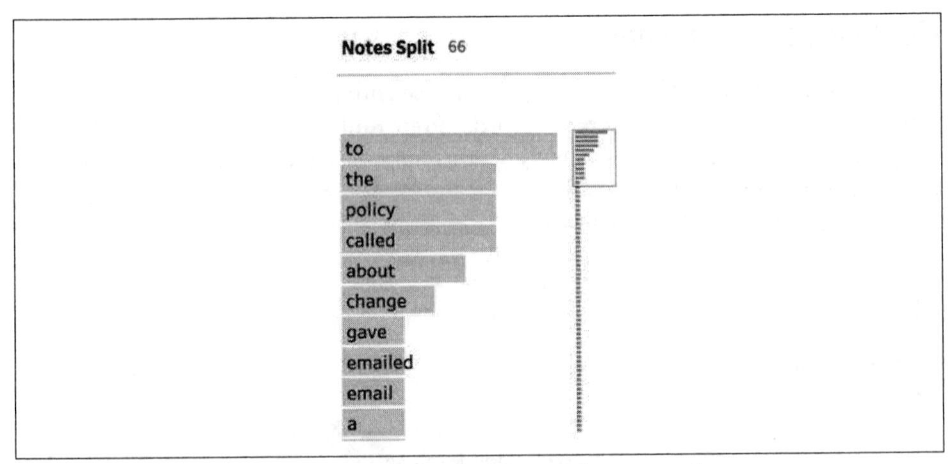

Figure 36-7. Top 10 words resulting from the pivot of the tweet split

It's easy to find a list of common words online. Having such a list will guide your removal efforts. Figure 36-8 shows a sample common words data set.

Rank	Word
1	the
2	of
3	to
4	and
5	a
6	in
7	is
8	it
9	you
10	that

Figure 36-8. A common words data set

Setting up the join properly is important to ensure the common words are removed. First, you'll need to join the list of common words to the list of single words you have split out from the free-text field, and then you can remove any words that are found in both lists. To do this, click on the intersecting part of the join condition. Your join setup should look like Figure 36-9.

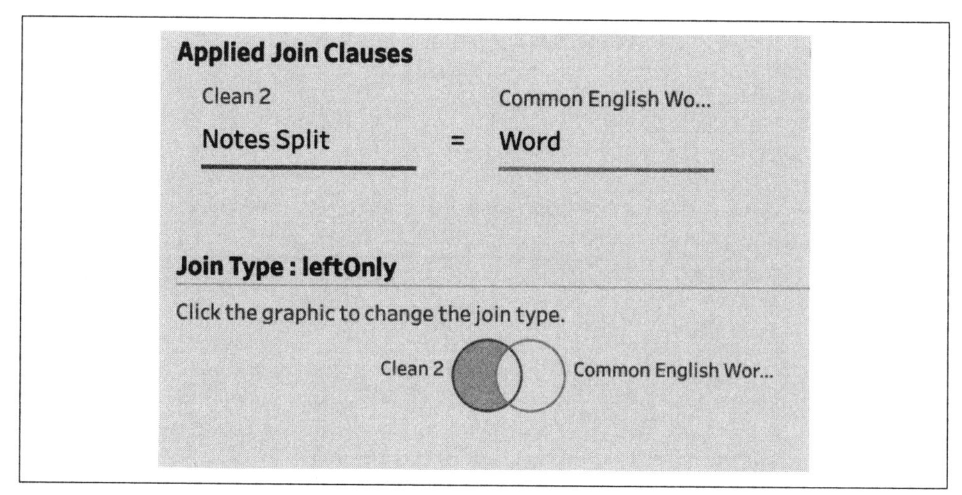

Figure 36-9. Join setup to remove common words found in the tweets

Group the Remaining Values

As Chapter 20 discussed, Prep Builder's grouping functionality is very flexible. This last step is optional and based on the analyst's opinion. On one hand, if you group values that shouldn't be grouped, you will get inaccurate answers to your questions. On the other hand, if you have similar values that can be grouped and aren't, you are going to struggle to find the signal in the noise. You can make some of the grouping algorithms more or less sensitive to better fit your requirements (Figure 36-10).

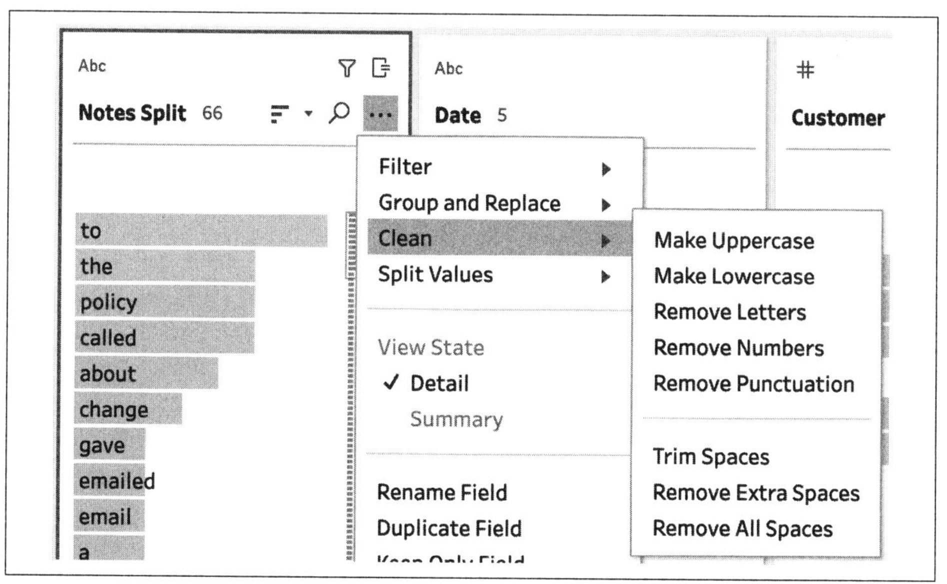

Figure 36-10. The Group and Replace options are one way to clean through grouping

The resulting data set of all five steps is shown in Figure 36-11.

Week	Policy Number	Notes Split	Date	Customer ID
17th June 19	#4899	Called	Monday	29,439
17th June 19	#4030	Called	Tuesday	39,822
17th June 19	#3001	Called	Tuesday	27,316
17th June 19	#2001	Email	Wednesday	12,219
17th June 19	#4030	Called	Thursday	39,822
17th June 19	#9220	Email	Thursday	49,291
17th June 19	#6090	Emailed	Friday	40,201
24th June 19	#2080	Emailed	Monday	72,617
24th June 19	#4030	Call	Monday	39,822
24th June 19	#4899	Called	Monday	29,439

Figure 36-11. Resulting data set from the steps in this chapter

This data set allows end users to look for common words and terms in the Notes field to help understand the feedback captured. Figure 36-12 shows the data visualized in Tableau Desktop.

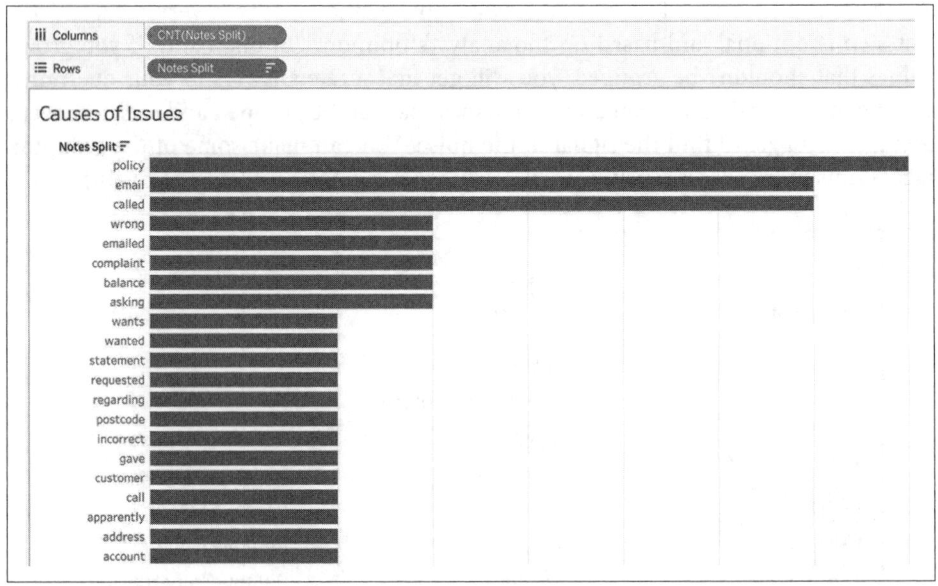

Figure 36-12. Common words in the Notes data field

Summary

Using free text to provide insight into your customers' opinions and feedback can add much more value than hours of number crunching, so mastering the techniques in this chapter can be a big time saver and improve your analysis. Due to the range of possible values entered in free-text fields, the number of steps involved in cleaning the resulting data can be considerable, but the effort is well worth it for the analytical benefits you'll gain.

CHAPTER 37
Using Smarter Filtering

The basics of filtering were covered in Chapter 23, but filtering doesn't have to mean just manual clicking or setting date ranges to specify what data should be included in your output. Calculations, joins, and measuring percentage variance can all make filters smarter and more automated. Removing manual work during the data preparation stage reduces the risk of mistakes and makes it easier to share the data set with others once it is published.

Calculations

Any calculation can be used as a way to trigger filters. However, there are certain types of calculations that are more common than others for use as filters.

Boolean Calculations

Testing whether or not a value in a data field meets a condition—for example, if a value exceeds a logical limit—is a great way to clean data and remove the records that you do not want. In the example shown in Figure 37-1, we know that Percent Complete cannot exceed 100%, so we use a Boolean filter calculation in Prep to ensure the value is 1 or less by removing any values that exceed 1.

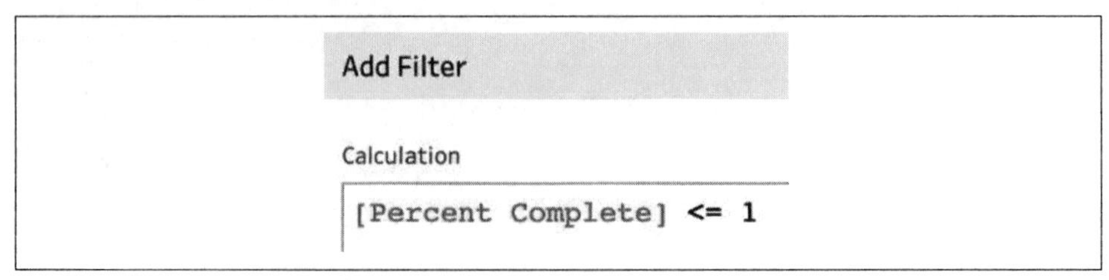

Figure 37-1. Limiting Percent Complete to 100%

Remember, percentages are often set as a value between 0 and 1 rather than 0 and 100, as this helps formatting in Tableau Desktop.

Logical Calculations

`IF` and `CASE` calculations can contain quite complex logic that makes filtering easier by matching certain scenarios across multiple data fields. In the sports data set in Figure 37-2, we want to return only games where one team went from winning at half-time to losing the game, or vice versa. Here are the fields the calculation is using:

`[HTf]`
: Half-time points for the team being assessed

`[HTa]`
: Half-time points for the opposition

`[Result]`
: Final result in the game

Field Name

Upset after Half Time?

```
if [HTf] - [HTa] > 0 and [Result] = 'loss' then
'Upset'
elseif [HTf] - [HTa] < 0 and [Result] = 'won' then
'Upset'
else 'Expected'
end
```

Figure 37-2. An `IF` statement in Prep

This calculation returns both game scenarios in the same result, in this case `'Upset'`, so we can filter `'Expected'` results out of the data set.

`IF` statements work through the conditions from first to last. As soon as one condition or set of conditions is met, the calculation returns the value that is set. The conditions set can be very complicated and detailed, so `IF` statements can be very large, complex calculations, but sometimes you need that complexity to determine whether or not the data should be filtered out.

Regex Calculations

For more complex filtering calculations, regular expressions (regexes) can be a massive help. As covered in Chapter 31, regexes allow you to set a pattern to test. For example, if a letter is present in an expected numerical field, then it can be handled separately with a regex (Figure 37-3).

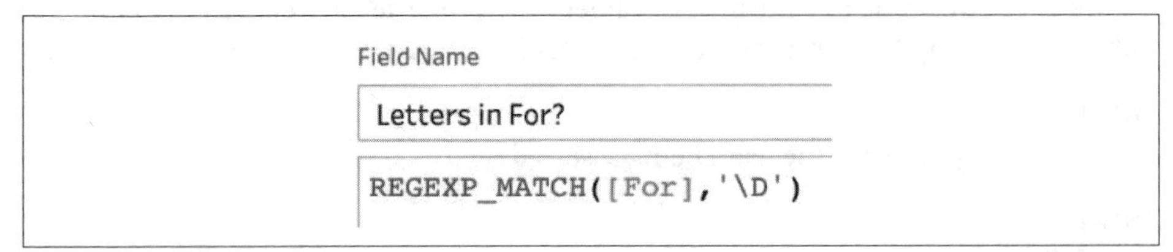

Figure 37-3. Regex function being used in Prep

This calculation tests whether a letter is present in the For column, which should contain only points scored, and will produce a `True`/`False` response (Figure 37-4).

Letters in For?	Team	Result	For	Aga
True	England	draw	0G	0G
True	England	draw	0G	0G
True	England	lost	0G	2G
True	England	draw	0G	0G
False	England	won	7	0
False	England	lost	0	1

Figure 37-4. Result of the calculation in Figure 37-3

A filter can then be configured to return just those values in the column that don't contain any letters (Figure 37-5).

Figure 37-5. Filter to remove values

After converting the data field For to "Number (whole)," we get the data set we needed, containing only the numerical point values (Figure 37-6).

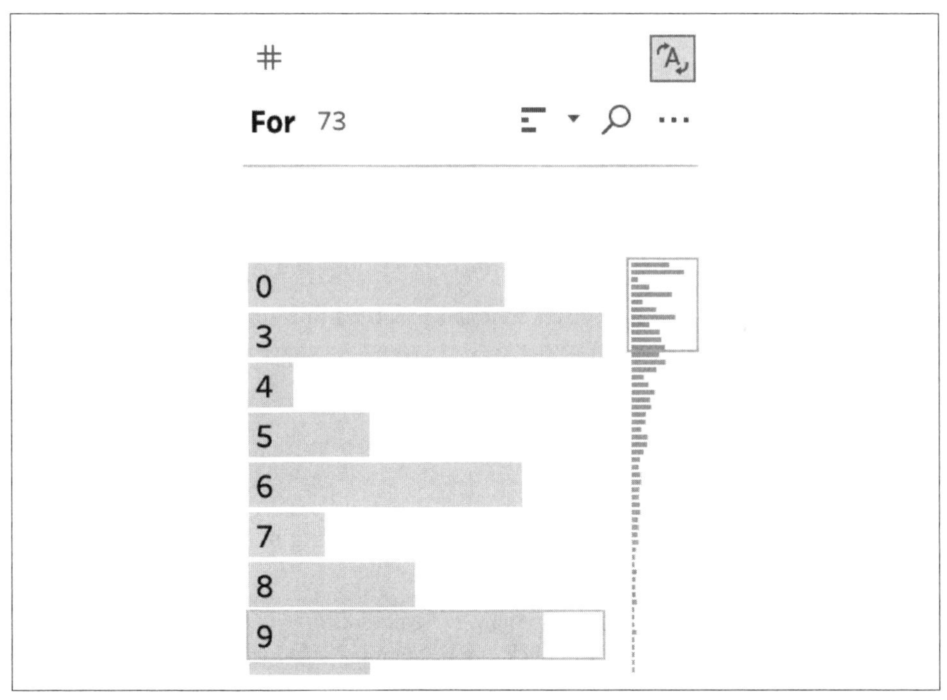

Figure 37-6. Results of For data field conversion after regex calculation and filter

Join Ranges

Using join conditions that intentionally exclude data prevents you from loading data that you would need to filter out of the data set later. One technique for doing this, a *join range*, allows you to join two data sets together based on join conditions using less than or greater than instead of equal to (the most common join condition).

Chapter 32 covers advanced join scenarios in a lot more detail. Chapter 39 covers scaffolding, another advanced join technique that uses join ranges to filter data.

By preventing rows from being formed during joins that you'd only have to filter out later, join ranges reduce the amount of data processed and require less computing power. Since Prep Builder is used primarily by "normal business users" on laptop and desktop computers, where processing power can be limited, this is a huge advantage.

Percentage Variance

Another form of smarter filtering is testing key measures to ensure they are within a tolerance level. These tolerance levels will often depend on business logic and rules. Finding significant variances in the previous minimum, maximum, or average is a strong indication that something within the data is amiss and requires addressing.

Percentage variances are often good checks to place into a flow where the data has been manually entered, increasing the risk of typos, or where the data will be reloaded. Let's look at each scenario in turn.

Manual Entry: Level of Detail Calculations

Because manual entry is likely to be OK most of the time with only the odd mistake, you can set the benchmark within the actual data set being used in the flow and test the values within it.

With the addition of Level of Detail (LOD) calculations in Tableau Prep Builder 2020.1.3, it became a lot easier to check the percentage variance because these calculations add a new column to the data set. We'll use the Soap Production Cost data set in Figure 37-7 to check for any outliers.

Type	Scent	Production Cost
Bar	Mint	1.25
Bar	Lemon	1.3
Bar	Lime	1.25
Bar	Strawberry	1.3
Bar	Blueberry	1.3
Bar	Heather	3
Liquid	Mint	1.55
Liquid	Lemon	1.4
Liquid	Lime	1.55
Liquid	Strawberry	1.4
Liquid	Blueberry	1.4
Liquid	Heather	1.5

Figure 37-7. Soap Production Cost data set

To identify outliers in the data set, first you need to test the "normal" level. For this data set, let's assume we know that there is normally a different production cost for the different types of product. We can build this calculation using the Visual Editor (Figure 37-8).

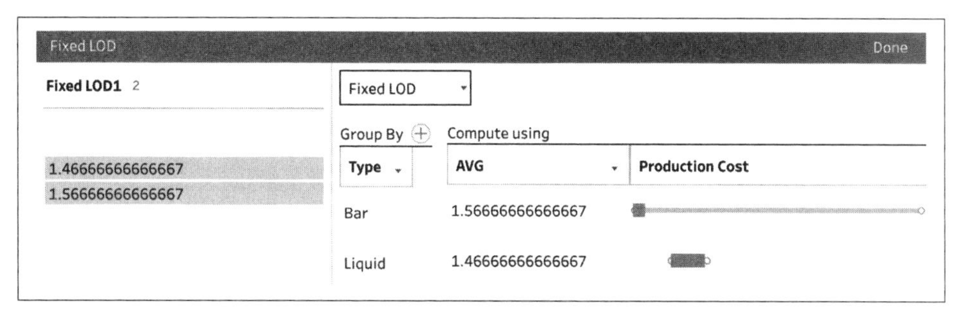

Figure 37-8. Visual Editor for LOD calculations

This gives us two values to test the production costs against. Creating two calculations that add and subtract the percentage variance will allow us to test each value. For this data set, the percent tolerance level is 30% above and below the average production cost per product. We can create the Upper Check field with the calculation shown in Figure 37-9.

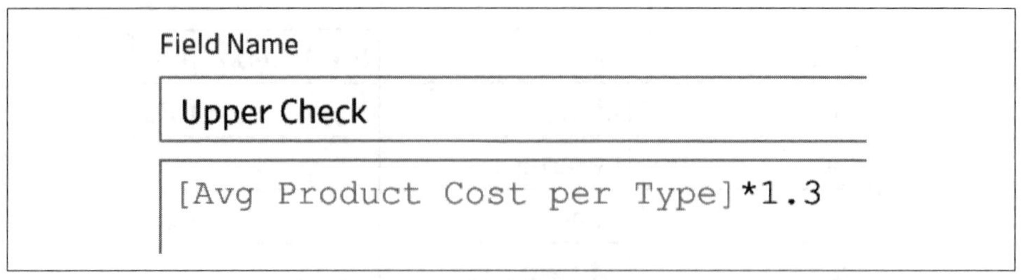

Figure 37-9. Calculation for the Upper Check field

We create the Lower Check field with the calculation shown in Figure 37-10.

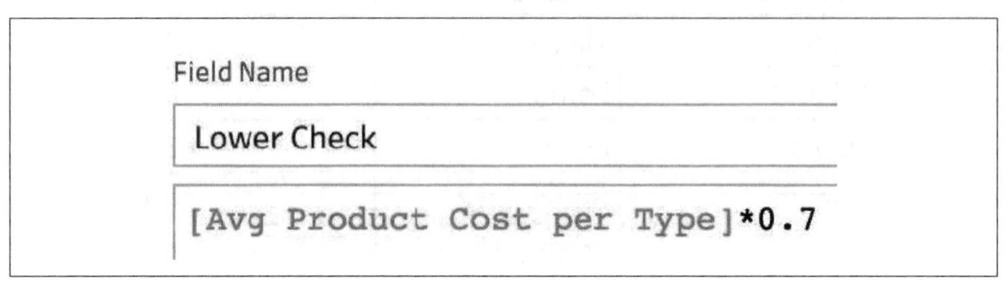

Figure 37-10. Calculation for the Lower Check field

We can then check the tolerance by assessing whether the Production Cost value lies between the Upper and Lower Check bounds. To do this, we use the `IF` statement shown in Figure 37-11 in a new calculation.

Field Name

Within Tolerance?

```
if [Production Cost] >= [Lower Check] and
[Production Cost] <= [Upper Check] then 'ok'
else 'outside'
end
```

Figure 37-11. Using the Upper and Lower Check bounds to set a tolerance level

The flow can then be split based on whether the values are within the specified tolerance level. The records containing values that sit outside the Upper and Lower Check bounds can be branched off for further checks and validation (Figure 37-12).

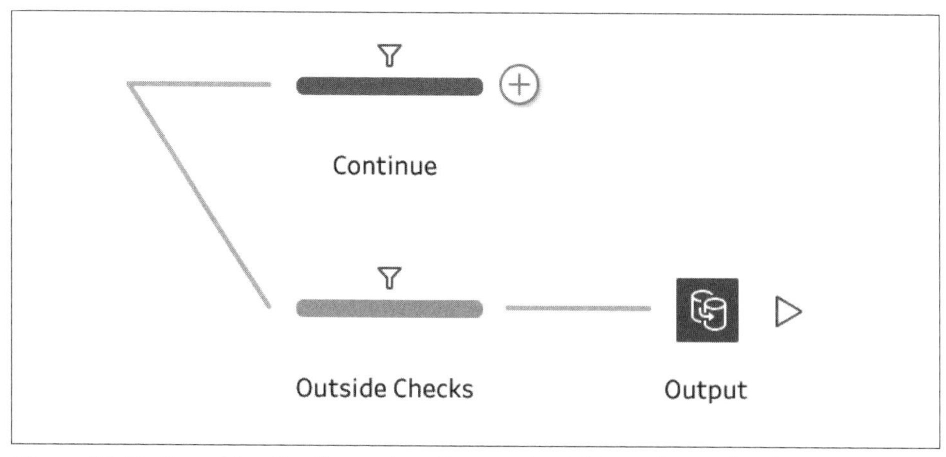

Figure 37-12. Branching the flow to handle records within and outside tolerance

Reloaded Data: Join to Previous Output

When a system goes wrong or not as expected, it often goes *very* wrong. If all of a data field's values have gone very wrong, assessing the tolerance of that data is not going to work. If all of the values fundamentally change, then the average technique we just used will not catch the issues, as the average values are driven by the changed column. To resolve this, we need to test the previous data set against the newly loaded data.

Let's structure a flow to recognize the issue in the data set shown in Figure 37-13, where the decimal points have suddenly disappeared from the Production Cost field.

Type	Scent	Production Cost
Bar	Mint	125
Bar	Lemon	130
Liquid	Mint	155
Liquid	Lemon	140

Figure 37-13. Data set resulting from a system error

We will use many of the same steps as in the earlier technique, but with some key differences.

Aggregating the Average Production Cost per Type

This Aggregate step changes the granularity of the data to the level of the tolerance check. We do this calculation on the original data, not the new data set being loaded. In Figure 37-14, we aggregate up to a single record per product type.

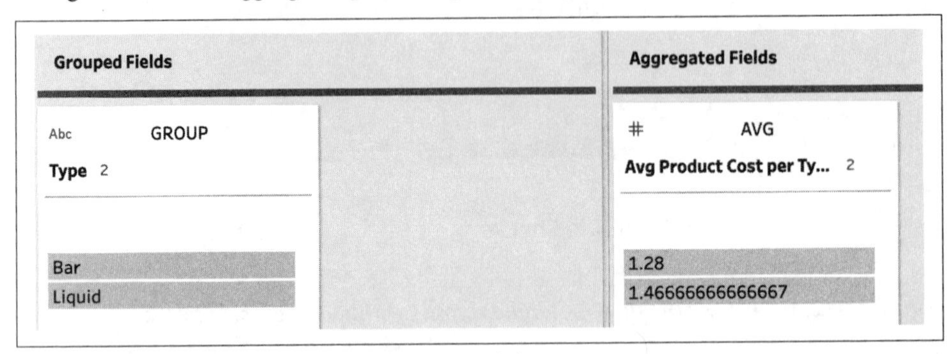

Figure 37-14. Aggregate step using the original data set

The benefit of using an Aggregate step is that it reduces the granularity to one record per element you want to join by. You will still need to calculate the Upper and Lower Checks based on the results from this step.

Joining the Data Sets Together

Now we'll join the new data set and original data set together to add the checks to the new data set (Figure 37-15).

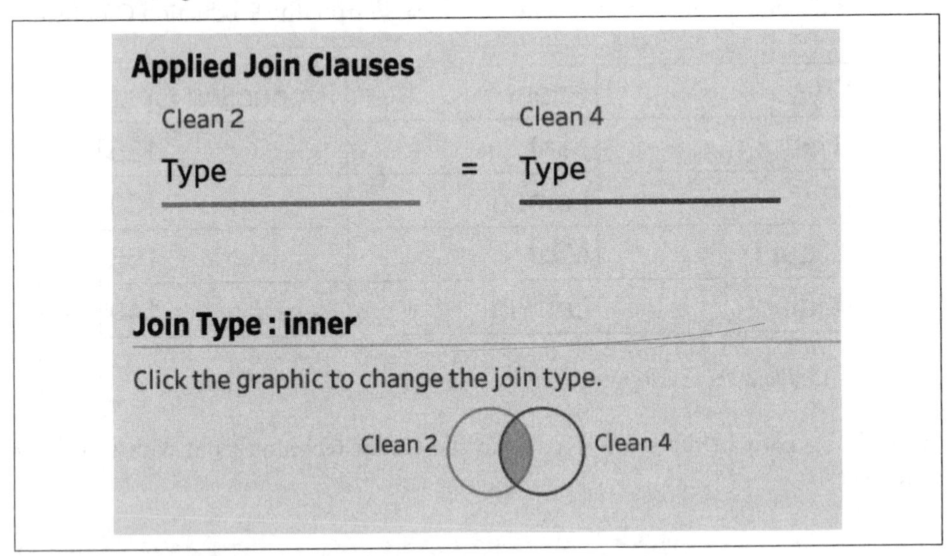

Figure 37-15. Join condition to join together the new and old data set

This results in a data set where we can use the same calculation from earlier to determine whether or not the Production Cost values are within the tolerance levels (Figure 37-16).

Within Tolerance	Lower Check	Upper Check	Avg Product Cost per Type	Type	Scent	Production Cost
outside	1.024	1.536	1.28	Bar	Mint	125
outside	1.024	1.536	1.28	Bar	Lemon	130
outside	1.17333333333333	1.76	1.46666666666667	Liquid	Mint	155
outside	1.17333333333333	1.76	1.46666666666667	Liquid	Lemon	140

Figure 37-16. Resulting data set of the Join in Figure 37-15

We can then make our filtering choices the same way as in the earlier technique.

Combining Techniques

Using the join range and percentage variance techniques together can also automate the filtering to leave only the results you require and flag potential issues in your data. In the previous example, we could set the join condition shown in Figure 37-17 to remove any values outside of the specified tolerance levels.

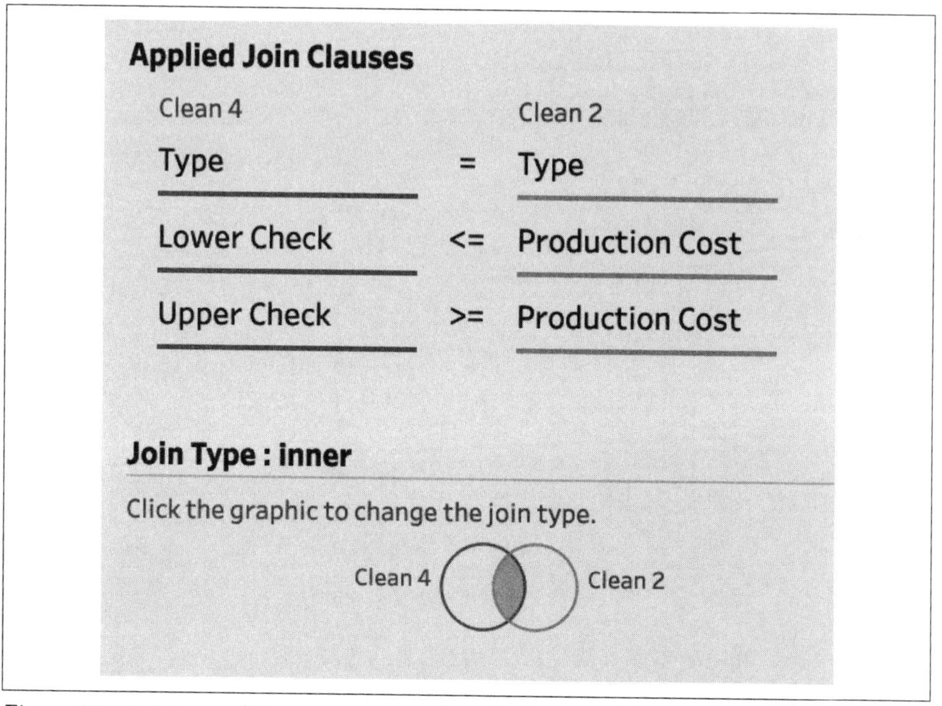

Figure 37-17. Join condition set up to combine both smart filtering techniques

This prevents any data that could lead to misinformed decisions from being processed and shared with end users. If there is any doubt that the filtered values may be real records—even if they are outliers—make sure to add checks to your process so you aren't removing data that could compromise the accuracy of your analysis.

Summary

The techniques described in this chapter can provide safeguards when things go wrong, either through data entry mistakes or because systems that produce data sets go wrong. Calculations, join ranges, and percentage variations take a rules-based approach to help you filter your data sets.

CHAPTER 38

Managing Conversion Rates

The world is getting smaller, with businesses and organizations working across national borders more than ever before. With this globalization comes data challenges that can be solved through a number of data preparation strategies. One typical use case for multinational organizations is determining currency conversion rates in order to ensure the financial data being analyzed is consistent and being accurately compared/aggregated.

Challenges of Conversion Rates

Converting rates such as currency seems like such a simple task. Multiply the currency rate by the original value, and you have the currency you require. Easy, right? Not so fast. Depending on the level of accuracy you want, using a single, constant exchange rate might not be sufficient. Currency exchange rates fluctuate throughout the day, every day.

Which rate you use may make a significant difference even in smaller transactions; added up over time, the small differences can accrue, resulting in over- or understated results or profits. The frequency of the conversion rates (hourly, daily, weekly, etc.) is another important decision to make. Greater frequency assists with accuracy but makes the conversion rate tables more difficult to apply, especially to different time periods.

Applying conversion rates consistently is also key, especially when you are working across a number of different business units. Let's look at how to do this within Prep.

Applying Conversion Rates in Prep

Using the 2020: Week 6 Preppin' Data challenge (*https://oreil.ly/dMwG3*), let's work through a currency conversion from British Pounds (GBP) to US Dollars (USD).

Step 1: Create a Consistent Granularity of Data for the Conversion

First, you need to import the input data (Figure 38-1 and Figure 38-2).

Date	British Pound to US ...
31/01/2020	1 GBP = 1.3205 USD
30/01/2020	1 GBP = 1.3086 USD
29/01/2020	1 GBP = 1.3021 USD
28/01/2020	1 GBP = 1.3023 USD
27/01/2020	1 GBP = 1.3058 USD
26/01/2020	1 GBP = 1.3063 USD

Figure 38-1. Conversion Rates data set

Week	Year	Sales Value	US Stock sold (%)
wk 27 2019	2,019	53,025	42
wk 28 2019	2,019	49,994	44
wk 29 2019	2,019	55,236	29
wk 30 2019	2,019	76,013	33
wk 31 2019	2,019	25,544	40

Figure 38-2. Sales Values data set

These two data sets are not initially ready for conversion, as you don't have a field to join them on. For these data sets, you need to "roll up" the individual dates at the day level to the week level to match the Date column in the conversion rate table in Figure 38-1. This is a judgment call; ultimately, you need to decide with what frequency you want to do the conversion, then find the rate within that period and ensure the correct conversion rate is held for that level.

This example requires you to find the best- and worst-case conversion scenario. In other words, the data preparation task is to find the minimum and maximum value of US dollars that would be exchanged for one British pound per week. Other scenarios to consider include:

- Finding the average value for a time period
- Finding the "opening" or "closing" value as an indicator of the likely value of an exchange in that time period

The context for the question being posed is what will ultimately determine which technique you need to deploy.

Step 2: Join the Data Sets Together

Using a Join step in Prep Builder, you can now join together the two different data sets (Figure 38-3). If everything is in one table already in your use case, then you can skip this step.

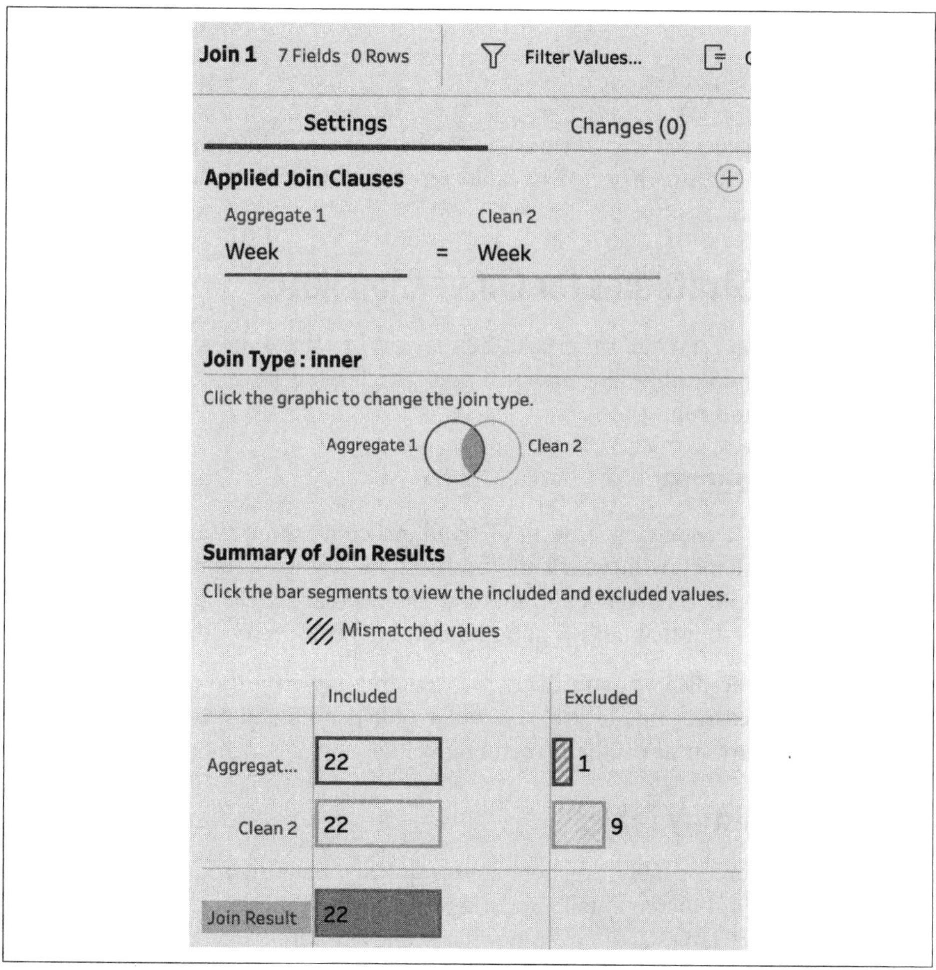

Figure 38-3. Join setup to pull together the Sales Values and Conversion Rates data sets

Step 3: Apply the Conversion Rate

Now that everything is in one table, you can apply the conversion rate to the value you are converting (Figure 38-4).

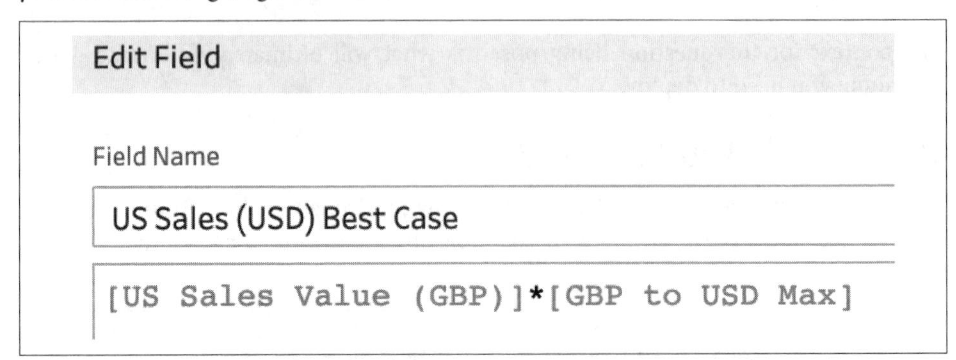

Figure 38-4. The conversion rate calculation

This applies the specified conversion rate to the value, making analysis for the end users a lot easier than if they had to build separate calculations themselves in each Desktop workbook.

Long-Term Strategies for Conversion Rates

Creating a solution for a one-off preparation flow is fine, but if the flow will be reused multiple times and/or the data updates, then you'll need longer-term strategies to ensure accuracy and robustness.

Managing Frequency

One of the most challenging aspects of handling conversion rates like currency is knowing the frequency with which the business will want to apply those rates. If a single value per hour, day, week, month, quarter, and year is held as a reference table, this is very useful—but that table is going to get pretty big, pretty quickly.

By storing only the data you are likely to need and ensuring the rules are clear for determining the correct rate to use, you will avoid spending a lot of time fixing mismatched values further down the reporting pipeline.

Maintaining History Tables

As just noted, reference tables can get big very quickly. Building robust history tables, with the latest values incrementally updating over time, is a worthwhile effort. Ensuring that the history tables are kept somewhere central where every "data prepper" can access them is very important. How far back the users' questions go will determine how long you'll need to maintain the history tables.

Summary

With more organizations trading across the world, it's especially important to be consistent in your approach to converting currencies. Using Prep Builder to apply the techniques covered in this chapter makes converting values very easy to do. Setting up methods for storing the rates used and keeping those methods consistent will help you more accurately report everything from earnings to costs.

CHAPTER 39

Scaffolding Your Data

Scaffolding is a term that you may not come across often, but the challenges that it resolves are increasingly found in modern data problems.

What Is Scaffolding?

Scaffolding is the process of filling missing rows of data within a data set to assist analysis. A data set may appear complete—with no nulls and a record for each individual entity—but still not be suited for the analysis you wish to conduct. Consider a mobile phone operator that wishes to analyze its monthly revenue from contracted customers. Figure 39-1 shows the data the operator is likely to have.

Customer ID	Start Date	Monthly Price	Contract Length
ABC123	05/01/2019	10	24
DEF456	27/02/2019	15	12
GHJ789	19/03/2019	18	18

Figure 39-1. Data set requiring scaffolding to assist analysis

As you can see, the operator has a record of the customer, the contract start date, the contract length, and the monthly price for the service. However, there is no date to determine the value we're seeking to analyze, monthly revenue. The only dates to use for analysis are the contract start date and end date. If the contract is for two years, we would need 24 records to gain full insight. As Figure 39-2 demonstrates, there is only a single month of revenue per Customer ID, and we can't see the revenue being collected over time.

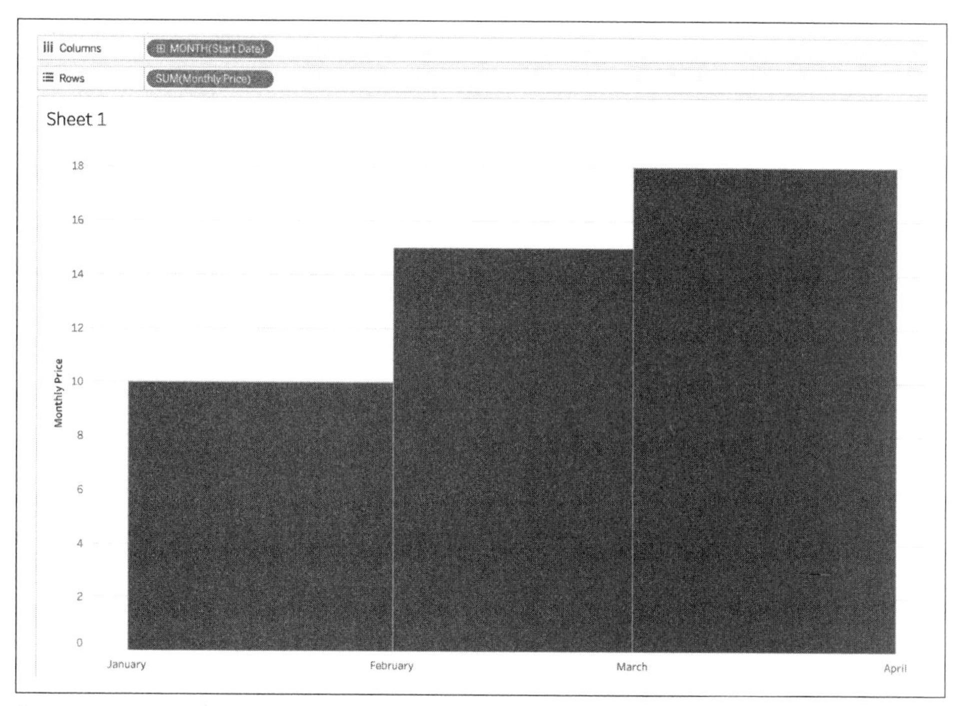

Figure 39-2. Resulting visualization from data set in Figure 39-1

Scaffolding is the answer to this problem. The technique adds a row for each of the "missing" records that are required for the analysis. In the mobile phone operator example, a scaffold would create the additional rows so each monthly payment would have its own record. The resulting data set would have a single row per contract, per month, for the cost of the service. This means there will be a row for each customer with an active contract per month. After 12 months, one contract is no longer active, and after 18 months, another ends. The scaffolding creates the data profile shown in Figure 39-3.

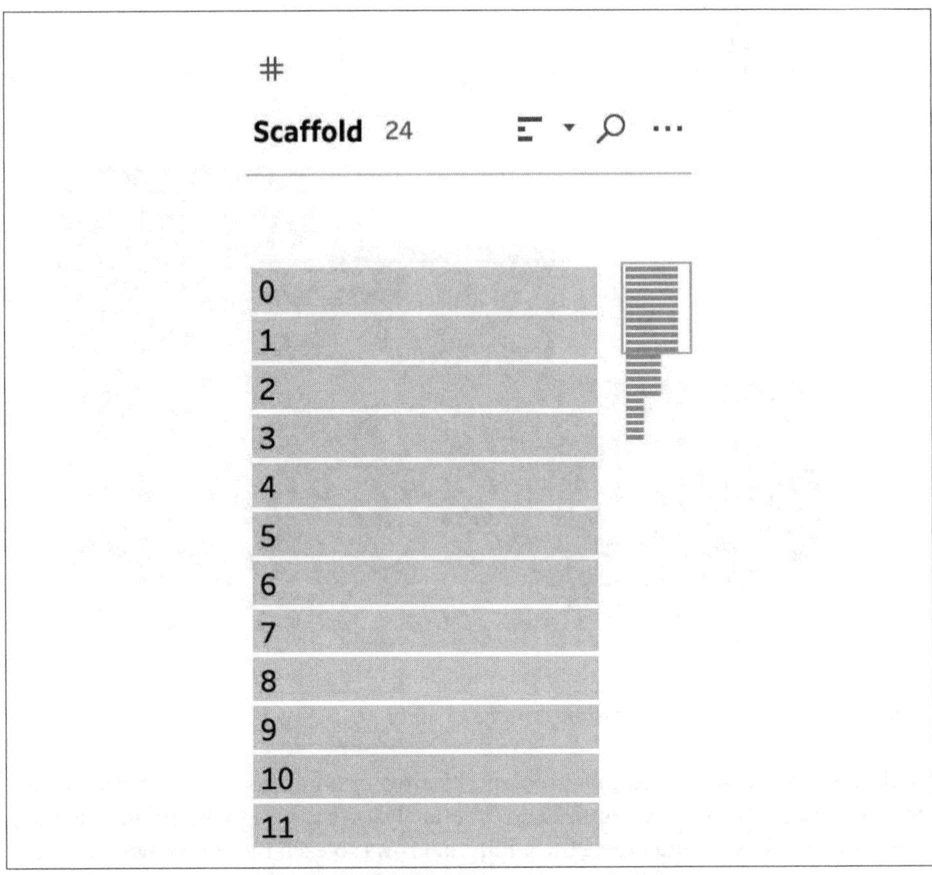

Figure 39-3. Profile of the scaffolded data in Prep Builder's Profile pane

Giving all of the individual payments a row of their own in the data set allows analytical software, like Tableau Desktop, to show the full revenue over time. Figure 39-4 shows a sample of the resulting data set.

Reporting Date	Customer ID	Start Date	Monthly Price	Contract Length	Scaffold
19/03/2019, 00:00:00	GHJ789	19/03/2019	18	18	0
27/02/2019, 00:00:00	DEF456	27/02/2019	15	12	0
05/01/2019, 00:00:00	ABC123	05/01/2019	10	24	0
19/04/2019, 00:00:00	GHJ789	19/03/2019	18	18	1
27/03/2019, 00:00:00	DEF456	27/02/2019	15	12	1
05/02/2019, 00:00:00	ABC123	05/01/2019	10	24	1
19/05/2019, 00:00:00	GHJ789	19/03/2019	18	18	2
27/04/2019, 00:00:00	DEF456	27/02/2019	15	12	2
05/03/2019, 00:00:00	ABC123	05/01/2019	10	24	2

Figure 39-4. Sample of the scaffolded data set

Challenges Addressed by Scaffolding

Scaffolding techniques address several challenges, many of which are rapidly becoming more common due to *subscription pricing*—a payment schedule wherein the consumer pays a smaller amount on a regular basis rather than making a large one-time purchase. Subscription music, software, and food services are increasing in popularity, creating a bigger need for data analysts to be able to work with subscription data. Scaffolding goes beyond addressing the challenge of subscription-based data sets, however. It can also help in these areas:

Retention
: Reporting on expected revenue can be presumptuous. Just because a customer signs up for a subscription-based product doesn't mean that they will keep up those payments. Scaffolding can help you remove rows of projected revenues to more accurately determine levels of customer retention.

Inflation/exchange rates
: Depending on the length of the subscription, you might need to account for factors like inflation or changing exchange rates in order to accurately represent earned revenue. Projecting these values is an art in its own right, but applying common forecasts with a scaffold can greatly improve the value of the analysis. Often projections will be stored in a secondary reference table and need to be joined onto the original data set along with the scaffold.

Data duplication
: If you have fields that are aggregated over the periods covered in a record that you are then scaffolding out, you'll need to divide these values over the number of periods to avoid duplicating the aggregations.

Challenges Created by Scaffolding

While the scaffolding technique does supply the additional rows of data required for analysis, it creates another challenge that should be considered.

Scaffolding will multiply the size of the data set you are analyzing considerably. This poses challenges for:

Data storage
: The larger the data set, the greater the amount of storage that you will need.

Computing power
: Not everyone has a brand new, powerful computer to use. Adding a huge number of rows to (potentially) an already large data set can really test your computer's working memory.

Filtering at the data source level can significantly reduce the impact of scaffolding on your data size.

The Traditional Scaffolding Technique

Scaffolding has been used for a long time, and the traditional technique uses a scaffold of all possible values. Returning to the mobile provider example, we need to create a new Date field containing each month from the earliest contract start date in the data set to the latest contract end date (Figure 39-5).

Date
01/01/2019
01/02/2019
01/03/2019
01/04/2019
01/05/2019
01/06/2019
01/07/2019
01/08/2019
01/09/2019
01/10/2019
01/11/2019
01/12/2019
01/01/2020
01/02/2020
01/03/2020
01/04/2020
01/05/2020
01/06/2020
01/07/2020
01/08/2020
01/09/2020
01/10/2020
01/11/2020
01/12/2020

Figure 39-5. Data set to be used for scaffolding

As you can see, the longest contract starts in January 2019 and runs 24 months, so the latest date we need is December 2020. Just having the date range required isn't sufficient, though. This scaffold needs to be joined to the customer data. To do this, we have two options:

- Create a calculation that defines the last month of the contract as End Date and then join using the conditions Start Date >= Scaffold Date and End Date <= Scaffold Date.
- Create a condition that will join each row in each data set together. This can then be filtered to remove any excess rows. The calculation that's easiest to join is just the number 1. Create this calculation in each data set and join on a condition where 1 = 1 (i.e., every row joins to every other row). This is another style of appending, covered in Chapter 33.

The second join condition using 1 is easier to follow, as you can see the results of the join and filter to ensure each aspect is correct. Let's follow that approach in the mobile provider example.

Step 1: Input the Data Sets

Input both data sets—Subscription Data and Scaffold (Figure 39-6).

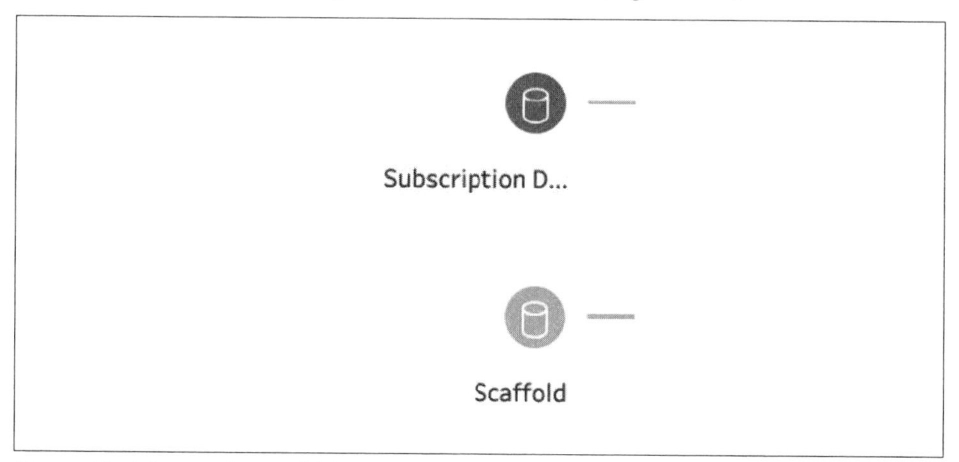

Figure 39-6. Inputs for scaffolding

Step 2: Build the Join Calculations

Build a calculation of the value 1 in a Clean step. For simplicity's sake, call the calculation "1" too (Figure 39-7).

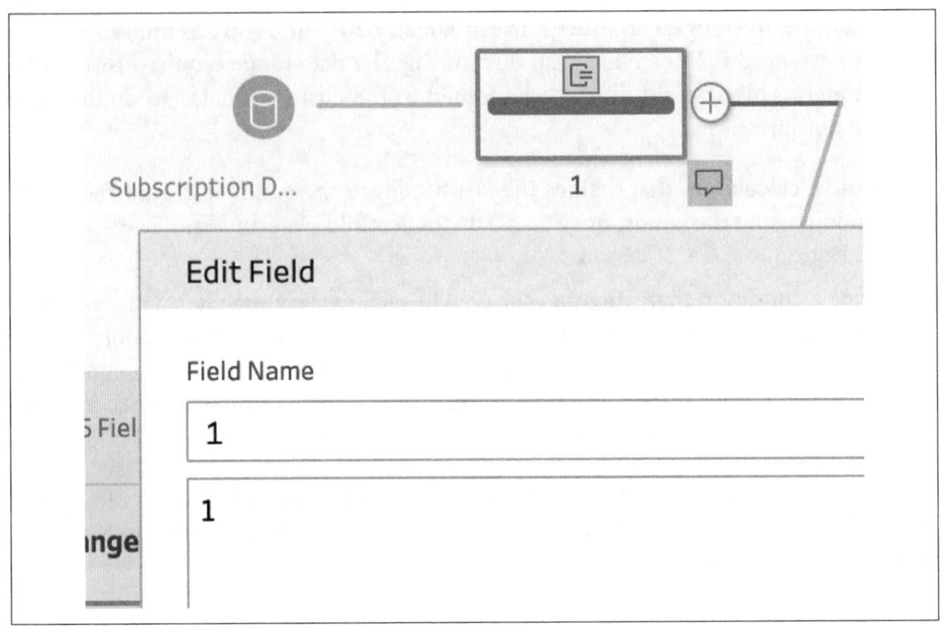

Figure 39-7. Creating a calculation to act as the join condition

Repeat this step for both inputs (Figure 39-8).

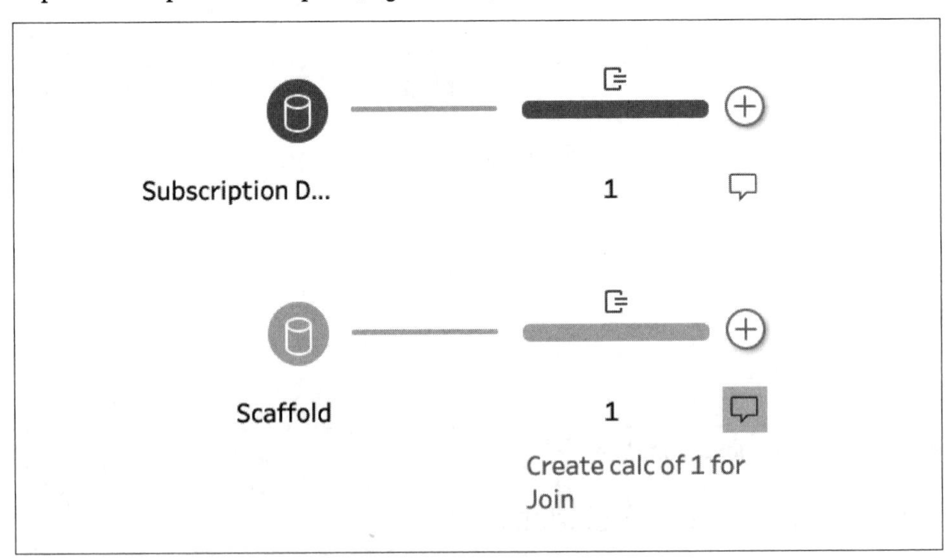

Figure 39-8. Flow resulting from creating calculations to act as the join condition

Step 3: Join the Two Data Sets Together

Using a Join step, connect the two flows together using the join condition of 1 = 1 for an inner join (Figure 39-9).

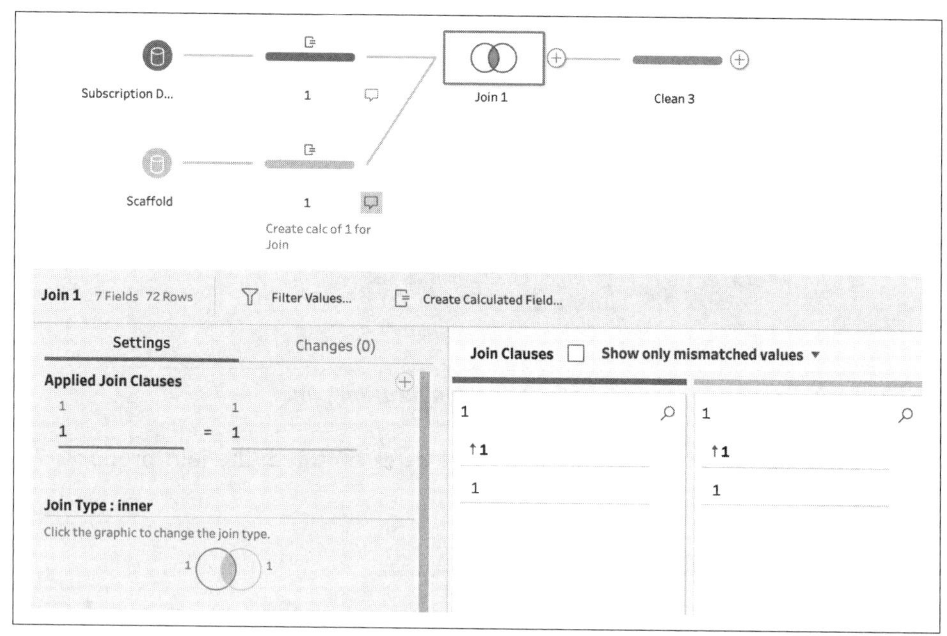

Figure 39-9. Flow resulting from joining the Subscription Data and Scaffold data sets

This adds each row of a customer record from the Subscription Data set to each date. Prep clearly displays the results of this join in the lower part of the Join configuration pane (Figure 39-10).

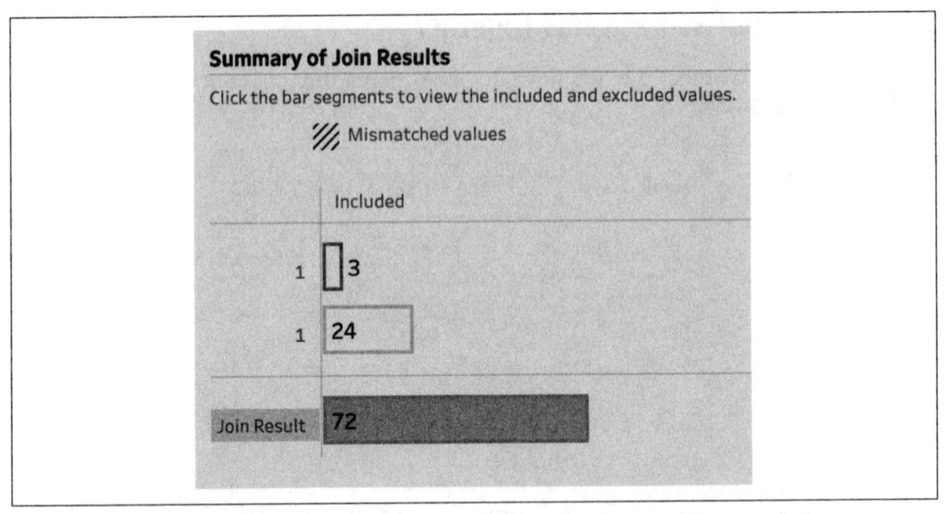

Figure 39-10. Join Results section of the Join configuration pane

As there are 3 subscriptions recorded and 24 dates captured, the join produces 72 rows of data.

Step 4: Filter Out Unnecessary Rows

To be able to filter out any rows containing dates before the contract start date or after the contract expires, we first need to create a calculation for End Date. Add a new Clean step and create the calculation shown in Figure 39-11.

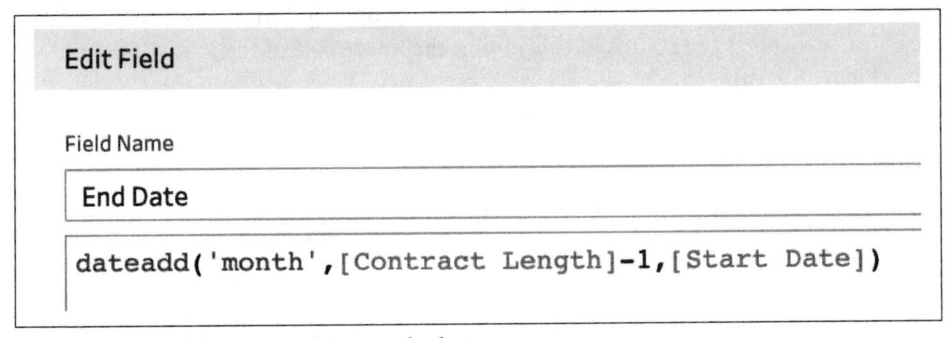

Figure 39-11. Adding an End Date calculation

This will add a new End Date field in your data set. You can easily check that it has worked as intended in Prep's Profile pane (Figure 39-12).

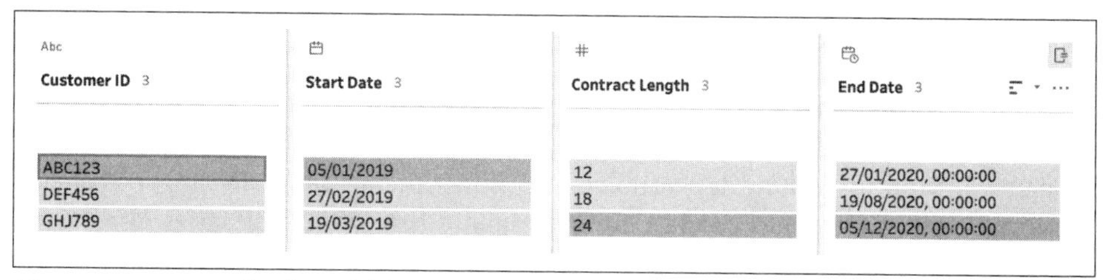

Customer ID	Start Date	Contract Length	End Date
ABC123	05/01/2019	12	27/01/2020, 00:00:00
DEF456	27/02/2019	18	19/08/2020, 00:00:00
GHJ789	19/03/2019	24	05/12/2020, 00:00:00

Figure 39-12. Result of scaffold

Once you have the end date of the subscription, you can filter out the dates that have been added by the scaffold that fall outside the contract start and end dates. The calculation in Figure 39-13 uses the `datetrunc()` function to set the frequency of the scaffold. This data set is monthly, so we can ignore the day of the month as you will accrue revenue for the entire month when the contract starts and each month until it ends.

Edit Filter

Calculation

```
datetrunc('month',[Date]) >=
datetrunc('month', [Start Date])
and
datetrunc('month',[Date]) <=
datetrunc('month',[End Date])
```

Figure 39-13. Calculation to retain only dates within the contract dates

A filter within a Clean step in Prep Builder must return a Boolean result. The records that meet the condition set—that is, those returning a `True` value—will remain in the data set. Those that do not meet the condition set in the filter—those returning `False` —will be removed from the flow, leaving one row per month of revenue per contract (Figure 39-14).

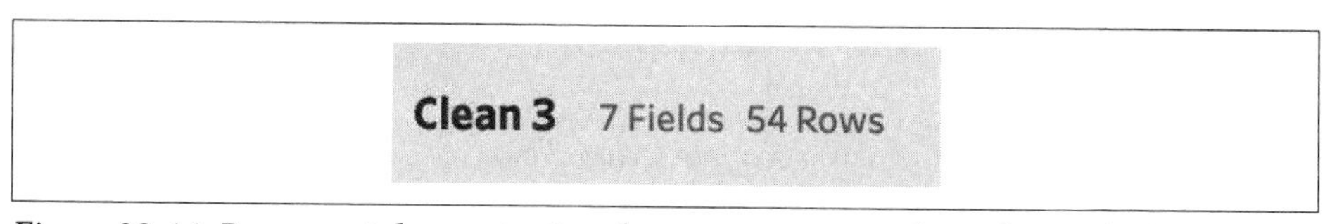

Figure 39-14. Row count demonstrating the correct outcome from the scaffolding

The Newer Scaffolding Technique

While the traditional technique works, it does have a number of shortcomings. The most notable of these is that the list of dates in the scaffold has to be maintained and updated. This doesn't sound too problematic, but it means remembering where the file is kept, remembering to update it, and making sure your colleagues also know where and how to do this (in case you leave).

This all changed when Bethany Lyons (*https://twitter.com/tablyze*) presented a new technique in Tableau Desktop at a Tableau conference a few years ago. Her approach resolved many of these issues by challenging the logic that a data set that needs to be scaffolded to create extra records for each date. Like the traditional method just detailed, the new technique utilizes different date calculations, but it differs in that they are all created from simple integers. In the mobile provider example, the longest contract length is 24 months, so the scaffold would require a column of integers from 0 to 23 (Figure 39-15).

Scaffold
0
1
2
3
4
5
6
7
8
9
10
11
12
13
14
15
16
17
18
19
20
21
22
23

Figure 39-15. Data set for the newer scaffold technique

Let's look in more detail at the differences in this technique and how it addresses the shortcomings of the traditional scaffolding approach.

Step 1: Input the Data Sets

Input both data sets as in the traditional technique.

Step 2: Join the Data Sets

This time, instead of 1 = 1, we set the join condition as Scaffold < Contract Length (Figure 39-16). This means that a scaffold row will be assigned to each subscription up to the point where the scaffold number is the same as the contract length. This works because the values in the scaffold start at 0 (we'll get to that shortly).

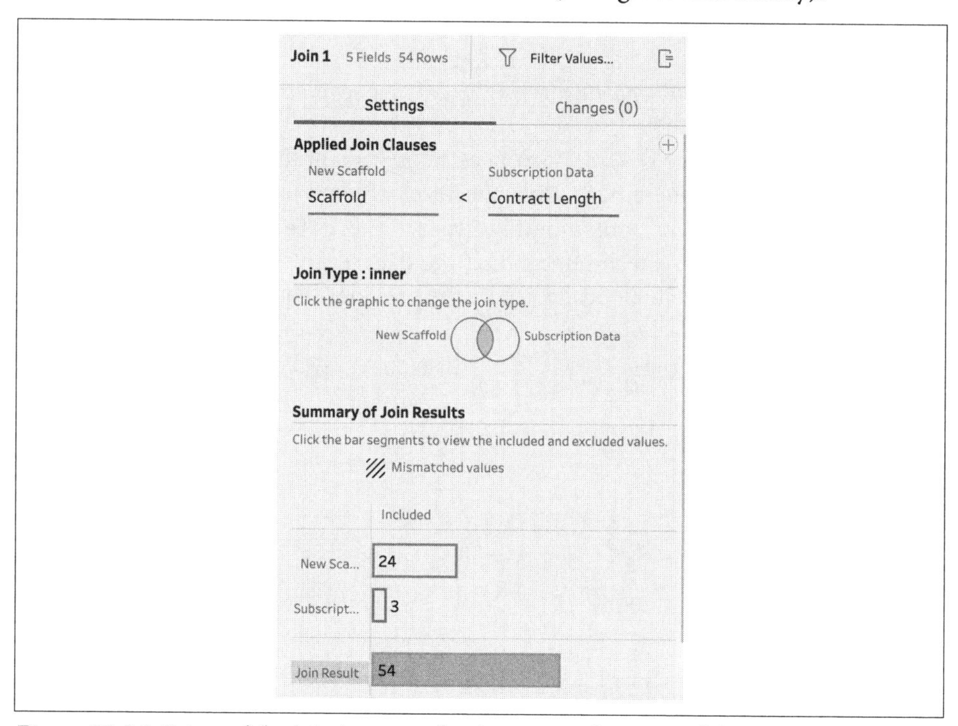

Figure 39-16. Setup of the join between the data set and new scaffold

Notice how the number of rows required is already correct, so there is no need to filter later in the flow. This is a massive advantage when you have thousands or millions of customers, since you aren't adding millions of rows only to remove them straightaway.

Step 3: Add the Reporting Date

In the previous technique, the scaffold date became the date used for reporting. As no such date exists this time, we need to create it. We use the `dateadd()` function, setting Start Date as the contract start date and Scaffold as the increment on top of this (Figure 39-17).

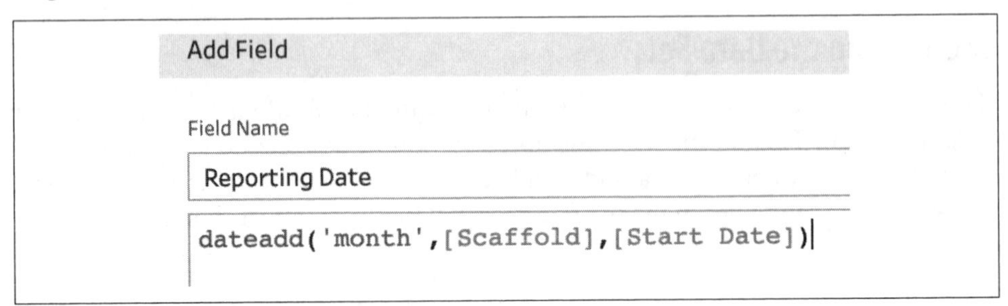

Figure 39-17. Creating the reporting date with the new scaffold

As the first scaffold value is `0`, we don't need to increment the contract start date further. Any subsequent increment adds the level of date part applied. Because the mobile contract payments are monthly, we set the `dateadd()` level as `'month'`. Another advantage of this technique is that the date returned is much more likely to be when the revenue is actually collected by the provider (Figure 39-18).

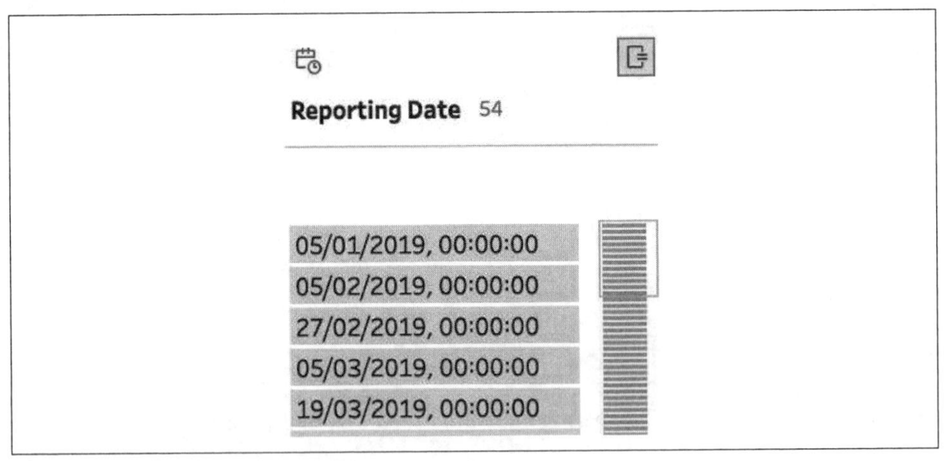

Figure 39-18. Result of the scaffold

Step 4: Remove the Scaffold Value

Deleting the scaffold value by removing that data field from the data set after you've created the reporting date ensures that it won't be mistakenly used by an end user of the data set.

The Result

You can generate the visualization of all monthly revenues in Tableau Desktop after outputting the data. In Figure 39-19, the Customer ID has been added so it's easier to see the effect of the multiple rows (one per month) of the scaffolding.

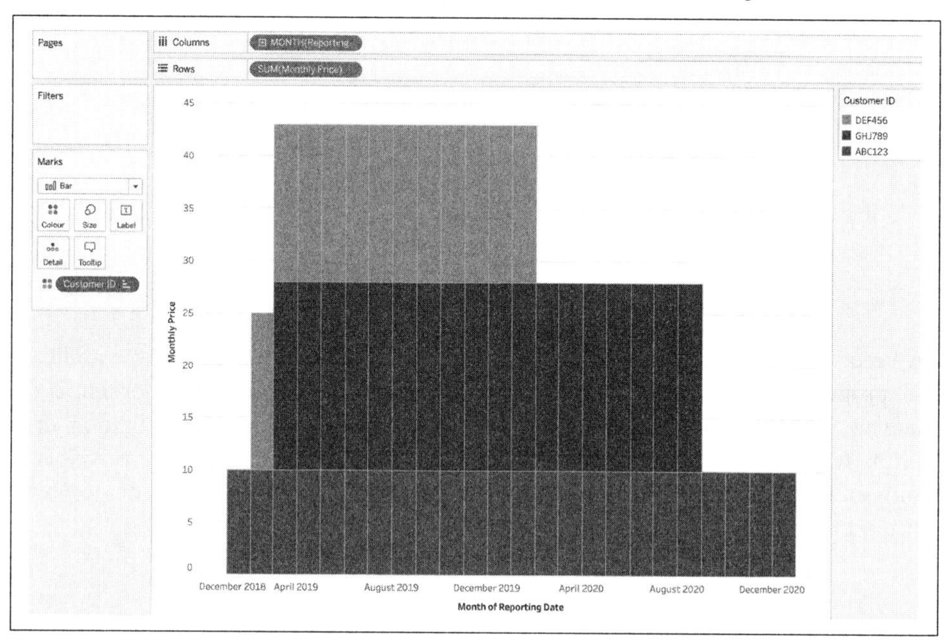

Figure 39-19. Updated visualization resulting from scaffolded data set

You should now be able to easily apply scaffolding to any data set that lacks full monthly records. This technique has enabled a lot of analytical solutions that would not have been possible otherwise.

Summary

For Desktop users who need to add all possible dates into a data set in order to analyze the data correctly, scaffolding is a good solution. Using Prep to avoid applying this technique in Desktop can simplify the analytical process for end users. Further use cases for scaffolding include dividing targets across multiple dates or carefully filling in missing values in the data (you'll need to disclose to end users that these are estimates). Now that you know the core techniques, you'll be able to approach these additional challenges with confidence.

CHAPTER 40

Connecting to Programming Scripts

In Prep Builder version 2019.3, the product's development team added the ability to call programming scripts, so now we can use functionality in Prep Builder that is not natively available. The Script step (Figure 40-1) allows you to enter Python or R scripts written by others or those you create yourself. You can use these scripts to connect to website APIs or conduct data preparation tasks that are not possible within Prep Builder's built-in functionality.

Figure 40-1. The Script step icon

This chapter will look at situations where you might want to connect a script into Prep, explain how to set up your computer to use the script functionality, and walk you through a short example.

When to Use the Script Step in Prep

The simple answer to when you should use the Script step is whenever Prep does not have the functionality you need to achieve your data preparation goals. Programming is a very flexible way to instruct computers to complete the tasks you require. Python and R are taught in many universities due to the number of packages available in the two languages. A *package* utilizes a programming language to complete common tasks, saving the user from having to write all the instructions from scratch.

Tableau Prep is designed to be easy to use, but the introduction of the Script step sits at the other end of the easy-to-use spectrum for many users. So how might nonprogrammers take advantage of the feature? Find a programmer who can help! This person might be a colleague or a helpful member of the Tableau community. Of course, you need to trust the source before using the script in order to ensure that you are not going to cause any harm with malicious code. The script file can be created once and then reused multiple times. The Tableau community loves to share, so Prep users will be able to take advantage of an ever-growing volume of useful scripts.

One constraint of the Script step is that you will need to install a version of TabPy or Rserve on your computer or a server you can connect to. This may not be possible within your organization, as some companies are protective of what programs can be run on their computers. If you don't have this constraint, you will be able to use scripts that can potentially save you a lot of time and effort within flows that require advanced techniques.

Setting Up Your Computer to Use Scripts in Prep

If you have the permissions on your computer to set up TabPy or Rserve, follow the instructions in this section to get started. Rather than providing step-by-step instructions for both languages, this chapter will focus on Python, which is becoming the most prevalent scripting language for data preparation in the business world.

To use Python scripts in Prep, first you need to download Python to your computer. Anaconda is free and easy to install from anaconda.com (Figure 40-2).

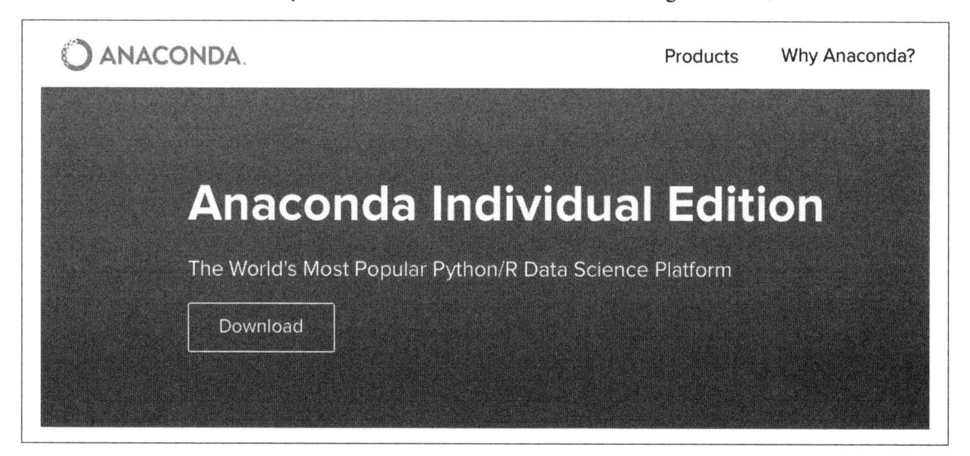

Figure 40-2. Anaconda home screen

Once Anaconda is installed, you will need to download TabPy. TabPy is hosted on GitHub (*https://oreil.ly/yrlWK*) and shown in Figure 40-3.

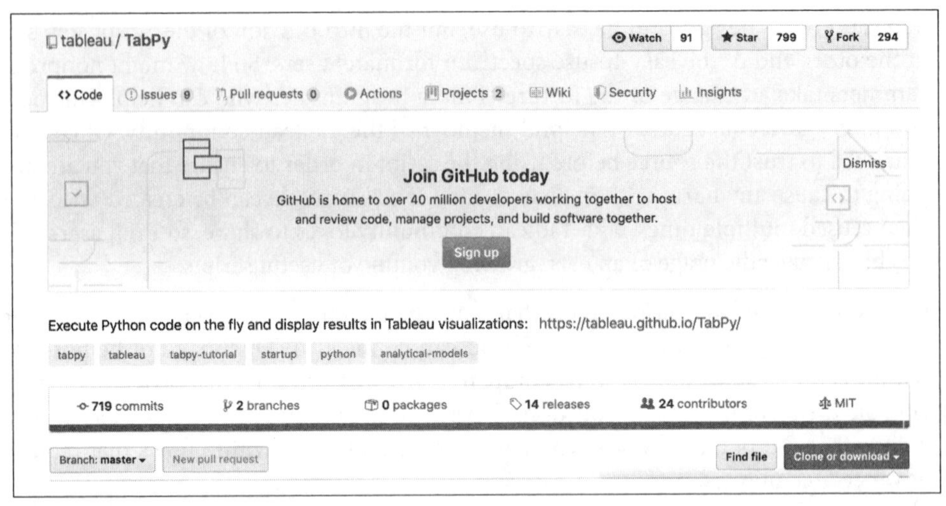

Figure 40-3. TabPy on GitHub

To install TabPy, run the **`install`** command shown in Figure 40-4 in either Terminal (Mac) or Command Prompt (Windows).

```
(base) Carls-MacBook-Pro:TabPy-master carlallchin$ pip install tabpy
Collecting tabpy
  Downloading tabpy-1.0.0-py2.py3-none-any.whl (81 kB)
     |████████████████████████████████| 81 kB 2.7 MB/s
```

Figure 40-4. Downloading TabPy in Terminal

If installation is successful, TabPy will be available on localhost (your local computer's address) and you should see the message `Web service listening on port 9004` in Terminal/Command Prompt (Figure 40-5).

```
2020-03-28,14:27:40 [DEBUG] (state.py:state:616): Returning value '[]'
2020-03-28,14:27:40 [DEBUG] (state.py:state:148): Collected endpoints: {}
2020-03-28,14:27:40 [INFO] (app.py:app:107): Web service listening on port 9004
```

Figure 40-5. View in Terminal once TabPy is installed

When TabPy is running, you will be able to connect to the instance in Prep Builder using the Script step. The Script step cannot be the first step in your flow, as it relies on reading in a data set from an Input or other preparation step. The Script step involves quite a bit of configuration. Again, we'll use Python for this example (Figure 40-6).

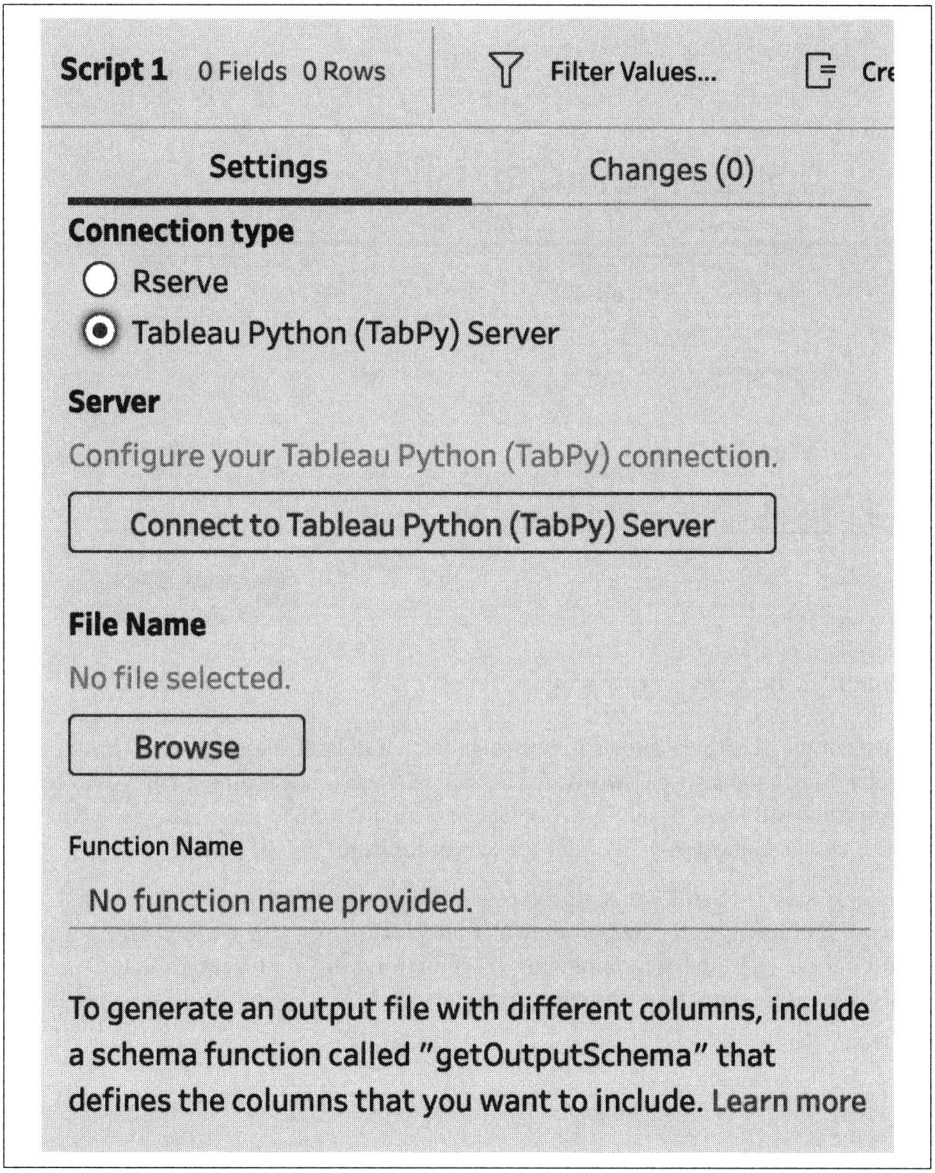

Figure 40-6. The Script configuration pane

When you initially add the Script step, you'll see an error. This is because you need to point the Script step to the programming server you want to use and configure the tool correctly. To configure the step, first select the connection type you want to use (TabPy, in this example). Once you've made the selection, you will need to connect to the server running that language (Figure 40-7).

Tableau Python (TabPy) Server

Tableau Python (TabPy) Server

Server: localhost Port: 9004

Username:

Password:

Require SSL (recommended)

Sign In

Figure 40-7. TabPy server connection

The example used here shows the settings from a default install. You do not need to enter a username and password if your instance doesn't require them. A successful connection will show a light gray message reading "Connection to *<server name>*"—in this case, localhost:9004—under the Server heading (Figure 40-8).

The next step is to link the script you want to run in Prep Builder. To do this, click Browse and navigate to the *.py* file you want to run. You can write *.py* files in many text editors. The final step is picking the function you want Prep Builder to return. This is set in the Function Name section.

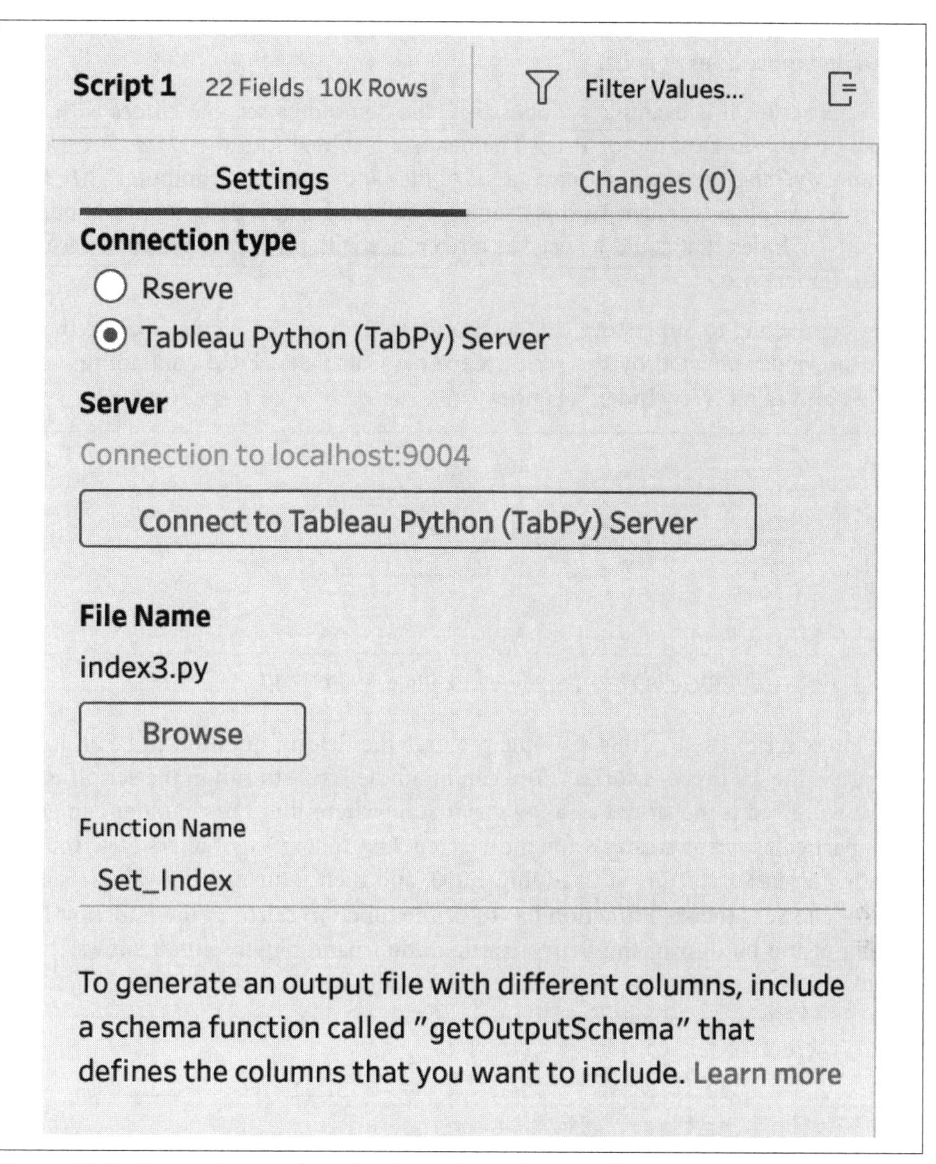

Figure 40-8. Connection made to localhost through port 9004

Using a Script Step

As this chapter is targeted to Prep Builder novices, let's go over a simple way to create a row ID that can be used as an ID for analysis. In Prep Builder 2020.1, Prep Builder added the ability to use the `ROW_NUMBER()` function, but before that, the technique we'll cover here was the easiest way to add a row ID to a data set where it didn't exist.

You'll see that adding a data field to an input is easier to do with the Python tool than by adding more lines of code.

The data set for this example is Superstore, the demo data set that comes with every install of Tableau Desktop or Prep. The data set is added into the *Datasources* folder of your *My Tableau Prep Repository*, a set of files added to your computer's *My Documents* or *Documents* folder. In this example, we'll add a new data field to Superstore called New Index that could be used as a reference with an ID per product, per order, for the Superstore.

After connecting to Superstore, add a Clean step to create a Calculated Field that will host the values affected by the script. Create a Calculated Field containing just the value 1 and call it "New Index" (Figure 40-9).

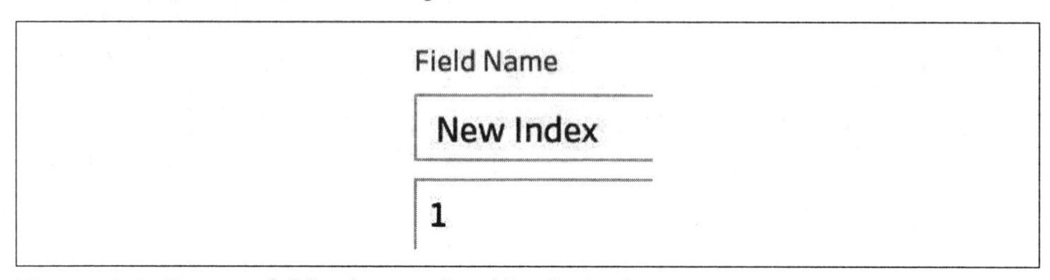

Figure 40-9. Dummy field to be populated by the Python script

The index script used in this example requires the field in question to be an integer, so ensure the data type is correct. You can build the script to run in the scripting tool, but it will need to be stored as a *.py* script somewhere that Prep Builder can access. This particular script creates a function called `Set_Index()` that alters New Index by creating an index starting at 0, adding 1,000, and then returning the values as a data frame. The `Set_Index()` function has to be the function called in the Function Name setting at the bottom of the Script configuration pane. Figure 40-10 shows the full script.

```
def Set_Index (df):
    df['New Index'] = df.index + 1000
    return df
```

Figure 40-10. Python script to form the New Index field

Data frames are the values returned from the script into the Prep flow. If you are creating new data fields, you will need to define the data frame being returned. This takes additional coding, which we avoid in this example by creating the dummy New Index field to be overwritten by the Script step. If you do require additional data fields, you will need to add the schema function `getOutputSchema()` to your script.

In this example, the output of the Script step is an updated New Index field that starts from 1,000 (Figure 40-11).

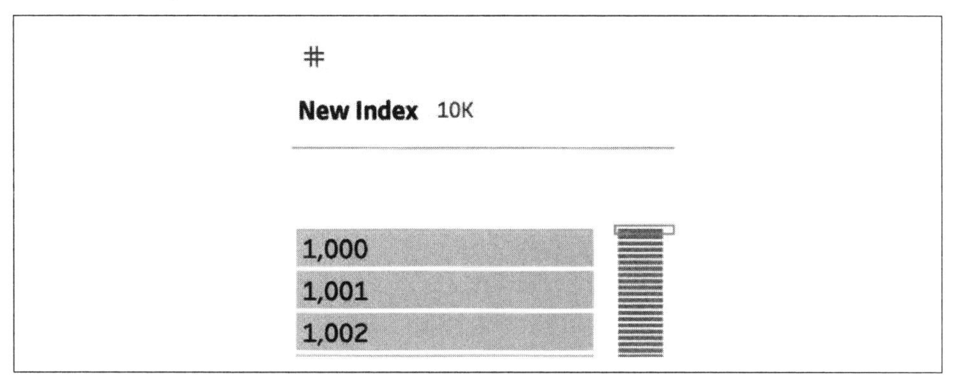

Figure 40-11. Field updated by the Python script

Summary

Prep Builder opened up its capabilities dramatically with the introduction of the Script step. Despite the feature's greater programming complexity, it offers significant flexibility in the data preparation process. Still, using programming techniques will not be suitable for every user of Prep Builder. Like any skill, scripting takes time to learn and develop, but it can supercharge your data preparation abilities.

CHAPTER 41
Handling Prep Builder Errors

Errors in software suck. Who hasn't spent time crying into their keyboard or banging their head against their screen asking the computer gods to solve their problems? Agreed—no one! As I have used Prep Builder a lot, I have developed a good sense of the common mistakes people make with the tool. This chapter will address the following trouble spots and pose some potential solutions:

- Parameter errors
- Blank Profile and Data panes
- Errors within a Calculated Field

Parameter Errors

You may often see an error message including the word *parameters*. For example, the error message in Figure 41-1 reads `Function 'DATEADD' can't accept parameters: (string, integer, string)`.

In this case, parameters are the data fields, numbers, and strings you are entering into your calculations. The calculation expects a certain data type in these fields and can't accept a different one. In this specific case, my colleague was using a date in a string field. The `DATEADD()` function specifically needs a valid date for the calculation to increment.

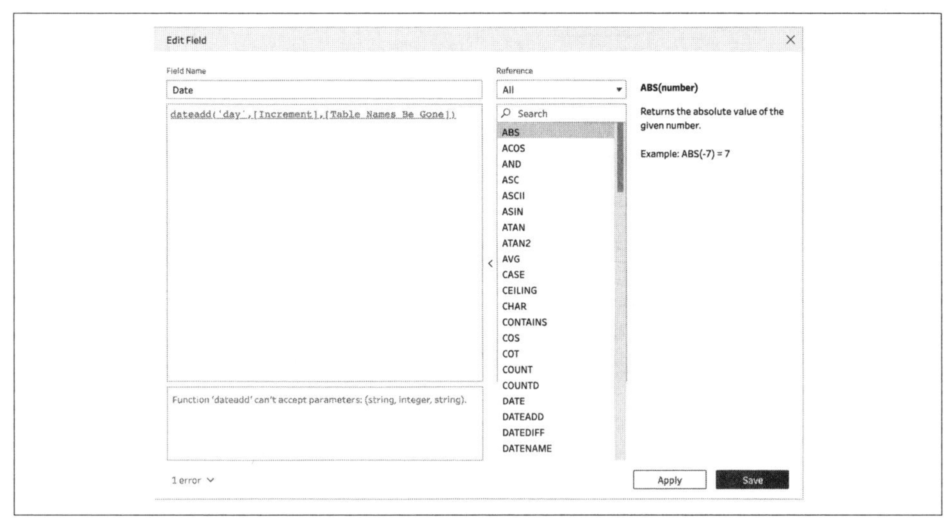

Figure 41-1. A common error message regarding parameters in Prep

Be sure to check the data types that you are using in the calculation against what the syntax expects. Remember, you can see an example of each function's syntax to the right of the Reference list in the Edit Field dialog.

Blank Profile Panes or Data Panes

There are multiple causes for blank Profile or Data panes, but I've found a few common culprits.

Changing a Calculation or Removing a Data Field Downstream

In Figure 41-2 I have created a duplicate NPS field, but I removed the original field in an earlier step.

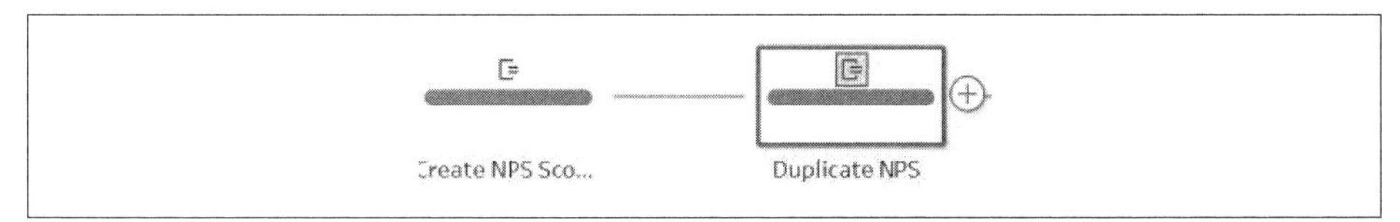

Figure 41-2. Example flow to show the effect of changing fields

Obviously, this breaks the flow, but I've done it to show the source of the issue. In Figure 41-3, the red exclamation point above the duplicate NPS field shows where the error prevents the flow from running, but this isn't where you would actually fix the flow; you need to do that in the earlier step where the original field was removed.

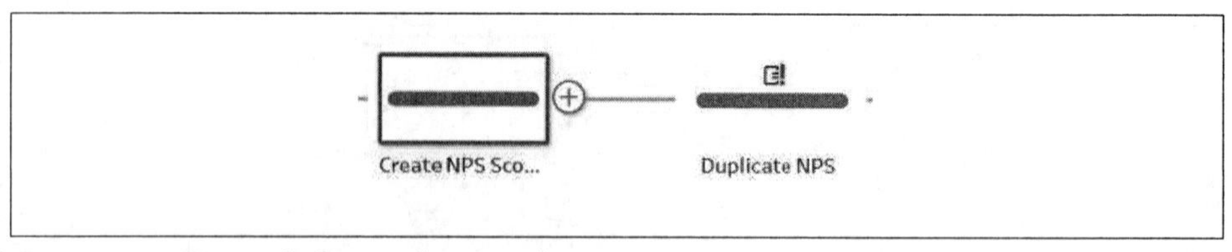

Figure 41-3. Example flow indicating the error

Use the Changes pane in Prep to work your way back through the flow to find where that original field was deleted. The error icon in the top-right corner of your flow allows you to read the error description, but the "View in flow" option in the bottom corner of this dialog box takes you to the error (Figure 41-4).

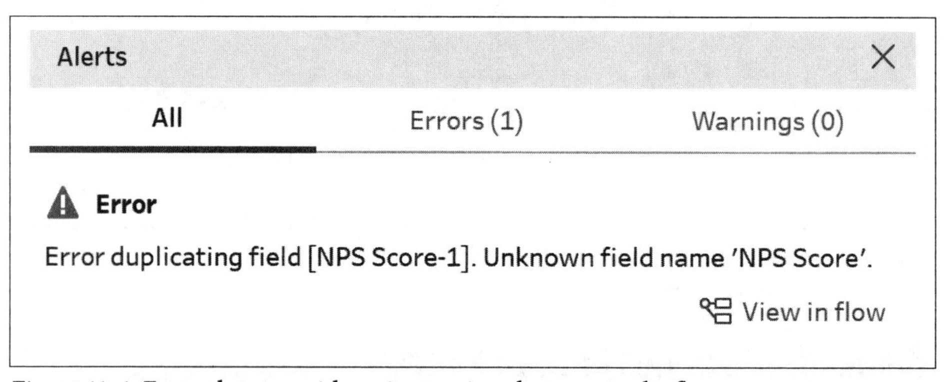

Figure 41-4. Example error with option to view the error in the flow

There's no easy solution here, but at least this might help you find where to start working your way backward.

The Data Source Has Changed

When you are using any data tool, if you change the data source in any way other than adding more rows of data, you are likely to get errors. Prep Builder is no exception. Common changes to check for are:

- The data source being used has been moved.
- The data source has been deleted.
- The columns used in calculations have been removed.
- The columns used in calculations have been renamed.

All of these changes will likely result in Prep Builder returning a blank Profile pane, as no data load will be possible, or data fields will appear in the Clean step with only null or blank values (Figure 41-5).

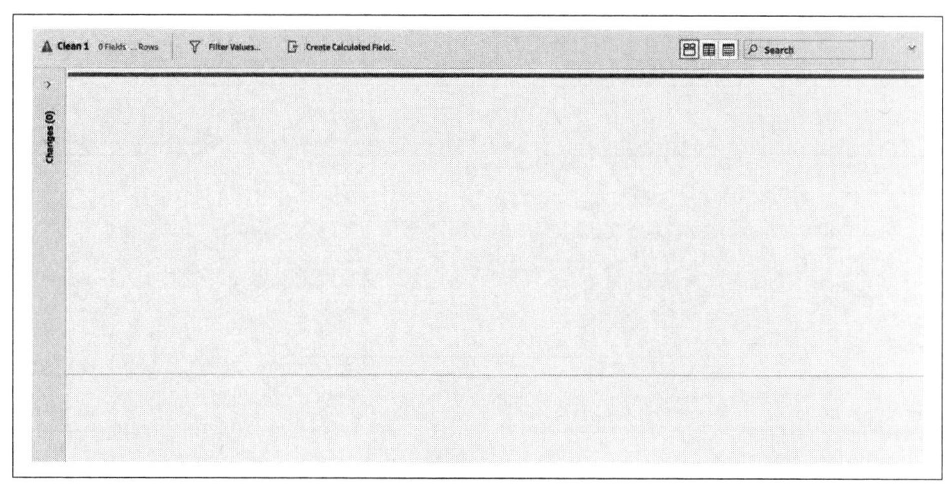

Figure 41-5. Blank Clean step resulting from an error

Errors Within a Calculated Field

When errors occur in Prep Builder, they may not be as catastrophic as losing all of your data as in the case just shown. Calculated Field errors are the most common mistakes you will make, but generally they are very easy to fix when you know how to spot them.

Incomplete Calculations

When writing calculations, you will very quickly see when the calculation is not in the format that Prep Builder is expecting. For example, the calculation in Figure 41-6 is erroring because the `IF` statement does not have an `END` condition yet.

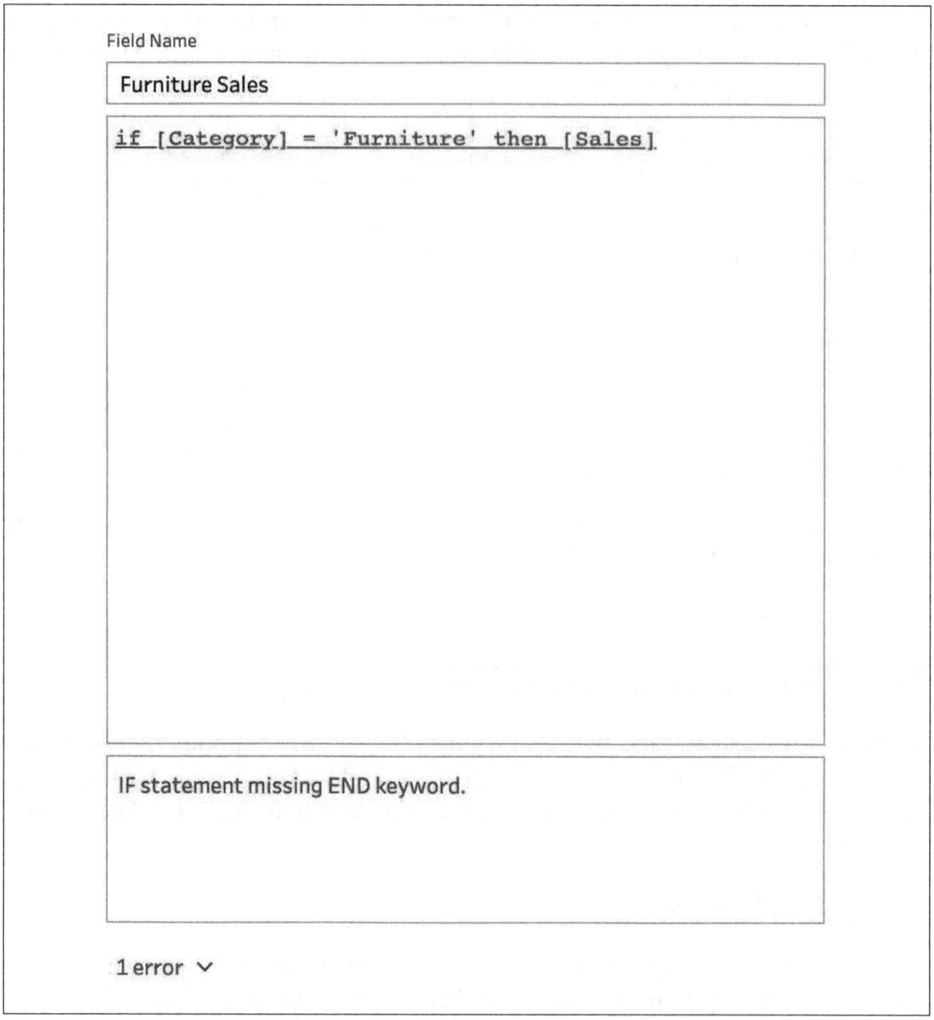

Field Name

Furniture Sales

```
if [Category] = 'Furniture' then [Sales]
```

IF statement missing END keyword.

1 error

Figure 41-6. Error caused by a calculation not matching the expected syntax

By clicking the "1 error" arrow at the bottom of the Calculation Editor, you can quickly see what is missing in your calculation. When building `IF` statements, I often add the `END` condition early and then edit the calculation between the words `IF` and `END` so I can see when each condition is complete as the error disappears (Figure 41-7).

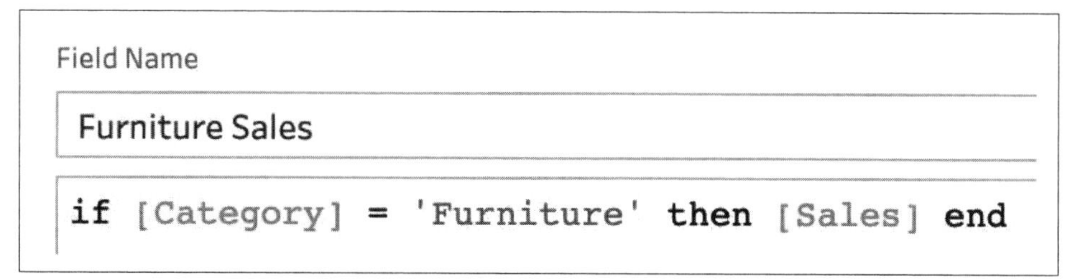

Figure 41-7. Non-erroring calculation in Prep Builder

Unsupported Functions

If you have worked with other data software before, sometimes you find yourself using tried-and-tested methods elsewhere. This is especially true for Tableau Desktop users who use Prep Builder. Using functions that are supported elsewhere but unsupported by Prep Builder—either in a certain area or altogether—can cause you a lot of frustration, even for simple functions like `sum()`, as shown in Figure 41-8.

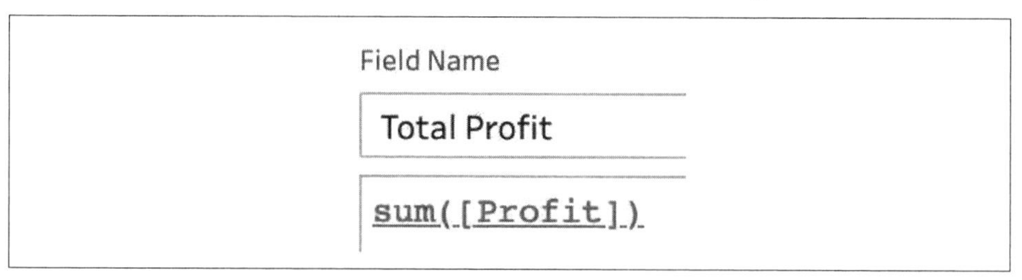

Figure 41-8. Error caused by an unsupported function

To resolve this error, you would need to use either an Aggregate step or a Level of Detail (LOD) calculation, as these are the areas where Prep Builder *does* support calculating totals.

Summary

Errors occurring in software when you least expect them can be very frustrating. The time you spend decoding them is time lost from the analysis and production of other pieces of work. Blank Data panes are a sign that something is amiss within Prep, and the warning symbols can get you only so far, but hopefully this chapter has given you a starting point to approach these challenges.

CHAPTER 42

Documenting Your Data Preparation

Jobs are no longer for life; these days, people look for variety and challenges in their careers. Data roles are no different, as the skills are highly transferable between companies and industries. Therefore, the data sets you prepare are likely to be passed along to others. To ensure the work continues to be understandable, up-to-date, and valuable, it needs to be well documented so you can walk other preppers through your logic once you move on or get promoted. In this chapter, we will cover the fundamentals of documenting your data preparation work, Prep Builder's built-in functionality to aid basic documentation, and the considerations to keep in mind for each data preparation step in Prep Builder.

Basic Documentation

No matter what tool you use to prepare your data sets, there are a number of aspects you should consider.

Folder Structure

All documentation within the data preparation file is useless if you can't find the flow file in the first place. Keeping a folder that is available to you and your team for all the flows will help tremendously. My organization uses Google Drive, as it has the benefit of not just controlled sharing but also being available on any computer I log in to. Setting up a standardized structure for those files is also key. In large organizations with more complex data flows, you should consider setting up the following folders:

In production
: The holy grail of folders, this file should be under strict control to avoid changes that break data sets your organization relies on. Flow files should be added to this folder only once they are fully tested.

Development flows
: This is the work in progress folder, or "sandbox."

Testing
: Once you've developed a flow, this is where you lock the version being tested.

Archive
: Having a history of key versions can help you track output changes.

Filenames

The naming convention you use for files is also key. If you can't locate the correct document in a folder of flow files, there is no point to spending time documenting the work in the first place. As a consultant, I have probably seen every naming convention imaginable. There's no single optimal solution; the key criterion is, *Does everyone understand it?* If they don't, then there is no reason to have a naming convention, as your colleagues will soon start to break it or chase you down to find out where a certain file is. You have data to prepare; you don't have time for that!

Data Sources

Within the file, the key piece of documentation that must be crystal clear is where that input data comes from (Figure 42-1). This may sound obvious, but those source files/tables are likely to move over time or change structure. Tableau tools can read only what is in the underlying source. As the source changes, so will the resulting data being used by the flow and, therefore, the output.

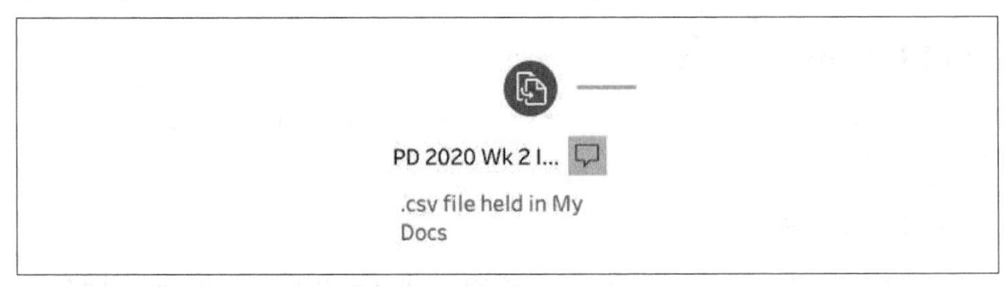

Figure 42-1. Documentation of the Input step

Recording the original data source, location, filename or table name, and the frequency of updates will help you and your colleagues understand what should, and should not, change when the flow is next run.

Output

Knowing which files it is safe to overwrite is critical in data preparation, as you may not be able to reverse the changes if you need to. The file location should be clear, and the output location shouldn't be changed, unless the existing output file is also moved. Clear documentation of what the output is and where it is held will help with this (Figure 42-2).

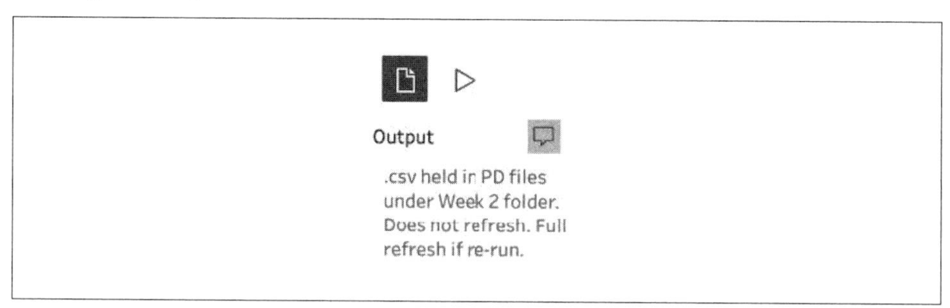

Figure 42-2. Documentation of the Output step

There are a number of other stages in Prep Builder where documentation can make the difference between work being easy to access and fix if necessary and it being an absolute nightmare.

Step Names

There are currently only eight different types of steps that can be set up, but clearly documenting them will help with handing over and maintaining the work.

Clean Step

The Clean step is the Swiss Army knife of Prep Builder, as one step can comprise hundreds of different preparation techniques. This step can include calculations, filters, cleaning string values, splitting fields, and renaming or even deleting fields. The Clean step can also contain any combination of these actions. Renaming the step with a description of what you are doing will ensure that the users of the flow—and your future self—can follow along (Figure 42-3).

Figure 42-3. The Clean step icon

The Prep developers have provided icons that show some of the top-level actions happening within the Clean step. For example, the step shown in Figure 42-4 includes both a filter and a calculation, and a field has been removed.

Figure 42-4. Clean step with icons demonstrating different clean operations

These icons give the user a very good quick overview of what is happening or give you a reminder of what happened if you return to the data at a later time. The step names need to be very concise, as only a limited number of characters will show in the Flow pane.

Step Descriptions

Step descriptions, unlike step names, can be much longer—200 characters. They also have the significant benefit that they can be toggled between visible and invisible. In the union shown in Figure 42-5, clicking the gray quotation icon will hide or show the description in the Flow pane.

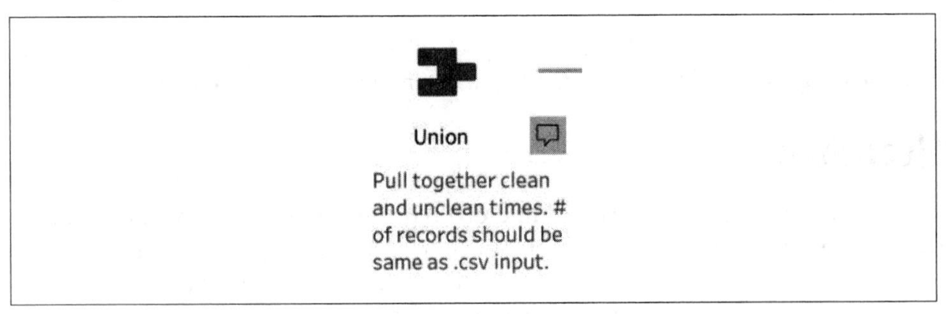

Figure 42-5. Documented Union step with description

The description allows you (the flow author) to add much more detail about what is happening at each step in the flow while preparing the data for analysis.

Color

One feature that was added very early on in the development of Tableau Prep was the ability to assign colors to steps to add visual documentation to your flow. This is useful both to the person building the flow and to someone picking up that flow for maintenance or further development. There are two key steps where color particularly makes a difference in Prep Builder.

Joins

When joining data sets together, or self-joining data together as in Figure 42-6, it's helpful to use color to show there are two different data sets coming together. Not only is this very useful when you are picking up someone else's flow, but it also helps with ensuring you have used the correct data fields from the incoming data sets.

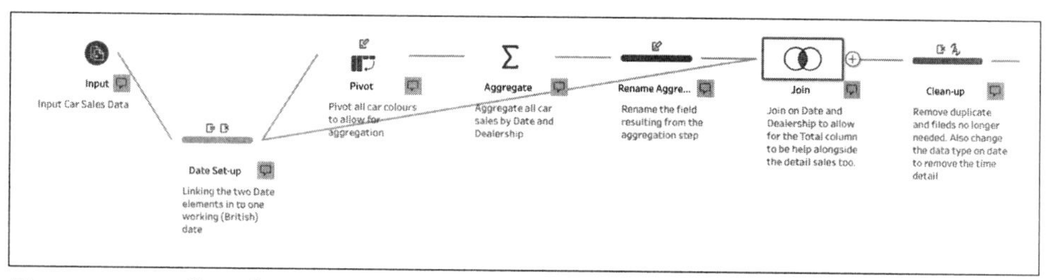

Figure 42-6. A flow showing a well-documented join including color logic

The color logic I like to apply to joins mixes together the yellow and blue inputs to create a green output. This way, I can instantly see the two data fields that are being joined (Figure 42-7). This is useful especially on inner joins, where the fields are normally identical; it's much easier to see where values have not been joined if you can see which input source has the mismatched fields.

Join Clauses
Show only mismatched values
Join Results
Rename Aggregagte output
Date Set-up
Dealership
Date
A 01/01/201
A 01/02/201
A 01/03/201
A 01/04/201
A 01/05/201
Date(Date-1) 24
01/01/2018, 0...
01/01/2020, 0...
Total Cars Sold 48
600
800
1,000
Date 24
01/01/2018,
01/01/2020,

Figure 42-7. Coloring the Join step assists with the setup

The green line running above the Profile pane in Figure 42-7 helps highlight to the flow developer the data fields that the join will be creating.

Unions

In a Union step, you can use color to demonstrate the two flows being stacked together. Unlike my approach to joins, my preference here is that the two flows are two shades of the same color rather than a mix, since the data structure is the same or similar (Figure 42-8).

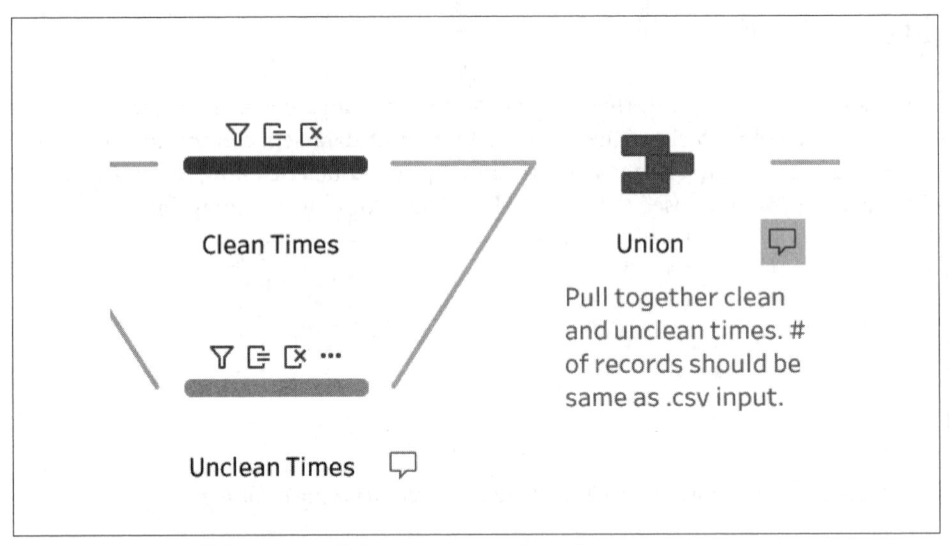

Figure 42-8. A well-documented Union step within a flow

In the setup of the Union step, the input flows' colors also represent (as with joins) where they have come from (Figure 42-9).

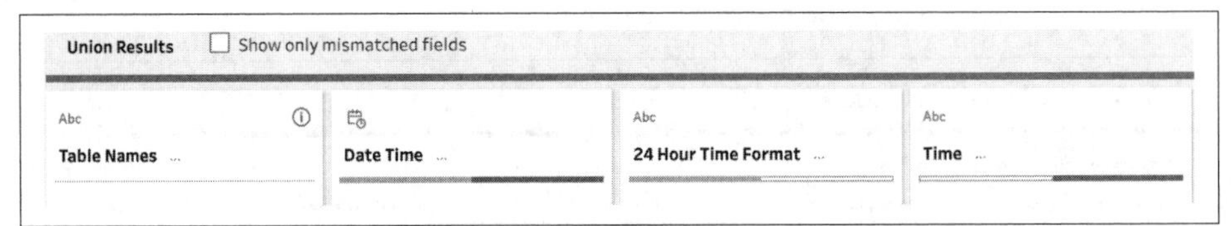

Figure 42-9. As with a Join step, the coloring of the flow assists with the Union step setup

In this example, the Time field has come from the Clean Times input, and the 24 Hour Time Format has come from the Unclean Times input. This helps you consider the different field names and where you may want to go back "upstream" in your preparation flow to either amend the names or investigate why they differ. The absence of color indicates nulls, which occur because they lack a corresponding field in the other data set (Figure 42-10).

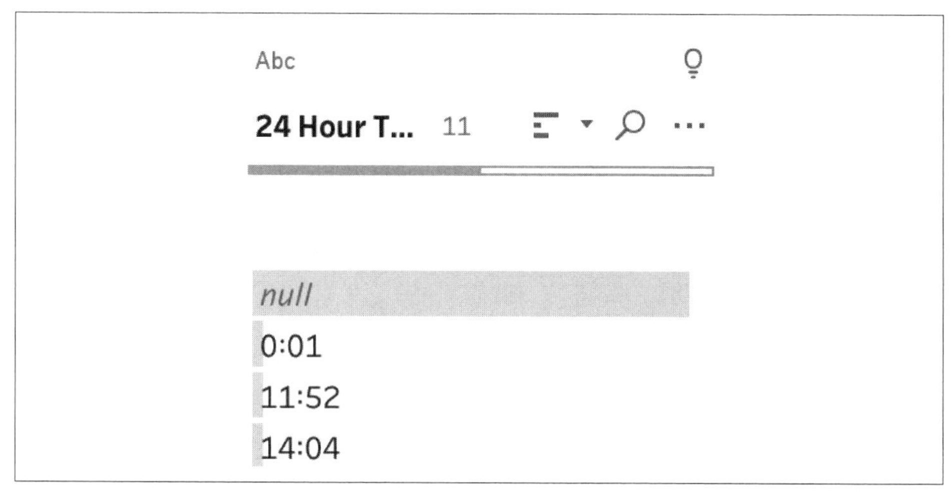

Figure 42-10. Absence of color represents nulls in a union

Summary

Although documentation might sound laborious and time-consuming, this isn't the case in Prep Builder. Editing step names, adding short descriptions, or changing the color of the preparation steps can make development less error-prone and hand over much easier.

CHAPTER 43

Deciding Where to Prepare Your Data

Like many software providers in the data landscape, Tableau doesn't just have one tool where you do every task. Frankly, such a tool would be either crazily complex to work with or incomplete in terms of functionality the user needs. Having multiple tools available within the Tableau platform does pose another question, though: Where should you complete certain processes?

Processes to Consider

Data preparation comes down to a number of key steps:

1. Inputting data
2. Joining/unioning multiple data sets
3. Pivoting
4. Cleaning
5. Aggregating
6. Outputting data

Each step could be appropriate for lots of hypothetical situations, but in reality, the majority of them should take place in the data preparation tool. Joins, unions, and pivots are common tasks at the data prep stage, and end users of the data set should be spared that complexity in Desktop. While flexibility might be required in some cases (specific visualization styles, for example, demand a different data structure), the majority of data sets have a relatively standard setup for analysis.

This leaves cleaning (including calculations) and aggregations as two processes that may fit in either the data preparation tool or the visualization tool. With small data sets and simple calculations, which approach is "best" is more ambiguous. However, as the size of the data set or the complexity of the calculations grows, your decision here might determine how successfully your organization can utilize the data.

Data Preparation Versus Visual Analytics

Balancing agility and functionality is a key consideration when you are evaluating which tool to use to complete each task. If you sacrifice agility by separately preparing data in a tool like Prep, you eliminate the option for each user to do this individually. Removing that flexibility might actually be useful, however, as it will potentially prevent mistakes, optimize performance, and enable tasks that would be impossible otherwise.

Considering what tasks should be completed in each tool can help you allocate the work of data preparation. To determine how work should be distributed, you will need to evaluate your organization's sophistication on a number of factors.

Data Literacy

Data literacy—or how well one understands data products like graphs, results, and the like—is a key determinant of where you will conduct the cleaning and aggregation. Making data easy to work with is important, but ensuring the accuracy of the answers derived from your data sets is even more important. If your peers do not have the data literacy to take on these tasks, you'll need to complete this work before making the data available to them. This means that you will need to understand and/or anticipate what questions those users want to answer and prepare your data set accordingly.

Organization Size

Having a team that is competent and able to complete the tasks is a major asset, but you'll also need to evaluate the volume of work required to repeat those tasks multiple times across the organization. If you are asking one person to complete a task once, it doesn't matter much where that task is completed. If that same task would need to be repeated hundreds or thousands of times by multiple individuals across an organization, however, then this task should be performed in the data preparation tool to reduce the amount of duplicated effort. Data preparation tools are designed to automate such tasks once they are set up.

Quality of Technological Hardware

The hardware on which the tasks are processed has a strong impact on the time it takes to complete them. Companies across the world pay people high salaries but then equip them with older laptops or underpowered computers. This situation can hinder people from being able to work with data, and the problem only gets worse with increasing volumes of data. If the data sets for analysis are small, then any basic data preparation task may still perform fine on the individual's computer. If the data sets are large, though, a data preparation tool might be a better option. Data prep tools can often work with just a sample data set (as Tableau Prep does automatically for large data sets) and process the full data set only when required. This full processing likely takes place on a server (which has more processing power) once the full end-to-end data flow and logic is established.

History of Data Investment

If organizations have historically invested in data solutions and continue to do so as technology advances, the likelihood is that their databases will contain clean, ready-to-use information. When conducting the analysis, you can add any necessary fields to the database for future use. If this isn't the case, then it's likely you'll be wrestling with messier data from multiple sporadic sources. There's no clear answer as to where you should do your data preparation; you will likely need to switch between the data visualization tool (to find out which data is useful) and the data preparation tool (to set up more strategic data sources for future use).

All of these contextual factors will help guide you to a decision, but it's only when beginning the actual work that you'll determine where is best to complete it.

Software Performance

As you'll see in this section, Prep is specifically designed to optimize the process of building the data preparation flow and then executing it.

Sampling

When you import a data set in Prep during an Input step, the software runs a sampling algorithm that shows you a suitable profile of the data without having to process all the rows of the full data set (Figure 43-1).

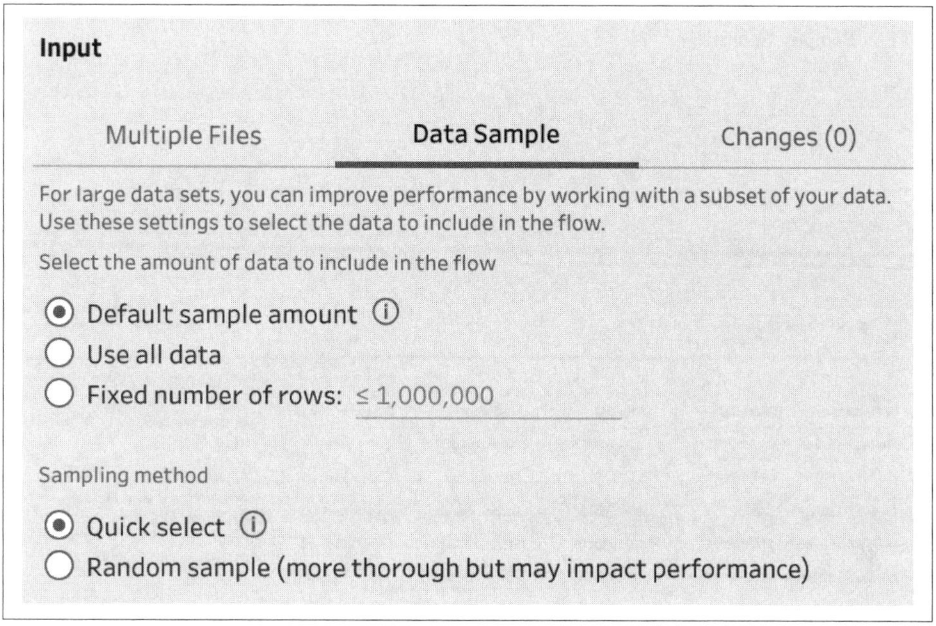

Input

Multiple Files | **Data Sample** | Changes (0)

For large data sets, you can improve performance by working with a subset of your data. Use these settings to select the data to include in the flow.

Select the amount of data to include in the flow

- (•) Default sample amount ⓘ
- ( ) Use all data
- ( ) Fixed number of rows: ≤ 1,000,000

Sampling method

- (•) Quick select ⓘ
- ( ) Random sample (more thorough but may impact performance)

Figure 43-1. The default sample in Prep for each Input step

The sample is designed to represent what steps your data preparation will need to include. Tableau Desktop also shows a small sample of data, but that sample is based only on a certain *N* number of rows. For many data sources, this will be just the first 1,000 or 10,000 rows of data in the table you're importing. Therefore, if there are issues in the last rows of a table, you might not see them until much later in the analytical process.

Functionality

Data preparation functionality was initially built into Tableau Desktop in the Data Connection window (Figure 43-2), but the need for additional features coupled with a desire to keep this screen uncluttered led to Prep.

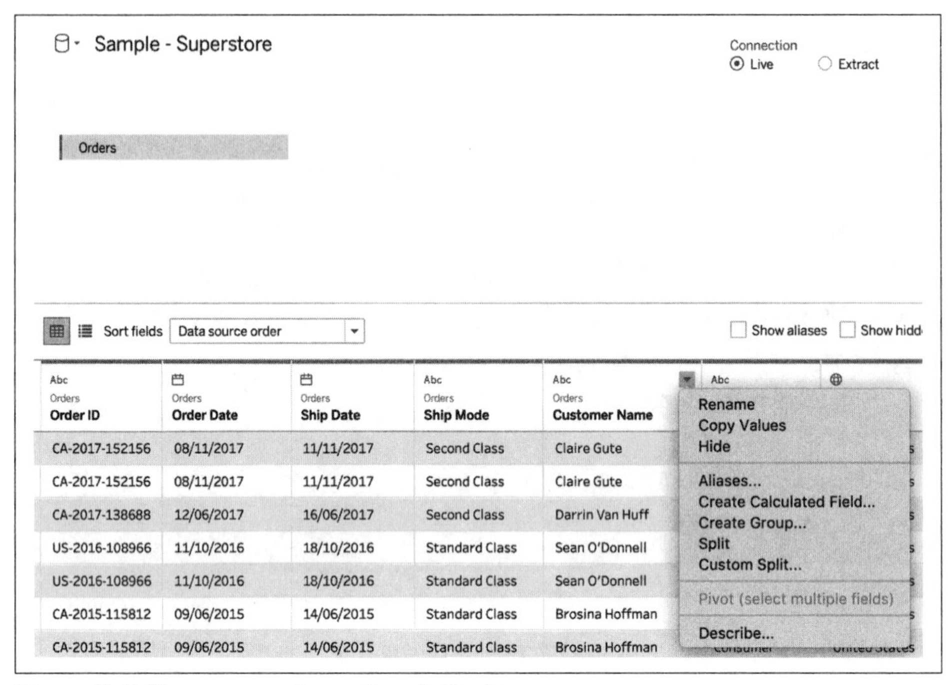

Figure 43-2. Data preparation options in Desktop

While basic tasks can still be completed in Desktop, it has limitations where Prep does not. Planning the required data preparation steps will often eliminate Desktop as an option for completing them. Multiple pivots, unioning data sets from different sources, and preaggregating tables before joining are just a few of the tasks that you will need to complete in Tableau Prep rather than the visualization tool.

Documentation

Being able to apply a solution not just to the problem at hand but also for future scenarios is a significant reason to do your data preparation in a Prep tool. Documenting your process—from naming the logical steps to describing what happens within them—makes it much more robust and maintainable (Figure 43-3).

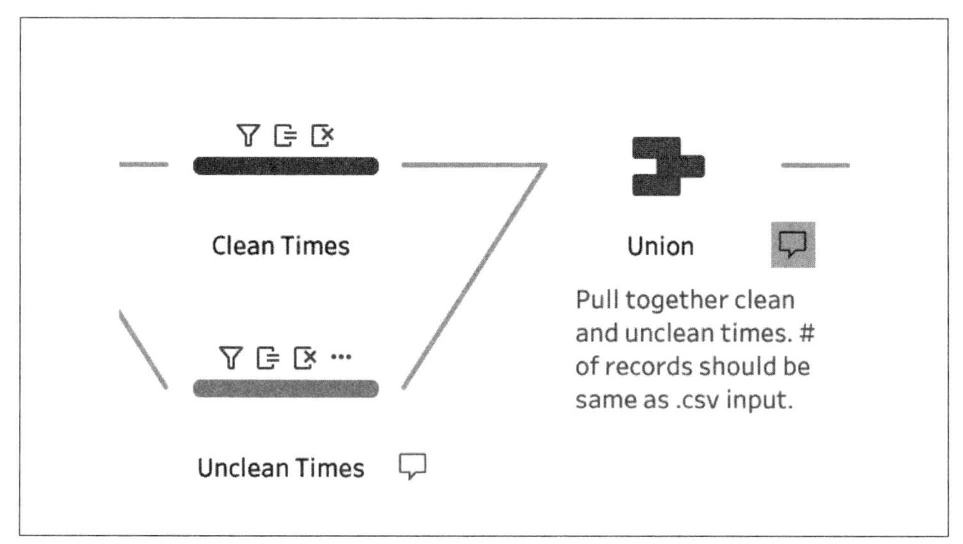

Figure 43-3. Step names and descriptions in Prep

To rename a step, double-click its current name. Once you have entered the new name, you will also be given the option to add a description. You can show or hide this option by clicking the speech bubble icon that appears after you've written a description. This way, you can hide it to keep your screen cleaner, but it's still available for new users of the flow or for when you need to revise certain steps.

Summary

There is no single answer as to where to prepare your data, but simply considering this question will improve everyone's ability to use it well. Your individual situation and how your end users access the data set will determine a lot about where you should process your data. The computing power, frequency of data updates, and the size of the data set are all major factors that will inform which approach you take.

CHAPTER 44

Managing Data

Data management and information security are growing focus areas for consumers and companies. This chapter is not a guide on what the laws are (or are not), but rather will highlight the areas that most Prep Builder users should consider. After learning about the types of sensitive data you may come across, you'll be better equipped to handle and use the data correctly. This chapter will also cover when and why you may wish to delete data.

Throughout this book, you may have picked up on the fact that I support the use of data and data proliferation. This is because I have seen many organizations lock down data access so much that it becomes impossible to make data-driven decisions, either due to the lack of data or lack of practical experience working with it once it does become available. This is not to say that you should be reckless with data access, but preventing data use altogether can cause consumers to make poor, even harmful, "gut instinct" decisions.

What Is Sensitive Data?

Data sensitivity is measured in many different ways depending on the organization, the sector in which it operates, and the data's subject matter. Most data is categorized by its sensitivity—that is, the extent to which the accidental or purposeful release of this data would cause issues (legal or otherwise) for the organization that owns the data or the subjects of the data. Typically, there are about three to five levels of data security. This section will describe four of the most common levels.

Public

Public information has been pulled from publicly available sources and is probably free to use. The data may come from government sources or just be data that is used openly—for example, spatial data sets (like post codes/zip codes), census responses, or social media posts.

Confidential

Confidential data has probably been processed; that is, it has been refined from the raw data. After investing the time and effort in that refinement process, the data holders might not want to openly share the data with potential competitors. If this information were to leak to the public, there would be no consequences for the entities covered by the data.

Strictly Confidential

Strictly confidential data is likely to be information on your sales, customers, and products. This is data that you cannot share and that you do not want your competitors to see. This data may contain lists of your customers' details, but nothing that is so sensitive that it would impact the customer if the data were leaked. This level of sensitivity is likely to include intellectual property within the company that may cover data assets like financial models and projections.

Restricted

Restricted data is the most sensitive data the organization holds. For large organizations, this can cover a vast spectrum, from customer-sensitive data like banking details to demographic information that could potentially be harmful if leaked to the public. Not all organizations store information like political affiliations or sexual orientation, but the data can be a proxy for this (e.g., bank transactions that show donations to a political campaign).

Restricted data can also include company Price-Sensitive Information (PSI), which relates to company shares traded on the stock market. If PSI is not managed correctly, shareholders can potentially engage in insider trading (i.e., trading shares based on information that isn't publicly available, a practice that is not just unfair but also illegal). People with access to this information need to understand how to manage this data; they must be prevented from trading shares themselves or divulging the information to those who do. Breaking rules around restricted information poses a risk not just of damaging the organization's reputation but also of incurring fines or even facing imprisonment.

Managing Data Based on Sensitivity

Striking the balance between overprotection, which prevents people from being able to do their work, and underprotection, which risks misusing sensitive data, is a challenge but far from impossible. Working with the users of the data to understand what they intend to do with it is the key to achieving this balance.

Having data sources that are stored centrally but enable you to grant or gain access quickly is a good target to aim for. Removing the small data sets held individually on colleagues' laptops and personal drives can help ensure data is up-to-date and correctly removed once it is no longer relevant. Providing a centralized data source will ensure that data is accurate, as multiple people will be using the data sets and can make adjustments as they find errors.

The challenge with centralized data sets is getting access in a timely fashion so people can answer the questions they have as they arise. Centralized data sets often include data across the sensitivity spectrum, so permissions are tightly controlled. It's important to have a fast turnaround and a devolved permission approval process, where multiple people are authorized to grant access to the individual making the request. Without these measures, no one is able to gain access and requests can be held up by individuals who are away from the office or busy.

For data preppers, practicing in a "sandbox" environment similar to the centralized stores can help them master the following skills:

- Connecting to a data store
- Using naming conventions for software and data sources
- Writing to the data store
- Removing content from the data store once it is no longer needed

Practicing these skills helps ensure that the data prepper will use the right terms in the centralized production environment and have a process developed in Prep that mirrors what needs to happen in production.

Production Versus Development Environments

Production environment is not a term covered much in this book thus far. Production environments are very tightly controlled, as they store lots of the regulatory or otherwise important reporting of a company's performance. In contrast, a *development environment* is where a data set or query will be tested before being placed into the production environment, so it is more flexible and has fewer rules constraining the use of the data sets it holds.

Not all data is prepared perfectly the first time, in the same way that a report or analytical dashboard is rarely production-ready right away. You'll often need to iterate based on the feedback of others using the data. This is why you need development environments to test the data preparation flow as well as the resulting data set. Only once the asset has been tested and is approved for widespread use and in key reports should you move the flow into a production setup. Again, the production environment is more tightly controlled, so most people will not have permission to write content to it, nor should they, lest they make mistakes that may be very difficult to resolve.

Deleting Data

So, if you understand the sensitivity of the data and have tested the data set in a development environment before publishing the flow to a production environment, are you done? Well, not really. You also need to consider when to delete data. In this section, we'll cover the two most common reasons for doing so.

When Data Becomes Outdated or Irrelevant

Data can become less relevant and potentially less accurate over time. When creating a data source, you should think about how long to retain that data. Obviously, designating a date for deleting the table or records doesn't mean you have to do it then. You can always reassess the data for its relevance and accuracy, but setting a date will at least keep you from putting off the decision to keep or remove it.

When a Customer or Client Leaves

You should retain data only as long as you are legally allowed to. This rule has become much stricter with the EU's introduction of General Data Protection Regulation (GDPR) in 2016. Detailed customer data should be removed when the customer leaves. To be able to do that, you must know or be able to find out where all of that customer's data actually resides—in which tables in what systems. If data is distributed through a lot of different sources and files, this is a more difficult process. Having data sources in a centralized location means that when you remove customer data from that central source, the change will also be applied to the other data sets.

Summary

Overall, data management and information security aren't the most fun subjects, but understanding them can make working with data much easier and faster. Having novice data preppers learn and develop these skills within a controlled "sandbox" environment can help ensure the security and integrity of the data once they are granted access to the centralized production environment.

CHAPTER 45
Storing Your Data

One of the key considerations in data preparation is where to hold the output. After all, what is the point of doing all that hard work if you then put the data somewhere that is:

- Inaccessible to those who need to use the data
- Slow or unresponsive
- Not protected against accidental overwrites, risking permanent loss of the source data

Let's consider each of these scenarios in turn to determine what you should consider when writing your output to a location (Figure 45-1).

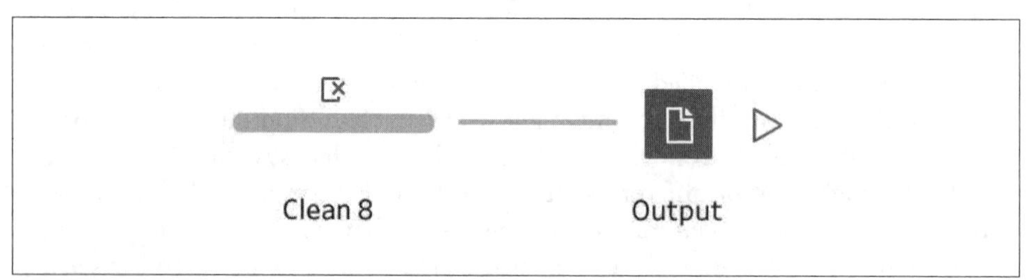

Figure 45-1. The Output step in Prep

Inaccessibility

As the previous chapter discussed, it can be challenging to find the right balance between data openness and data security. More restrictive data legislation is being passed across the world as the general public realizes the value of their personal data and the potential effects of a data breach on their lives. At the same time, *without breaking any rules or laws*, giving people freedom to work with data will lead to future innovation and better, more efficient decisions. So how can you strike this balance? Well, let's first consider the absolute *don'ts* of data access.

Don't Break the Law

There are some things that you just can't do, and breaking the law is one of them. The following are two important things to keep in mind.

Personally identifiable information

Personally identifiable information (PII) is any data that can identify an individual. For operational reasons, you may need to be able to identify someone (i.e., check the balance in their bank account), but for analytical purposes, this shouldn't be the case. This isn't a book on data security, so I won't go into too much detail here, but the point is that you should restrict access to any data that could be used to identify the individuals it involves.

The right to be forgotten

Numerous pieces of legislation have been created or restructured over the last few years protecting an individual's right to have their data removed from your organization's possession. In the EU, this policy is known as the *right to be forgotten*. To comply with these laws, be sure there is a clear trail of where data is used and what it is used for, and delete it once it has been used in your analysis or is no longer relevant.

Don't Delete Operational Data

Operational systems—the technology systems that allow you to make payments, take orders, or provide services—must not be affected by your analytical queries. All of these systems rely on data and often store it in databases. If you are querying these systems directly, you are one poor query away from causing a lot of damage by deleting operational data points or slowing down key operational processes. If it is legal and useful for your analysis, copy this data into a specialist analytical environment. This way, you are querying a database that, in the event of an issue, will not affect the key systems your organization relies on to operate.

So, with these aspects in mind, now let's look at the other side of the coin and still see data as an asset, not a liability. Making data accessible by following this list of *do's* is key for any organization to progress and develop.

Do Grant Access to Data for the Experts

Not being familiar with what is in the data can lead to poor decisions. This isn't a matter of technical skills but of understanding the context of the data. Giving subject-matter experts access to the data, and working with them to understand exactly what each column is doing, will ultimately ensure the data source gets documented and becomes useful. Otherwise, you are storing or potentially using data that could be misconstrued. Storing the data in a location the business experts can't access will result in "opinion-driven" rather than data-driven decisions.

Do Document Your Sources

Storing data in such a way that it isn't clear and obvious what it is won't lead to any success. Curating data sources so that end users can understand them is more important than saving on small amounts of storage space. Clearly naming column headers, creating views on top of tables to "humanize" the language, and publishing data sources on user-friendly platforms like Tableau Server are all ways to document a data source and make it more accessible.

Slow/Unresponsive Performance

One major consideration when you are deciding where to store data is the response time. In an age where people all over the world are able to ask questions online and get answers in seconds, a data set that takes 20 seconds to load can feel positively glacial. Ensuring data sets are responsive to the queries being made is key to putting data at the heart of your organization. Not all data resides in this state—lots of data sets are stored on slow, archaic databases—but as data preppers, our task is not just to clean input data sets but also to ensure the output data set will be responsive. Ultimately, if it isn't, your users will tell you by not using the data set.

Overwriting Risks

Although Tableau Desktop is a read-only tool, the same can't be said for Prep Builder. Any alterations a user makes in Tableau Desktop will never change the underlying data source. This frees users to experiment and try out new techniques or queries without fear of damaging anything. If you enable the same level of freedom in Prep Builder, however, you run the risk of users overwriting data that you might not be able to recover.

This risk is nothing new for running data infrastructure. For decades, database environments have been a balancing act between managing limited resources and meeting the needs of the users. If you reduce the administrator's workload, you open up less experienced users to more responsibility. However, tilt that balance in the other direction, and everyone is stuck waiting for the administrator to be able to do anything. Creating an equilibrium is clearly impossible, but there are some approaches that can help.

Grant Read-Only Access

Giving people access to the raw data can help them find what they need without putting huge processing loads on the data storage environment. At a large bank, the idea of giving users Tableau Desktop was initially met with resistance for fear that the increased demand would overwhelm an already stretched processing environment. The opposite actually happened, thanks to the quality of the drivers that Tableau uses. Queries became more optimized in the majority of cases, and running the environment dramatically increased the value that users were getting from the data assets. With Desktop, this was an easy move, as the tool is read-only and therefore couldn't affect the underlying data.

Still, you should look at Prep Builder in a similar way. The only difference is that instead of producing visual analysis, users will be producing cleaner, more streamlined data sets. Giving the users a set location to publish these to (a "sandbox" or "playground") not only will give them the confidence to try to gain more value from the data but also can improve specifications for future developments, since users will know what they need to do to empower themselves and others. However, remember that with Prep, you can write the data back to its source with the Output to Database option (Chapter 20). To prevent any issues with accidental overwrites, give data preppers access solely to read-only information.

Train Before Publishing

The Prep Builder flow itself doesn't have to be run as an output. The process of cleaning data can be beneficial both for users, who learn what they would like to achieve and how to do so, and for those who hardcode the results (i.e., write them to the database), who get to see the details of the cleaning process. This means the users' requirements, which may have been spurious or iterative, are now clarified before the administrator spends time implementing this work. Over time, the skills the users develop will empower them to do the publishing work themselves, as they will know how to use the tool and the environment correctly.

So, Where Do You Write That Output?

The short answer to where to write your output is...it depends. It depends on the user, the type of data involved, the responsiveness of the database, the organization's investment in the data platform, and the key roles supporting it. Empowering end users and allowing them to learn over time will promote data-driven decision-making in your organization. Ensuring that they can't go too far wrong is good for them and the platform administrator both, as fixing mistakes can be time-consuming. Where possible, writing data sources to a centrally managed location will help protect it, especially as data privacy laws and corporate policies become more stringent.

Central locations, like databases and shared spaces, means potential data users know where to look for data sets that may provide answers to their questions. Potential users need to be clear on what they can and cannot use, and what is in the process of being tested. Organizations will need to create data repositories for validated data sets, as well as repositories to hold data sets that are in development. These environments are dramatically different, so be sure to check with the data experts in your organization about how to store the data you have prepared.

Good documentation will ease the migration from a development environment to a "productionalized" data set and help reduce the time it takes to fix issues that may arise.

Documentation in Prep Builder was covered in Chapter 42.

Summary

Creating an analytical data store that allows users to work with their data is a worthwhile investment that has huge benefits for the organization. Allowing users to access Tableau Prep, which is focused on making data preparation skills easier to learn, is another plus in an organization's approach to data work. Combining the two is a strategy that has a lot of upsides but may take some time to achieve. Setting this as a target is a great starting point if data is locked down or potential users don't yet have the skills to access the data sources they need. Those skills will develop over time, and navigating this learning curve is far better than a situation where data is unclearly documented or locked away for fear of overwriting or misuse.

CHAPTER 46

Using Identifiers and Keys in Data

As mentioned in multiple previous chapters, databases are designed as powerful, performant, and secure locations to store and work with data. Two concepts at the heart of most databases you will use in your organization are *keys* and *identifiers* (IDs). Databases that utilize these concepts are known as *relational* databases, because keys and ID fields create relationships between the various data tables. In this chapter, we will look at ID fields, how they can be used as keys between tables, and how to create them in Prep Builder.

What Is an Identifier?

As in most computing software, numbers in databases are processed more efficiently than other characters. When working with data, you'll often need to process large volumes of data, so any method you can use to make processing more efficient can make a significant difference in the time it takes to prepare your data or conduct your analysis. This is where identifier fields come into play for databases. Rather than repeatedly storing names, addresses, or other long string values, a relational database architect will use separate *look-up tables* to store those values just once and associate an ID to the value. This purchases table shown in Figure 46-1 has a number of ID fields that we can use to link together other reference tables with the names of customers or products. Tables containing lists of transactions or similar metrics are often referred to as *fact tables*.

Order ID	Customer ID	Product ID	Quantity	Order Dispatched
1000	103	10002834	2	1
1000	103	10003821	1	1
1000	103	10001237	10	1
1001	101	10002834	5	0
1002	119	10001237	7	0
1002	119	10003821	3	0
1003	145	10002834	25	1

Figure 46-1. The Purchases data set, an example of a fact table

Identifiers can also be used in a Boolean way, indicating whether something is true or not. For example, "yes" (or "true") is often encoded as `1` and "no" or ("false") as `0`. This type of logic often does not have an associated reference table, but some data architects will insist on one for clarity's sake.

What Is a Key in a Database?

Keys help us identify how to join tables together in databases. Traditionally, there are two types of keys in database terminology:

Primary key
: A unique identifier that cannot be replicated in a table

Foreign key
: An identifier for a unique row of data stored in a different table

Figure 46-2 shows a reference table that has a unique key for each customer.

Customer ID	Name	First Order
101	Caroline Bell	04/01/2020
103	Helen Yates	14/01/2020
119	Steve Francis	16/02/2020
145	Kirsty Frank	30/03/2020

Figure 46-2. The Customer data set, an example of a reference table

We could use this table to add the customer's name to the Purchases table shown in Figure 46-1. The Customer ID in this table must be unique, as we wouldn't want to attribute orders to the wrong customer. In the Purchases table, the Customer ID could be classified as a foreign key because the values in the column are duplicated but are used to join the Customer table.

Using Keys and Identifiers in Prep

Keys in the context of Prep Builder are very similar to those in a database; they guide users on how to join tables together. Information about which data fields in a database table are keys is likely stored in separate database documentation. Prep Builder does not display keys any differently than other data fields, and you can use them like any other data field in join conditions. This is where the Join configuration pane comes in particularly handy; you can see what values are matched as well as the resulting number of rows being created by the Join step (Figure 46-3).

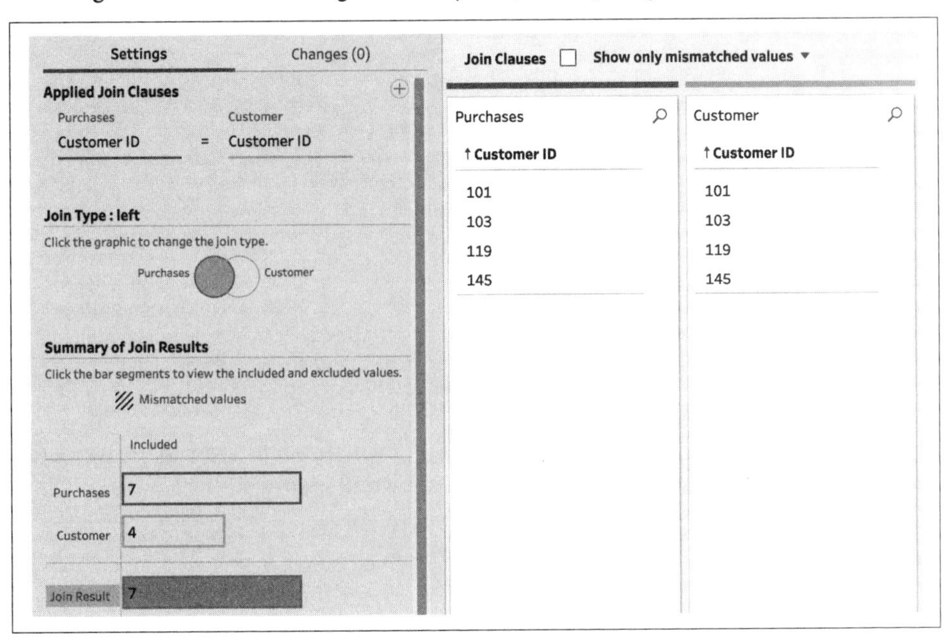

Figure 46-3. Setting up a join in Prep

When joining fact tables like the Purchases table, you would expect to retain the original number of rows. (Remember, fact tables are database tables containing measurements or metrics.) If the resulting number of rows doesn't match the original fact table, the possibilities are:

- Increased number of rows. You have duplicated the purchases, and simple counts or aggregations of values may overrepresent reality.
- Decreased number of rows. You have lost rows of data, meaning records of sales or the number of customers will be underrepresented.

The reason why this view in Prep Builder is so useful is because it clearly shows the number of rows both entering and resulting from the Join step. In this example, there are seven rows of data from the Purchases table and seven rows of data are returned. This is a likely sign that the correct key has been joined up, but verify the results for all joins even if the number of rows matches what you expect. The Join Clauses pane demonstrates that all Customer IDs have been matched, as they remain black text. Any unmatched values will turn red and show which source table they came from.

As mentioned earlier, identifiers don't have to be used as keys in join conditions. In the Purchases table, the Order Dispatched column is very likely an identifier field, as the only values it contains are `1`s and `0`s. You can easily clarify these identifiers if you are preparing the data for end users, by aliasing the values in a data field (Figure 46-4).

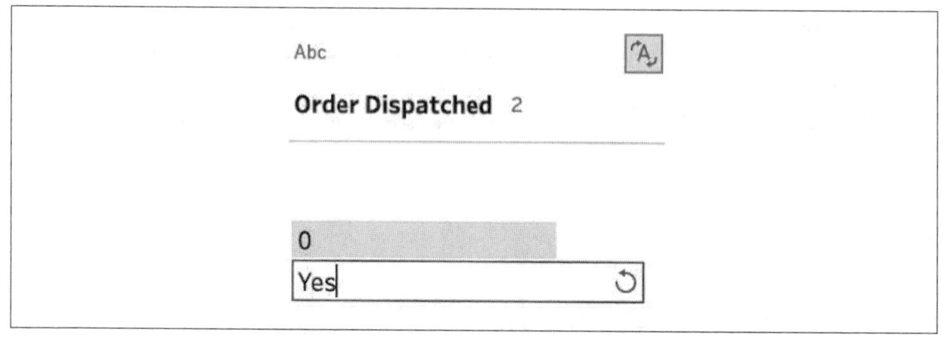

Figure 46-4. Setting up an alias in the Profile pane

As the data field is numeric, you'll need to update the data type to string before changing the value. The resulting data field is shown in Figure 46-5.

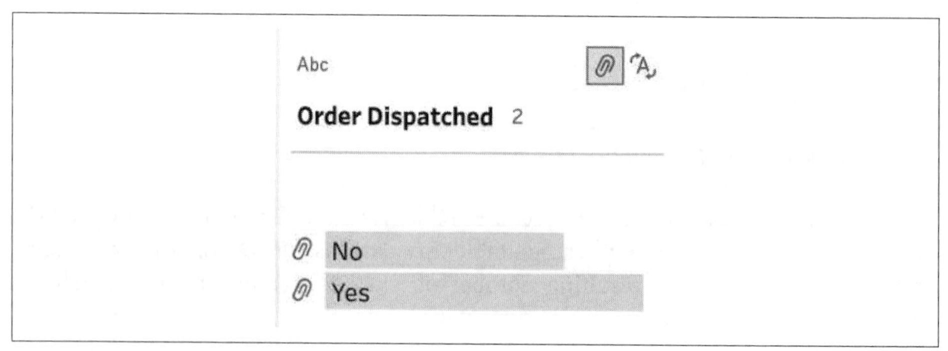

Figure 46-5. Results of changing the data type and aliasing

Creating Identifier Data Fields in Prep Builder

All of the previous points rely on an identifier or key being available in your data source. However, if you are preparing your data source for storage in a database, you may need to create a key from the data you have. Let's use Superstore, the sample data set that is downloaded as part of the Prep Builder or Desktop package, and create a category table that could act as a reference table.

As of Prep Builder version 2020.1, the tool includes the functionality to create a rank field automatically. In the example shown in Figure 46-6, you can remove some of the complexity by creating ranks based on the Category data field.

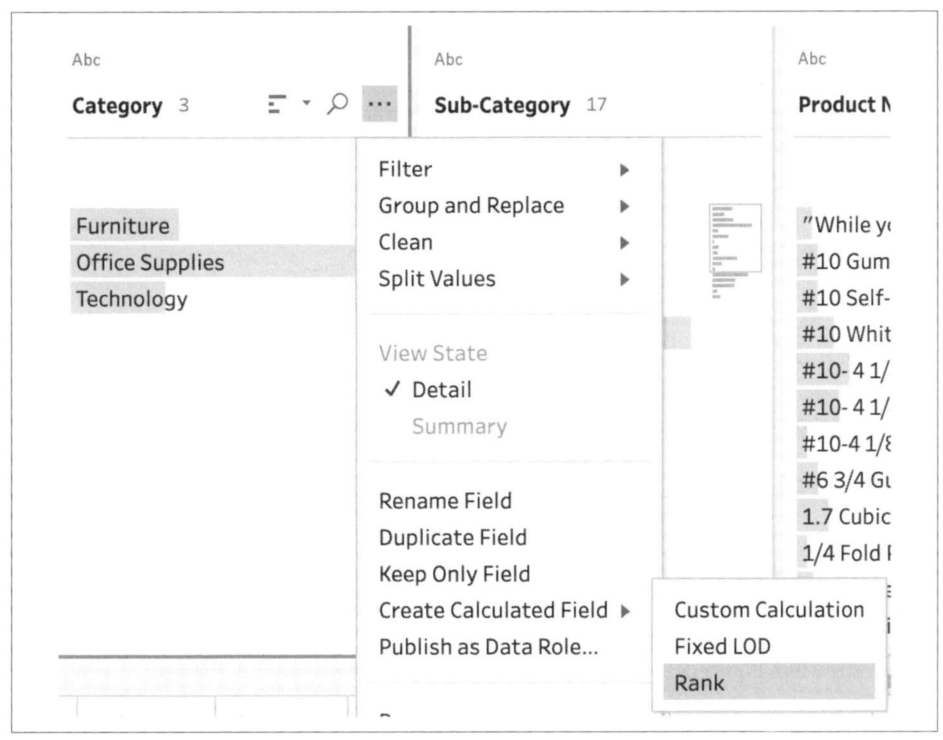

Figure 46-6. Creating a rank through the Profile pane data field menu

Selecting this option will launch the rank Visual Editor, which makes the process coding-free (Figure 46-7).

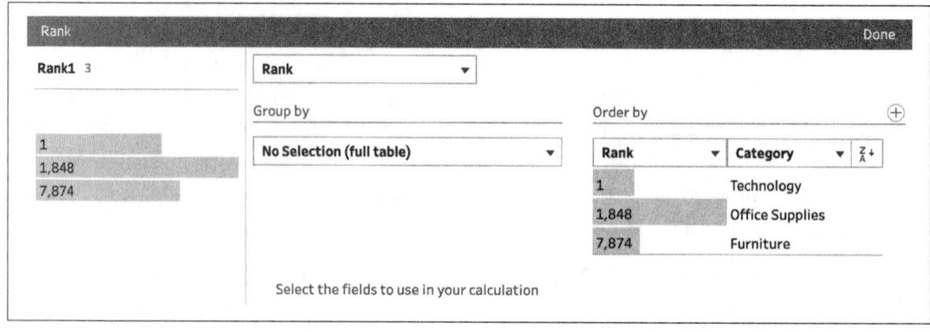

Figure 46-7. Setting up a rank in the Visual Editor

In the default rank setup, the value of the rank is ordered by the number of rows attributed to each category. To get past this and create a simple ID for category, change the rank option in "Order by" to Dense Rank (Figure 46-8).

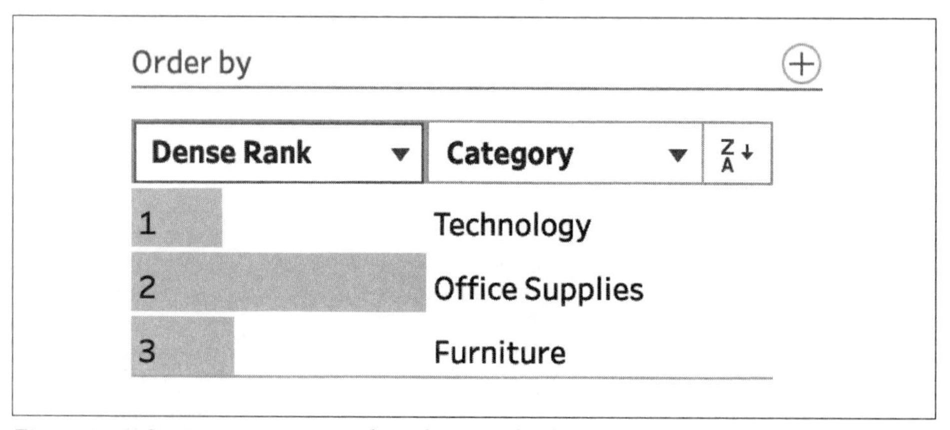

Figure 46-8. Setting up Dense Rank in the Visual Editor

To create the reference table to be able to look up these values, you can branch off an Aggregate step. Aggregate steps remove all data fields from your data set that are not used in the Aggregate step. In this step, group by category and average the rank to return a single ID per category. You can rename the Rank field to Category ID (Figure 46-9).

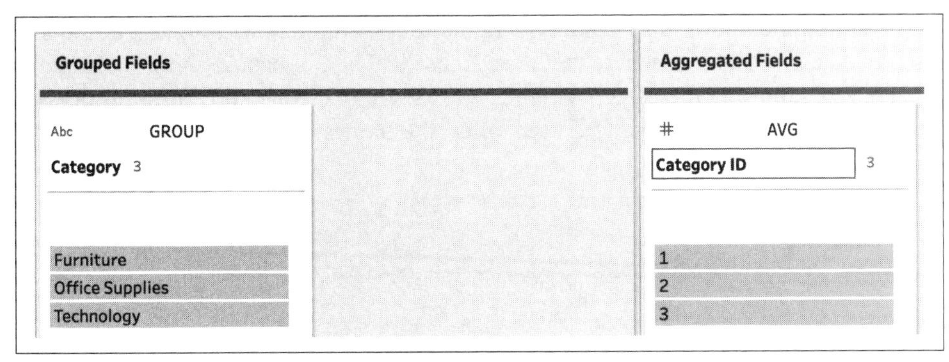

Figure 46-9. Changing the Rank field name to Category ID in an Aggregate step

The resulting table from the Aggregate step is very simple and can be output separately (Figure 46-10).

Category ID	Category
1	Technology
3	Furniture
2	Office Supplies

Figure 46-10. Result of the Aggregate step creating the Category ID field

The 1, 2, 3 result is perfect for creating the IDs we want in the data set. You can remove the Category field from the original data set now (Figure 46-11).

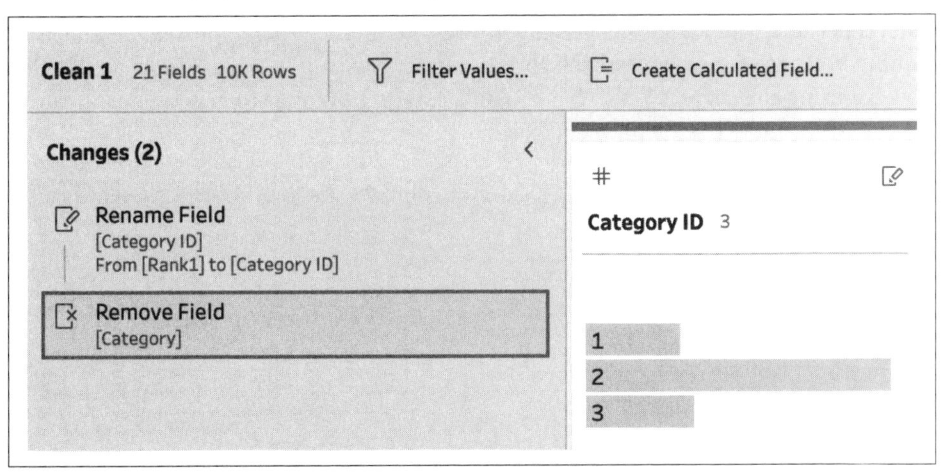

Figure 46-11. Changes pane showing the steps to remove the Category field

The resulting flow looks like Figure 46-12, although you may wish to add outputs to use the data or write it to where you need to use it. The full data set now has only the category ID and not the category name. This saves on space and can be processed faster. The reference table can be used when the names need to be added back into the data set if required.

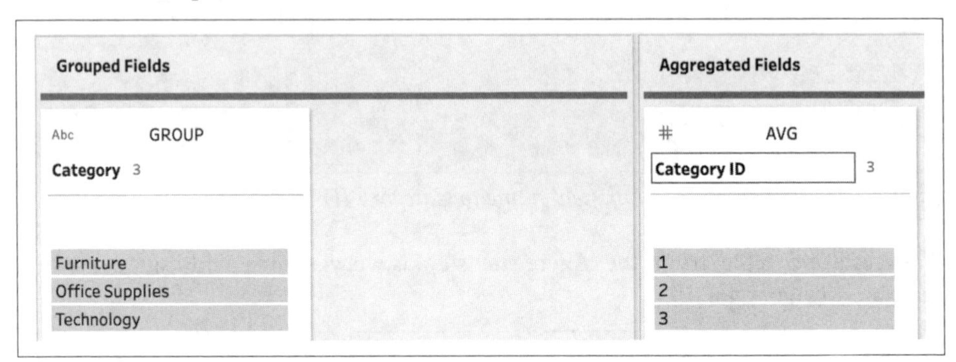

Figure 46-12. Using the Aggregate step on a separate branch of the flow to create the reference table

If you need to update the data set and increase the number of potential values in the Category field, be sure to keep the original ID set and assign the next rank value to the new values.

Summary

Keys and identifiers might seem to present unnecessary challenges when you are new to data preparation. However, by computing faster and requiring less storage space, you can reduce the time it takes to complete your analysis and the cost of storing the data, especially when data sets get larger. Prep Builder can help you see the effects of joining your reference tables back into the data, which is a good way to verify that you are using the correct keys. You can also use Prep Builder to create your own reference tables and write them to databases.

CHAPTER 47

Keeping Your Data Up-to-Date

Once you've prepared a data set for analysis, the next choice you want to make is whether the process needs to be repeated and, if so, how often. If the input data source is updated, in all likelihood your flow will need to be refreshed. Whether or not you have Prep Conductor will determine what options are available to you to maintain the data.

In this chapter, we will look at the two different approaches to refreshing data sets—full and incremental—and how they differ. For each technique, we will walk through how to set it up in Tableau Prep Builder and Conductor. I'll also offer a few key things to look out for when you're using these techniques.

Refreshing Data

Refreshing data means checking to see if the original input has new or altered data. If it does, you must decide whether the new or changed data should be pushed through the data preparation flow. One of the advantages of Prep Builder is that once you've built the preparation flow, you can reuse it simply by clicking the run icon or setting up a refresh schedule in Prep Conductor.

The refreshed data may have the following changes:

- New rows
- Overwritten values
- New columns

Each type of refresh poses its own challenges but can be handled with the different techniques in this chapter.

When you first import or receive a data set for analysis, you should ask whether that data will refresh. If the answer is yes, you'll likely need to rerun the data preparation flow each time new data is available. Understanding the frequency of the refresh is important; sometimes requests for "live" reporting will be impacted by the frequency of the updates to the original source. Also, you should assess how frequently the reporting requires fresh data, as in many cases the data may be constantly refreshing, so if the reporting is to be refreshed once a day, you'll need to determine a cut-off point.

Full Versus Incremental Refreshes

As their names suggest, a *full* refresh refers to a complete update of the data set, whereas an *incremental* refresh is only a partial refresh. With a full refresh, the original data set has all its records removed and replaced with the records in the refreshed data source. Therefore, a full refresh is usually irreversible, so you should check the data before running your flow to ensure data isn't removed or overwritten inadvertently.

An incremental refresh uses only the latest new records to update an existing data set. Normally, these new records are added to the end of an existing data set. There is less risk associated with this type of refresh, as those records being added can always be removed as long as no overwriting occurs.

In Tableau Prep Builder, the development team has added flexibility by allowing incremental refreshes to overwrite existing data points such that only the latest data points remain in the data source. This will be a new concept for most Tableau Desktop and Server users, as it wasn't an option previously. For a full refresh, rather than overwriting the original data set, you can add the new data to the end of the existing data set, forming a history table capturing snapshots of data over time.

Setting Up Different Types of Refresh

These new approaches to full and incremental refreshes allow you to set up refreshes for many different scenarios.

Full Refresh

A full refresh is set up in the Output step of Prep Builder. At the bottom of the Output configuration pane, you'll see an option to set a full refresh (Figure 47-1).

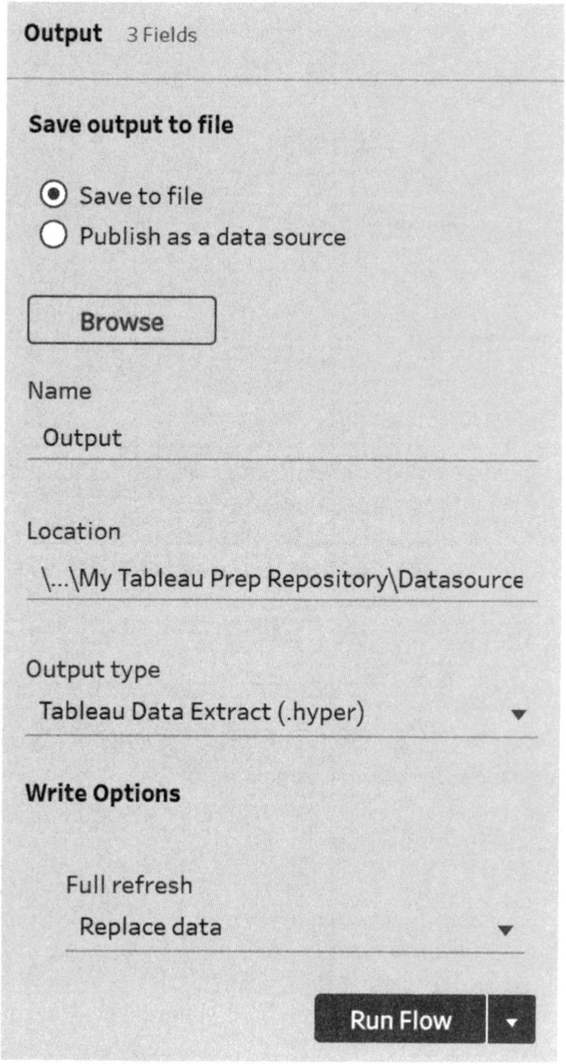

Figure 47-1. *Full refresh ("Replace data") option*

If you used Prep Builder before the new refresh options were added, any run of the Prep flow would have behaved like a full refresh. If the data set to which the Output step writes already existed, it would be overwritten; otherwise, it would be created.

Incremental Refresh

Setting up an incremental refresh begins with a new option in the Input step. As mentioned earlier, this functionality is a newer feature for Prep Builder (version

2020.2), so the Input step looks different from how it appears in previous chapters (Figure 47-2). Hyper files are the only output file type Prep Builder supports for incremental refreshes.

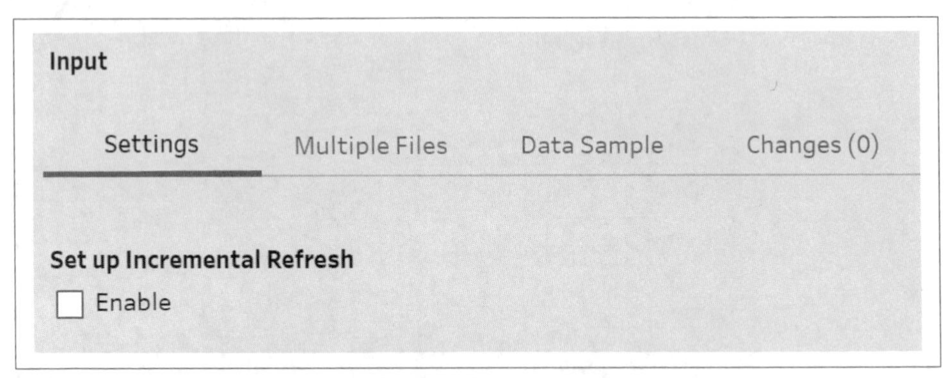

Figure 47-2. Enabling an incremental refresh option within the Input step

After selecting the "Set up Incremental Refresh" box, you'll need to configure some additional settings to avoid error messages appearing on your flow (Figure 47-3).

Figure 47-3. Configuring the incremental refresh

The first configuration option is to select which data field will indicate new data in the Input data source. In Figure 47-3, Date has been selected, as profit figures are recorded daily per store and product type. Thus, as the next day's data is available, a new date will appear in the data set and you'll need to incrementally add it to the output data set.

The second setting is which output to update. A Prep flow can result in multiple outputs, so this is why this setting needs to be specified. In Figure 47-3, I have a single output and haven't renamed it from the default, "Output."

Finally, you need to set the name of the field to match the one being accessed for new data. This is purely due to the likelihood of data fields being renamed as you prepare the data throughout the flow. In the example, the Date field hasn't been renamed, so the "Field name in output" setting is still Date.

When running the output, you can choose whether to run an incremental refresh or a full refresh within both Prep Builder and Prep Conductor. The options for processing the results of the full or incremental refresh are available from the drop-down menu on the Output step itself (Figure 47-4).

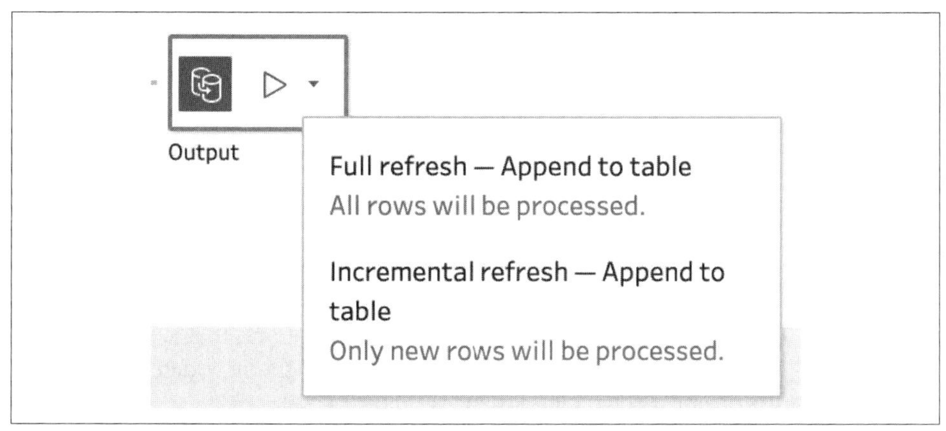

Figure 47-4. Selecting an incremental refresh in the Output step to append data

You can also control whether the data is replaced or added to the existing data set. You set this at the final stage of the Output configuration step (Figure 47-5).

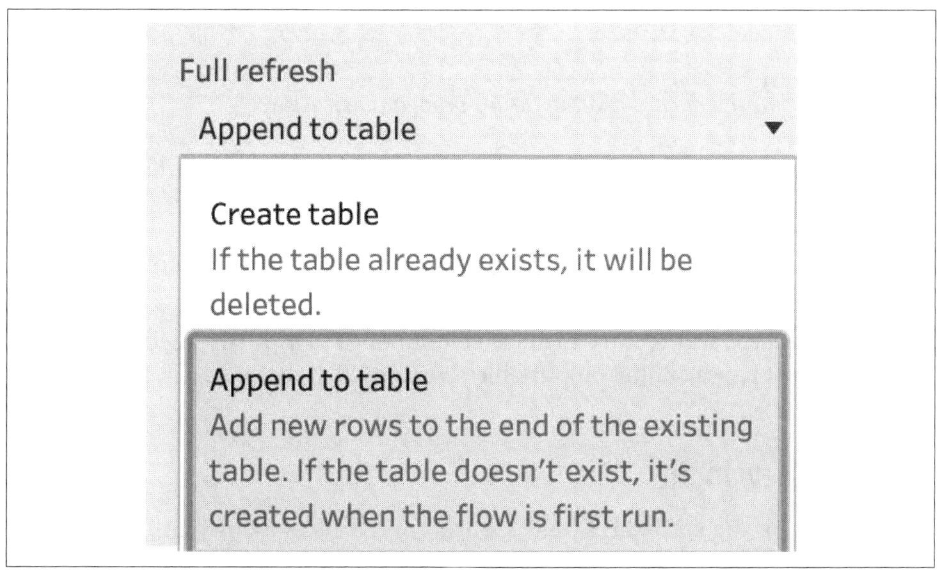

Figure 47-5. Option to append data during a full refresh

What to Watch Out for When Refreshing Data Sources

When setting up a refresh flow within Prep Builder, you'll need to evaluate a few areas to determine whether the results of the Prep flow are as intended.

Changing Data Values

When setting up an incremental refresh that adds new rows of data to an existing data set, you must consider the existing records too. If there is any chance that those records might be updated, you should periodically do a full refresh to ensure that any updates to existing records are also reflected in the data. Otherwise, because the new data rows were added, most users of the data set will incorrectly assume that all the data is current.

Altering the Structure of Sources

There is a difference between a new column being added that will be pivoted into a new record of data (e.g., a new month of data) and a new column that is not expected within a flow. You can apply the wildcard pivot option to the new date columns to prevent any errors or unexpected consequences (Figure 47-6).

Drop fields here to pivot them
Or
Use wildcard search to pivot

Figure 47-6. Using a wildcard pivot to future-proof data structure

With unexpected new columns, it's harder to predict how the output will be affected. Here are some possibilities:

- If there are no pivots, the data is still likely to output but might not include the new column.
- Aggregate steps will remove the new column of data, as it isn't used in the step, and it won't appear in the output unless you add that data set back in through a self-join.

New Data, New Input

If the new data being created is saved as a data source under a different name, it may not get picked up by the existing Prep flow. To ensure that it does, you can set up a wildcard union. The data structure of the file will need to remain consistent for the flow to work as intended.

Summary

When setting up data preparation flows, you must consider the updates made to the original source of the data. These will determine how you should design your flow to ensure that the output is as intended. In addition, you must specify the type of refresh and whether the original source should be amended or overwritten. Get all these factors correct, and you will be able to analyze the latest data on an ongoing basis with little additional effort.

CHAPTER 48

Using History Tables

Live data connections are a fantastic way to keep your analysis relevant and prevent repeated work each time a data set is updated, which may be frequently. Building data visualizations that connect to live data sets gives end users confidence that they are getting the latest view on the topic.

The challenge with frequently updating data is deciding what happens with records that are either updated or removed when no longer relevant. Building a history table allows you to record data points at a set interval without worrying that they will be overwritten and lost forever.

Why Are History Tables Required?

When analyzing data, users will ask many questions to get a comprehensive view of key trends, outliers, and unexpected results in the data set. However, once they have found these data points, users will want to know:

- Why are these trends or points showing what they do?
- Has it always been this way?

The first question relates to gathering context, and it can be tough to answer. Often, working with subject-matter experts can be helpful. The second question often boils down to having historical data to draw upon. Despite decreasing storage costs for data, however, the rapid growth of data sets can make it challenging to retain sufficient historical data points to answer the question.

Retaining too much information will swamp both the analyst and the tool. Too little history, though, and it will be impossible to judge whether a trend is present. This is where a well-built history table can make all the difference in your analysis. Retaining the right information about a situation can allow for meaningful analysis. Let's consider a few scenarios where history tables can be useful:

Staff progression
: Maintaining a view of who held which role when, along with how employees have progressed through the organization, can offer a lot of insight on how well your organization develops and retains its staff.

Customer retention
: As with staff retention, understanding what products or holdings a customer had with your organization before leaving gives you the opportunity to improve your offerings rather than looking only at the current snapshot.

Operational processes
: Customers often get frustrated when they are stuck going around in circles while trying to meet their needs (e.g., calling customer service and getting forwarded to multiple representatives to try to resolve an issue). If the organization doesn't have a clear view of the customer journey, it is more difficult to improve the process for the future. If a system holds only the latest status for a customer, it is impossible to judge where operational issues might lie. Operational data, including telephone and computer system logs, can be queried to provide insight into a customer's journey throughout an operational process.

History tables allow you to store data points for future analysis and offer insights that may otherwise have been lost.

What to Consider When Creating History Tables

As mentioned previously, there is a constant battle between storing everything and paying high storage costs, or potentially saving on storage and losing meaningful data. So, this section will discuss some considerations to keep in mind as you build your own history tables for analytical use.

Ability to Join to Live Data

History tables alone can provide valuable insight for an analyst. However, joining history tables to the live data can create even stronger, more pertinent results.

Relevance of Information

Unless you are a shoe retailer, storing your employee's or customer's shoe size isn't going to help with your analysis. Keeping information about the products customers purchased and when they joined and left the service or organization is useful for understanding patterns of consumer behavior. Too much information, on the other hand, makes it harder and more time-consuming to conduct the analysis. Keeping just the relevant data is key.

Frequency of Updates

Taking a historical snapshot of the data set is useful, but don't do it too frequently. In many cases, monthly views are sufficient for showing patterns of behavior if your customer is interacting with you daily. The more frequent the update, the less movement you will notice. On the other hand, if you don't capture your customer's traits with enough frequency, you will not have the data points to show trends in your customer's holdings and transactions.

Level of Granularity

One data point per group of customers? One data point per customer? You will need to decide what is relevant to the analysis you wish to conduct. Whatever you decide, you may need to aggregate the data further for your analysis. This is only possible when you are going from more granularity to less, as aggregating data means removing detail from the data set. When analyzing patterns over time, think about the comparison you may want to make: This month versus the same month last year? This quarter versus the same quarter last year? This decision will dictate the volume of data your history table will need to retain.

All of these choices may change over time, but by building the history tables, you are giving yourself the opportunity to conduct analysis you might not be able to otherwise.

Performance

Relevance, frequency, and granularity of data all factor into the performance while you're building the analysis as well as while you're applying it. Data software processes large data sets at an ever-increasing pace. But with history tables, keeping data sets small and concise enough to join to what might already be a large data set is challenging. Ensuring you have a clear join condition is important, but so is knowing how many rows your join will create.

If you have 10 million customers and hold a monthly snapshot for the last year, then all of a sudden you are analyzing *120 million records* of data instead. The size of that data set is likely to slow down your analysis when running calculations or rendering visualizations. When forming the analysis, consider aggregating the 110 million rows of data in the history table to a lower granularity by reducing the monthly frequency. An alternative approach might be to stop analyzing the data set on a per-customer basis and instead form clusters of customers with similar demographics or product holdings.

The benefit of maintaining a separate history table is that you can treat it differently and then join on only the relevant data. Removing data for analysis from your live data set dramatically increases the risk of making mistakes that will compromise your analysis compared to removing data from your history table and then joining that to the live data.

Data Regulations

The major consideration about history tables is to respect the laws on data use. When a customer or employee leaves, they have the right to have their data removed. However, a record that an anonymous customer had certain products or carried out certain transactions is useful. As long as that information is not personally identifiable, you can hold these kinds of records, but otherwise they should be deleted (some industries have differing regulations on this). This information becomes a lot safer to keep when the data is aggregated to a higher level than individual transactions.

An Example History Table

There are a number of considerations to take into account when writing a history table.

First, how will the data update? Let's use an example that is common across many companies—recording complaints. Figure 48-1 shows a table representative of how complaint records are often held in operational systems. The Extract Date field is used in many internal reports to show when the data was taken from the system.

Complaint ID	Status	Date	Extract Date
C001	Replied	10/01/2020	12/01/2020
C002	Open	08/01/2020	12/01/2020
C003	Open	09/01/2020	12/01/2020

Figure 48-1. Sample Complaints data set

This is a data set that seems simple and straightforward. The complexity is in how this table updates. Figure 48-2 shows the same table a week later.

Complaint ID	Status	Date	Extract Date
C001	Resolved	13/01/2020	19/01/2020
C002	Open	08/01/2020	19/01/2020
C003	Replied	14/01/2020	19/01/2020
C004	Resolved	16/01/2020	19/01/2020

Figure 48-2. Updated Sample Complaints data set

When analyzing complaints data in organizations, just knowing the number of complaints is rarely enough information. Here are some questions that are frequently asked:

- Is the number of complaints rising or falling?
- How long does a customer have to wait for a response?
- How long does a customer have to wait for a resolution?

Despite being seemingly easy-to-answer questions (as we have a Date field within the data set), sadly, they are not. The dates in the data set reflect only the latest state of the complaint. If a complaint has just been opened, for example, it shows the date when the complaint was received. As soon as the complaint has been responded to, the open date has now been lost. We also can't determine the time between response and resolution from this data set. If these questions are likely to be asked, why is the data overwritten as the complaint progresses toward resolution? Simply put, many operational systems are designed to optimize the flow of the process rather than focus on the analysis. Therefore, the onus is on the data analyst to collate the data frequently to capture and maintain each stage of the process.

To accomplish this, first we'll set up Prep Builder's incremental refresh option, new in version 2020.2, from the Input step. When the flow runs, we want to add only rows with an Extract Date value later than the one currently held in the file. As covered in Chapter 47, you also need to specify which Output step and subsequent file you will be assessing for additional data (Figure 48-3).

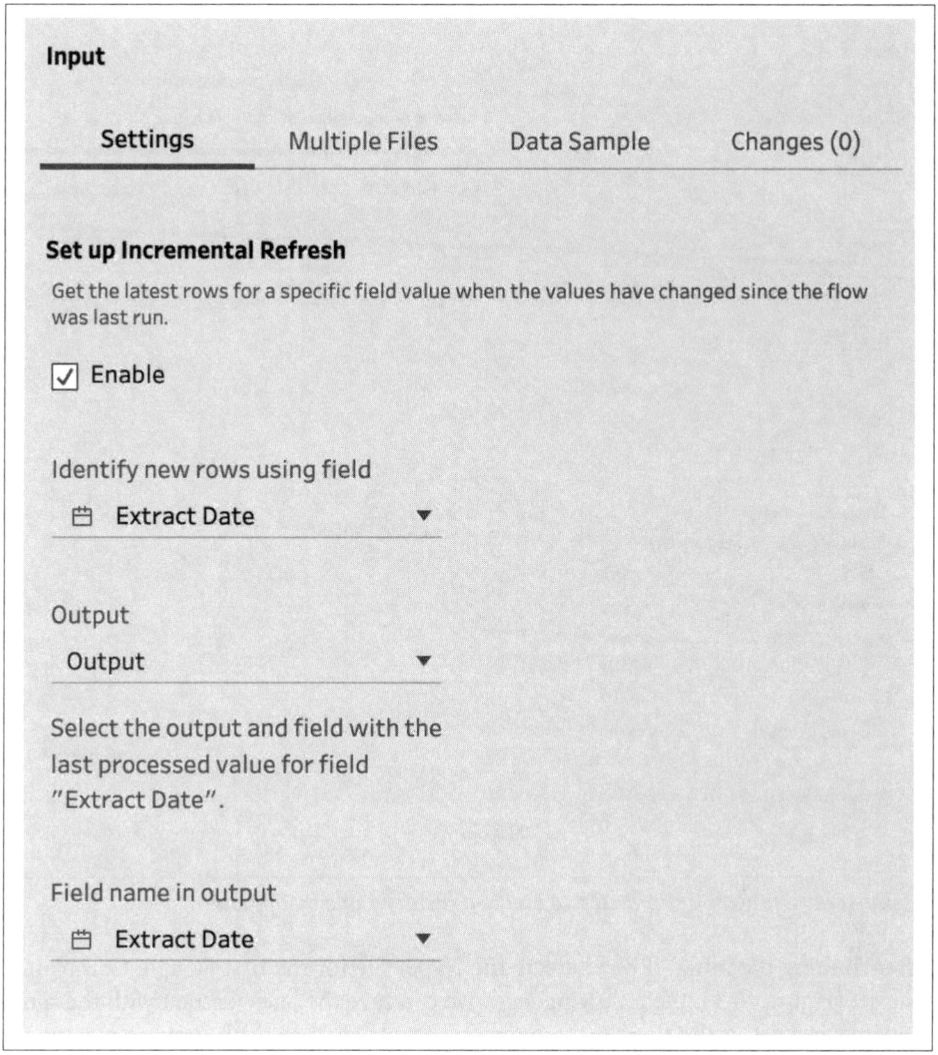

Input

Settings Multiple Files Data Sample Changes (0)

Set up Incremental Refresh

Get the latest rows for a specific field value when the values have changed since the flow was last run.

Enable

Identify new rows using field

Extract Date

Output

Output

Select the output and field with the last processed value for field "Extract Date".

Field name in output

Extract Date

Figure 48-3. Setting up an Input step to create a history table

In this example, the data needs to be appended and not overwritten at any point; otherwise, we'll lose the history of the data set since the data set we are referring to preserves only the latest status. Therefore, we need to output to a Hyper file, as that is the only file type that allows data to be appended (Figure 48-4).

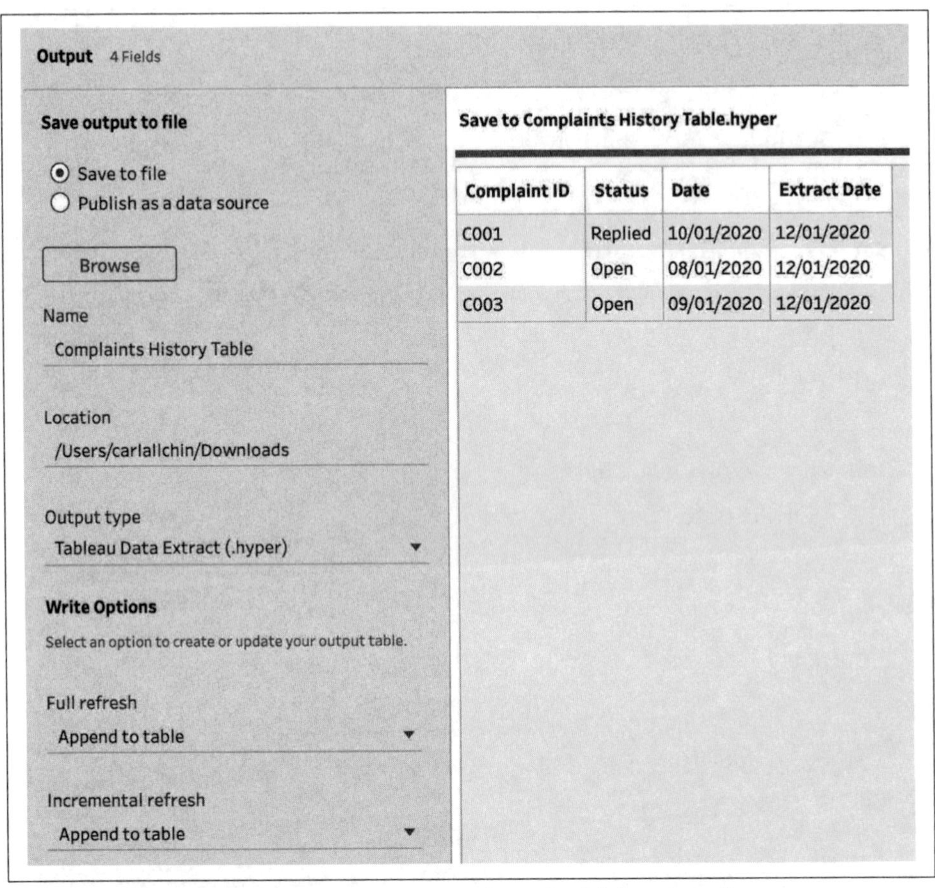

Figure 48-4. Using an Output step to create a history table by appending rows

After running the initial flow to create the Hyper file for the first Sample Complaints data set (Figure 48-1), the following week we can save the latest extract with the same filename and rerun the flow. If the name of the file changes with each extract, you'll need to create a wildcard union to absorb the new data. Once the flow is rerun with the new data in the input file, it looks like Figure 48-5, which has maintained the original data while adding the new data.

Abc Extract **Complaint ID**	Abc Extract **Status**	Extract **Date**	Extract **Extract Date**
C001	Replied	10/01/2020	12/01/2020
C002	Open	08/01/2020	12/01/2020
C003	Open	09/01/2020	12/01/2020
C001	Resolved	13/01/2020	19/01/2020
C002	Open	08/01/2020	19/01/2020
C003	Replied	14/01/2020	19/01/2020
C004	Resolved	16/01/2020	19/01/2020

Figure 48-5. Example Hyper file in Tableau Desktop

This history table will grow as each week's data is updated and the Prep Builder flow rerun.

Summary

Building and using history tables will allow you to conduct better and deeper analysis than you could otherwise. Refining the relevance, frequency, and granularity of the data stored in the history tables will ensure the analytical value you gain is well worth the effort it took to build the analysis. History tables can grow rapidly if they are not aggregated at a high level, so you will need to decide how much history is relevant before removing the data. You must also account for any legal limitations on how long data can be stored before it must be deleted.

CHAPTER 49

Evaluating Whether You Need Prep Builder at All

We all overthink things sometimes, and it happens a lot in data preparation. Once you have honed your skills, it can be easy to overthink the data and take steps you simply don't need. This chapter will cover when you can use Tableau Desktop to complete the data preparation process and situations where Prep Builder is the better choice.

A History of Data Preparation in Tableau

Back in 2010, when I was trying Tableau Public for the first time in version 5.2 and becoming a heavy user of Desktop version 7 onward, completing your data preparation in Tableau was tough unless you were doing very simple use cases. Prep Builder was just a glint in the eye of the Tableau developers at that point.

Common data prep tasks had to happen outside of Tableau Desktop for a number of reasons, like performance (slower processing speeds), but it was the lack of functionality that caused the most significant issues. Users often had to use external tools to complete tasks. The Tableau Excel add-in extension (see the top right of Figure 49-1) was the principal way to pivot data from columns to rows. This was very useful for dealing with survey data and dates held in separate columns (i.e., data that's not Tableau friendly, as covered in Chapter 4).

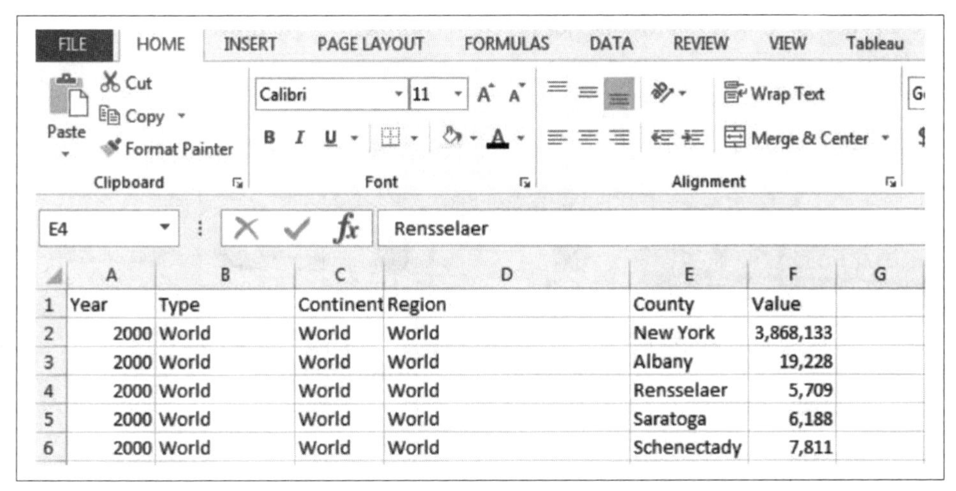

Figure 49-1. The Tableau Excel add-in extension

Complex joins, data removal, and unions were not part of the Data Connection window in Desktop as they are now. This meant that to use multiple data sources, you'd have to prepare your data outside of Tableau, often in SQL servers.

As Tableau's development aim has always been to keep the user in the flow of their analysis—rather than making them continually move out of Tableau to complete their tasks—the developers built more preparation functions into the core tool, Tableau Desktop. The issue with this was that all the data preparation happened within the Data Connection window and space was becoming limited. The complexity would rapidly clutter up Tableau's simple user interface. Version 2018 solved the problem by spinning out Tableau Prep Builder as a separate tool.

Where to Try Desktop First

There are instances where it might be easier to use Desktop instead of launching Prep Builder. However, not everything has a straightforward answer, so we'll go over some factors to consider. After discussing each technique you might attempt first in Desktop, we will cover when it might be more appropriate to move to Prep Builder.

Simple Joins

Joins have been a feature of data connections in Desktop for years (Figure 49-2). Tableau Prep can of course handle these, but keeping joins in Desktop allows you to change the join type and conditions whenever you want. With the addition of join calculations, you can handle most situations where you need to add columns from additional data sources right in Desktop. If you are unfamiliar with the data, or if you are unsure of the result of the joins, I'd recommend you move to Prep Builder instead.

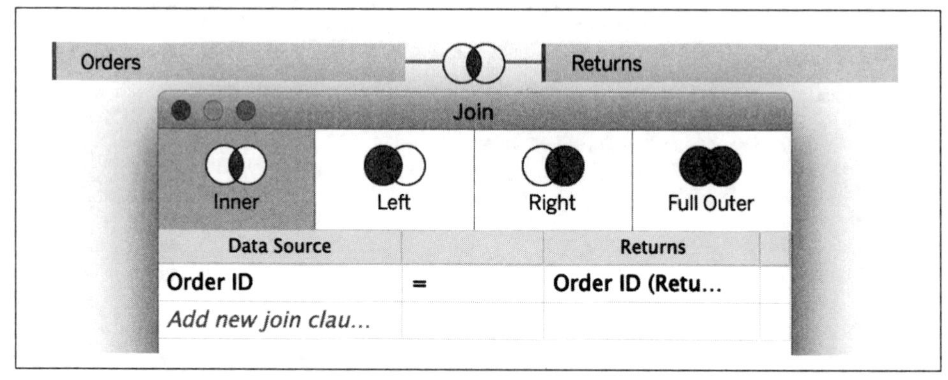

Figure 49-2. Setting up a join in Tableau Desktop

When to move simple joins to Prep

When you are making multiple joins or using multiple join conditions, it's easy to lose clarity on the output in Desktop. With Prep's Profile pane, it's much easier to see when join conditions have gone wrong or created a lot of nulls that you were not expecting. Also, look to use Prep when you need to change the level of aggregation of one of your data sets before joining it to another.

Joins are covered in more detail in Chapter 16 and Chapter 32.

Unions

Like joins, when unions were added to the Tableau Desktop Data Connection window (Figure 49-3), it dramatically reduced the number of times data prep had to be completed externally. Unions in Desktop are quite flexible—so much so that the options for unions within Prep are very similar.

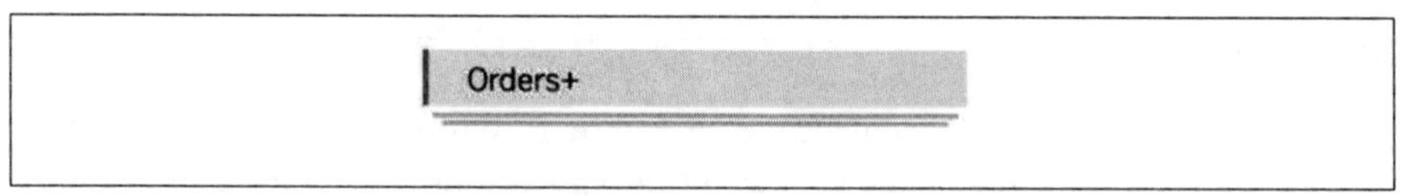

Figure 49-3. Union icon in Tableau Desktop

To add a basic union in Desktop, simply drag the additional data set underneath the original connection in the Data Connection window (Figure 49-4).

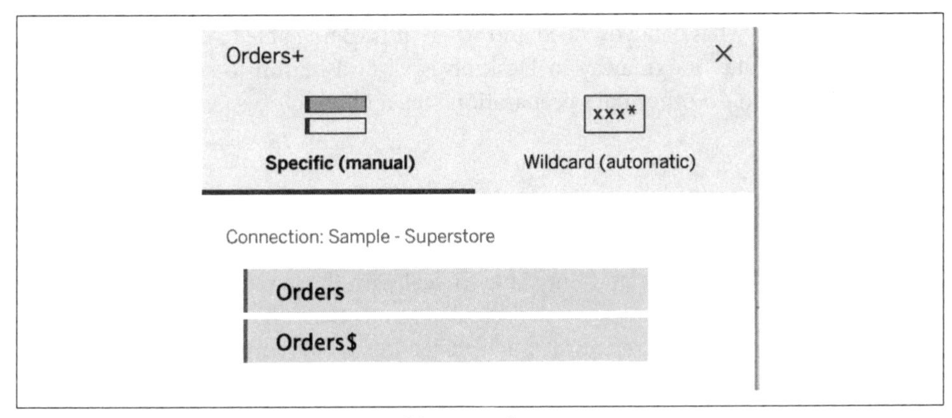

Figure 49-4. Setting up a union in Tableau Desktop

Once the union is formed, you have the option to edit it or add other data sources to it. More complex unions can be accomplished with wildcard unions (Figure 49-5), as we saw in Chapter 17.

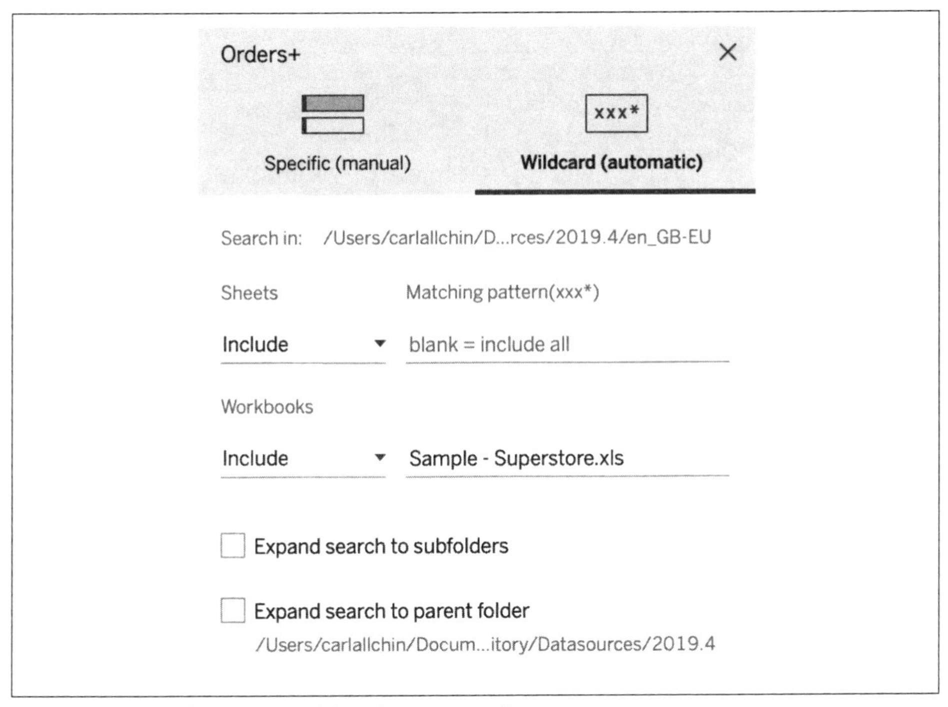

Figure 49-5. Tableau Prep wildcard union configuration pane

Similar to Prep, you can wildcard the union at the sheet or workbook level when connecting to Excel sheets. Different data types have different options at this point, so it's important to plan what data you need and where it is. Being able to explore the effects of the unioned data straightaway in Desktop is a good reason to start in the tool—unless you need to do other data preparation steps, in which case you might want to switch to Prep.

When to move unions to Prep

When you union data with different structures (column names), the resulting data set will contain a lot of null values. Being able to deal with those nulls before progressing to your analysis can be complex, so the calculations to clean this up must be accurate. Prep Builder's Profile pane makes it much easier to see these nulls, and the resulting changes, to check that you are proceeding correctly.

Single Pivots

When pivoting was added to Desktop (Figure 49-6), it significantly reduced the need to use external tools like the Excel add-in for Tableau. Survey data has long been a considerable challenge in Desktop, but being able to pivot a column or set of columns in Desktop made it much simpler to prepare and shape the data set.

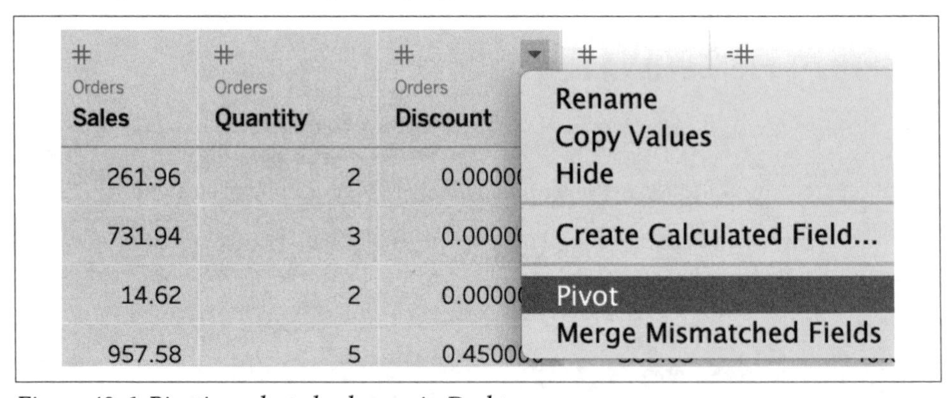

Figure 49-6. Pivoting selected columns in Desktop

The resulting data field names are far from useful, but they are easy to adjust (Figure 49-7).

Abc Pivot **Pivot Field Names**	# Pivot **Pivot Field Values**
Discount	0.000000
Discount	0.000000
Discount	0.000000

Figure 49-7. Data field names resulting from the pivot in Desktop

When to move single pivots to Prep

The limitation of pivoting within Desktop is that you can pivot only once. In addition, you cannot split and pivot at the same time or use Calculated Fields within pivots. Many data sets, especially that pesky survey data, are much more complicated than what is suitable for a simple pivot. Therefore, while planning your data preparation, if you see a need for multiple pivots, then move straight to Prep Builder.

Where to Start with Prep Builder

You've seen that even when it might make sense to start in Desktop, it's often a good idea to switch to Tableau Prep depending on the data you're working with. In the case of data review or handover, however, you will always want to head straight for Prep.

Tableau Prep allows for a lot of traceability. In other words, it is largely self-documenting as you build the data preparation flow. Desktop does not have the same capability. In Prep Builder, the use of iconography, the Changes pane, and the ability to step through the changes one-by-one allows anyone to retrace the journey from input to output and understand why each stage exists. Being able to rename steps with logical names and add descriptions (Figure 49-8) makes the work much easier to understand than the lengthy project handover Word documents that many of us have had to translate over the years because Desktop lacks that documentation capability.

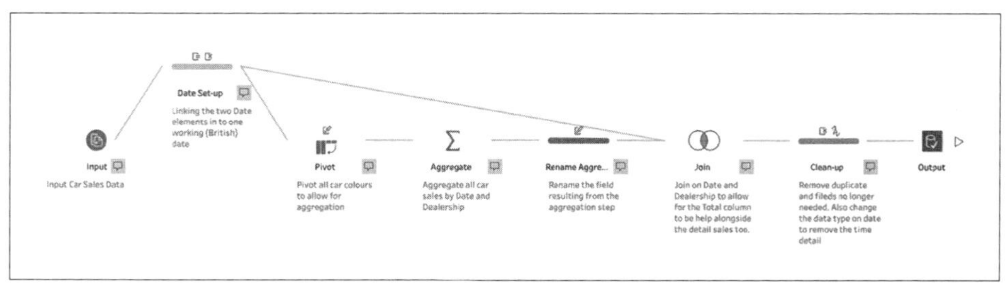

Figure 49-8. A documented flow within Prep Builder

In short, do the data review in Prep and save yourself from having to write the documentation for handover separately.

Summary

All these steps in isolation are possible within Desktop, but in combination they increased complexity and crowded the Desktop user interface, prompting the launch of Tableau Prep. Also, handing over complex workbooks with multiple stages of data preparation in Desktop is far from ideal for the recipient, so being able to document your flow succinctly and clearly within Prep Builder is one more advantage in addition to the more complex functionality featured in earlier chapters.

CHAPTER 50

Final Thoughts

Self-service data preparation enables those working with data to answer their own questions when they want to. Hopefully, after reading this book, you have discovered that Tableau Prep Builder is a great tool for preparing and working with data. In this chapter, I'd like to offer a few final thoughts from my experience of teaching people how to work with their data.

One aspect I mentioned early on was planning your preparation. As you gain experience, you might be tempted to skip this step, but planning allows you to periodically sanity-check what you are trying to achieve. Planning also gives you the opportunity to take your preparation further by adding beneficial data that you wouldn't have discovered otherwise. Inevitably, you will face some difficult challenges that need to be broken down and structured. A quick sketch can go a long way toward helping you formulate the best solution and make it a lasting one.

Prep Builder's simplicity makes it not just a great introduction to data preparation but also a product you can continue to use as you gain experience and take on even more challenges. Prep Builder has a lot of deep functionality that can help you tackle those tougher problems. Once a solution is in place, you can easily automate it by combining Prep Builder with Prep Conductor. With the time you save, you can work more on the actual analysis or on tackling additional data problems. Prep Builder's ease of documentation alone creates significant time savings.

Before reading this book, you might have viewed databases as an inaccessible option for working with data. With Prep Builder, you can set up a simple database connection and even write your clean data back to the database—without having to do any coding. If you do have coding skills, the scripting tool can deliver some great additional functionality. Handing over coded solutions to noncoding colleagues in the scripting tool is a much smoother process thanks to Prep Builder's user-friendly interface.

With each release of Tableau Prep Builder, the development team includes a richer and richer feature set. The platform is already an easy-to-use data preparation tool that removes unneeded complexity from the process, so I'm excited to see what comes next. These developers are the secret heroes and heroines in this book; without them, I wouldn't have had the functions to write about. I want to thank them for all their hard work and thank *you* for choosing to learn about data preparation with Prep Builder. I hope reading this book has given you a greater sense of data freedom and empowerment.

Index

D

E

F

G

Q

R

S

T

U

V

W

Z

About the Author

Carl Allchin is a Tableau Zen Master, multiple-time Tableau Ambassador, and "the other head coach" at one of the world's leading data analytics training programs at The Data School in London. After over a decade in financial services as a business intelligence analyst and manager, he's supported hundreds of companies through consulting, blogging, and teaching on market-leading data solutions. Carl is the cofounder of Preppin' Data, the only weekly data preparation challenge on Tableau and other data tools.

Colophon

The animal on the cover of *Tableau Prep: Up and Running* is a quokka (*setonix brachyurus*). The quokka is a small marsupial found in southwestern Australia and on nearby coastal islands.

With coarse brown fur and a somewhat lighter underbelly, these small wallabies are about the size of a domestic cat. Quokkas have black noses, relatively short arms and tails, and a pouch in which they carry their young. Their round ears and friendly-looking faces make them popular subjects for photographs. Male quokkas are often slightly larger than females, and baby quokkas reach maturity after about a year. Quokkas are nocturnal, and do most of the foraging at night. They eat a variety of plants including succulents, shrubs, grasses, and fruit, depending on the season. Quokkas typically stay in the same place all year round. Quokkas swallow their food whole, then regurgitate it and chew the cud.

The introduction of new predators from Europe (notably the red fox and domesticated dogs and cats) alongside habitat destruction and climate change has caused the quokka population to decline. Recent conservation efforts, however, have aimed to preserve the quokka's natural habitat, and further study has helped promote proper vegetation growth to help sustain larger populations. Nevertheless, the quokka's current conservation status is still "Vulnerable." Many of the animals on O'Reilly covers are endangered; all of them are important to the world.

The cover illustration is by Karen Montgomery. The cover fonts are Gilroy Semibold and Guardian Sans. The text font is Adobe Minion Pro; the heading font is Adobe Myriad Condensed; and the code font is Dalton Maag's Ubuntu Mono.

O'REILLY®

There's much more where this came from.

Experience books, videos, live online training courses, and more from O'Reilly and our 200+ partners—all in one place.

Learn more at oreilly.com/online-learning

©2019 O'Reilly Media, Inc. O'Reilly is a registered trademark of O'Reilly Media, Inc. | 175